ENGINEERING TECHNOLOGIES FOR RENEWABLE AND RECYCLABLE MATERIALS

Physical-Chemical Properties and Functional Aspects

Innovations in Physical Chemistry: Monograph Series

ENGINEERING TECHNOLOGIES FOR RENEWABLE AND RECYCLABLE MATERIALS

Physical-Chemical Properties and Functional Aspects

Edited by

Jithin Joy

Maciej Jaroszewski, PhD

Praveen K. M.

Sabu Thomas, PhD

Reza Haghi, PhD

AAP | APPLE ACADEMIC PRESS

Apple Academic Press Inc.
3333 Mistwell Crescent
Oakville, ON L6L 0A2
Canada

Apple Academic Press Inc.
9 Spinnaker Way
Waretown, NJ 08758
USA

International Standard Book Number-13: 978-1-77188-653-6 (Hardcover)
International Standard Book Number-13: 978-1-315-14715-4 (eBook)

Library and Archives Canada Cataloguing in Publication

Engineering technologies for renewable and recyclable materials:
physical-chemical properties and functional aspects / edited by Jithin Joy,
Maciej Jaroszewski, PhD, Praveen K.M., Sabu Thomas, PhD, Reza Haghi, PhD.

Includes bibliographical references and index.
Issued in print and electronic formats.
ISBN 978-1-77188-653-6 (hardcover).--ISBN 978-1-315-14715-4 (PDF)

1. Waste minimization. I. Joy, Jithin, editor II. Jaroszewski, Maciej, editor III. K. M., Praveen, editor IV. Thomas, Sabu, editor V. Haghi, Reza K., editor

TD793.9.E55 2018 628.4 C2018-902279-5 C2018-902280-9

CIP data on file with US Library of Congress

Apple Academic Press also publishes its books in a variety of electronic formats. Some content that appears in print may not be available in electronic format. For information about Apple Academic Press products, visit our website at **www.appleacademicpress.com** and the CRC Press website at **www.crcpress.com**

Printed and bound by CPI Group (UK) Ltd, Croydon, CR0 4YY

ABOUT THE EDITORS

Jithin Joy

Jithin Joy is Assistant Professor of Chemistry at Newman College, Thodupuzha, India. He is engaged in doctoral studies in the area of nanocellulose-based polymer nanocomposites. He has also conducted research work at Clemson University, South Carolina in the United States. He received his MSc degree in Chemistry from Mahatma Gandhi University. He is a coeditor of the books *Natural Rubber Materials: Volume 2: Composites and Nanocomposites,* published by the Royal Society of Chemistry, and *Micro- and Nanostructured Polymer Systems: From Synthesis to Applications*, published by Apple Academic Press. Mr. Jithin Joy received a prestigious research fellowship administered jointly by the Council of Scientific and Industrial Research and the University Grants Commission of the Government of India. He has publications in international journals and conference proceedings to his credit. He is a coauthor of several books chapters, peer-reviewed publications, and invited presentations at international forums.

Maciej Jaroszewski, PhD

Maciej Jaroszewski, PhD, is an Assistant Professor and Head of the High Voltage Laboratory at Wroclaw University of Technology in Wroclaw, Poland. He received his MS and PhD degrees in high voltage engineering from the same university in 1993 and 1999, respectively. Dr. Jaroszewski was a contractor/prime contractor of several grants and a head of grant project on "degradation processes and diagnosis methods for high-voltage ZnO arresters for distribution systems" and is currently a contractor of a key project cofinanced by the foundations of the European Regional Development Foundation within the framework of the Operational Programme Innovative Economy. His current research interests include high-voltage (HV) techniques, HV equipment diagnostics, HV test techniques, degradation of ZnO varistors, and dielectric spectroscopy.

Praveen K. M.

Praveen K. M. is an Assistant Professor of Mechanical Engineering at Saintgits College of Engineering, India. He is currently pursuing a PhD in Engineering Sciences at the University of South Brittany (Université de Bretagne Sud)—Laboratory IRDL PTR1, Research Center "Christiaan Huygens," Lorient, France, in the area of coir-based polypropylene micro composites and nanocomposites. He has published an international article in *Applied Surface Science* (Elsevier) and has presented poster and conference papers at national and international conferences. He also has worked with the Jozef Stefan Institute, Ljubljana, Slovenia; Mahatma Gandhi University, India; and the Technical University, Liberec, Czech Republic. His current research interests include plasma modification of polymers, polymer composites for neutron shielding applications, and nanocellulose.

Sabu Thomas, PhD

Sabu Thomas, PhD, is the Pro-Vice Chancellor of Mahatma Gandhi University and Founding Director of the International and Inter University Center for Nanoscience and Nanotechnology, Mahatma Gandhi University, Kottayam, Kerala, India. He is also a full professor of polymer science and engineering at the School of Chemical Sciences of the same university. He is a fellow of many professional bodies. Professor Thomas has (co-)authored many papers in international peer-reviewed journals in the area of polymer science and nanotechnology. He has organized several international conferences. Professor Thomas's research group has specialized in many areas of polymers, which includes polymer blends, fiber-filled polymer composites, particulate-filled polymer composites and their morphological characterization, ageing and degradation, pervaporation phenomena, sorption and diffusion, interpenetrating polymer systems, recyclability and reuse of waste plastics and rubbers, elastomeric crosslinking, dual porous nanocomposite scaffolds for tissue engineering, etc. Professor Thomas's research group has extensive exchange programs with different industries and research and academic institutions all over the world and is performing world-class collaborative research in various fields. Professor Thomas's Center is equipped with various sophisticated instruments and has established state-of-the-art experimental facilities, which cater to the needs of researchers within the country and abroad.

Professor Thomas has published over 750 peer-reviewed research papers, reviews, and book chapters and has a citation count of 31,574. The H index of Prof. Thomas is 81, and he has six patents to his credit. He has delivered over 300 plenary, inaugural, and invited lectures at national/ international meetings over 30 countries. He is a reviewer for many international journals. He has received MRSI, CRSI, nanotech medals for his outstanding work in nanotechnology. Recently Prof. Thomas has been conferred an Honoris Causa (DSc) by the University of South Brittany, France, and University Lorraine, Nancy, France.

Reza Haghi, PhD

Reza Haghi, PhD, is a Research Assistant at the Institute of Petroleum Engineering at Heriot-Watt University, Edinburgh, Scotland, United Kingdom. Dr. Haghi has published several papers in international peer-reviewed scientific journals and has published several papers in conference proceedings, technical reports, and lecture notes. Dr. Haghi is expert in the development and application of spectroscopy techniques for monitoring hydrate and corrosion risks and developed techniques for early detection of gas hydrate risks. He conducted integrated experimental modeling in his studies and extended his research to monitoring system to pH and risk of corrosion. During his PhD work at Heriot-Watt University, he has developed various novel flow assurance techniques based on spectroscopy as well as designed and operated test equipment. He received his MSc in Advanced Control Systems from the University of Salford, Manchester, England, United Kingdom.

ABOUT THE INNOVATIONS IN PHYSICAL CHEMISTRY: MONOGRAPH SERIES

This book series aims to offer a comprehensive collection of books on physical principles and mathematical techniques for majors, nonmajors, and chemical engineers. Because there are many exciting new areas of research involving computational chemistry, nanomaterials, smart materials, high-performance materials, and applications of the recently discovered graphene, there can be no doubt that physical chemistry is a vitally important field. Physical chemistry is considered a daunting branch of chemistry—it is grounded in physics and mathematics and draws on quantum mechanics, thermodynamics, and statistical thermodynamics.

Editors-in-Chief

A. K. Haghi, PhD
Editor-in-Chief, International Journal of Chemoinformatics and Chemical Engineering and Polymers Research Journal; Member, Canadian Research and Development Center of Sciences and Cultures (CRDCSC), Montreal, Quebec, Canada
E-mail: AKHaghi@Yahoo.com

Lionello Pogliani, PhD
University of Valencia-Burjassot, Spain
E-mail: lionello.pogliani@uv.es

Ana Cristina Faria Ribeiro, PhD
Researcher, Department of Chemistry, University of Coimbra, Portugal
E-mail: anacfrib@ci.uc.pt

Books in the Series

- **Applied Physical Chemistry with Multidisciplinary Approaches**
  Editors: A. K. Haghi, PhD, Devrim Balköse, PhD, and Sabu Thomas, PhD
- **Chemical Technology and Informatics in Chemistry with Applications**
  Editors: Alexander V. Vakhrushev, DSc, Omari V. Mukbaniani, DSc, and Heru Susanto, PhD

- **Engineering Technologies for Renewable and Recyclable Materials: Physical-Chemical Properties and Functional Aspects**
  Editors: Jithin Joy, Maciej Jaroszewski, PhD, Praveen K. M., Sabu Thomas, PhD, and Reza Haghi, PhD
- **Engineering Technology and Industrial Chemistry with Applications**
  Editors: Reza Haghi, PhD, and Francisco Torrens, PhD
- **High-Performance Materials and Engineered Chemistry**
  Editors: Francisco Torrens, PhD, Devrim Balköse, PhD, and Sabu Thomas, PhD
- **Methodologies and Applications for Analytical and Physical Chemistry**
  Editors: A. K. Haghi, PhD, Sabu Thomas, PhD, Sukanchan Palit, and Priyanka Main
- **Modern Physical Chemistry: Engineering Models, Materials, and Methods with Applications**
  Editors: Reza Haghi, PhD, Emili Besalú, PhD, Maciej Jaroszewski, PhD, Sabu Thomas, PhD, and Praveen K. M.
- **Physical Chemistry for Chemists and Chemical Engineers: Multidisciplinary Research Perspectives**
  Editors: Alexander V. Vakhrushev, DSc, Reza Haghi, PhD, and J. V. de Julián-Ortiz, PhD
- **Physical Chemistry for Engineering and Applied Sciences: Theoretical and Methodological Implication**
  Editors: A. K. Haghi, PhD, Cristóbal Noé Aguilar, PhD, Sabu Thomas, PhD, and Praveen K. M.
- **Theoretical Models and Experimental Approaches in Physical Chemistry: Research Methodology and Practical Methods**
  Editors: A. K. Haghi, PhD, Sabu Thomas, PhD, Praveen K. M., and Avinash R. Pai

CONTENTS

LIST OF CONTRIBUTORS

Srabanti Basu
Department of Biotechnology, Heritage Institute of Technology, Kolkata 700107, India.
E-mail: srabanti_b@yahoo.co.uk

Aparupa Bhattacharyya
Department of Chemical Engineering, National Institute of Technology Durgapur 713209, Durgapur, India

Udayan De
West Bengal Academy of Science and Technology, Kolkata, India. E-mail: ude2006@gmail.com

Ela M. Dedhia
Textiles and Fashion Technology, College of Home Science, Nirmala Niketan (Affiliated to Mumbai University), 49, New Marine Lines, Mumbai 400020, India. E-mail: elamanojdedhia@yahoo.com

Susmita Dutta
Department of Chemical Engineering, National Institute of Technology Durgapur 713209, Durgapur, India

David Morillón Gálvez
Institute of Engineering, Coordination of Environmental Engineering, National Autonomous University of Mexico, Post Box 70-472, Coyoacán 04510, D. F. Mexico, Mexico

Cincy George
Department of Chemistry, Newman College, Thodupuzha, Kerala, India

K. E. George
Albertian Institute of Science and Technology, Kalamassery, Kochi, Kerala, India

Neenu George
Research and Post Graduate Department of Chemistry, St. Georges College, Aruvithura, Kerala, India. E-mail: neenuathickal@gmail.com

Juma Haydary
Institute of Chemical and Environmental Engineering, Faculty of Chemical and Food Technology, Slovak University of Technology, Radlinského 9, 812 37 Bratislava, Slovakia.
E-mail: juma.haydary@stuba.sk

Sona John
Department of Chemistry, Newman College, Thodupuzha, Kerala, India

Cintil Jose
Department of Chemistry, Newman College, Thodupuzha, Kerala, India

Aji Joseph
Department of Chemistry, Bishop Kurialacherry College, Amalagiri, Kottayam, Kerala, India

Jithin Joy
Department of Chemistry, Newman College, Thodupuzha, Kerala, India

K. Kanny
Department of Mechanical Engineering, Durban University of Technology, Durban, South Africa

Luz Quintero López
Institute of Engineering, Coordination of Environmental Engineering, National Autonomous University of Mexico, Post Box 70-472, Coyoacán 04510, D. F. Mexico, Mexico

V. Madhurima
Department of Physics, Central University of Tamil Nadu, Neelakudi Campus, Thiruvarur 610101, Tamil Nadu, India. E-mail: madhurima@cutn.ac.in

Sergio Marín Maldonado
Institute of Engineering, Coordination of Environmental Engineering, National Autonomous University of Mexico, Post Box 70-472, Coyoacán 04510, D. F. Mexico, Mexico

Alfredo Galicia Martínez
Institute of Engineering, Coordination of Environmental Engineering, National Autonomous University of Mexico, Post Box 70-472, Coyoacán 04510, D. F. Mexico, Mexico

Ivy Mathew
Research and Post Graduate Department of Chemistry, St. Georges College, Aruvithura, Kerala, India

Suman D. Mundkur
Department of Textiles and Apparel Designing, SVT College of Home Science (Autonomous), S.N.D.T Women's University, Juhu Road, Mumbai 400049, India. E-mail suman_mundkur@yahoo.com

A. Nasimah
Faculty of Engineering, University Malaysia Sabah, Kota Kinabalu, Sabah, Malaysia

F. Picchioni
Dutch Polymer Institute (DPI), Eindhoven, The Netherlands. E-mail: f.picchioni@rug.nl

L. M. Polgar
Department of Chemical Engineering, University of Groningen, Nijenborgh 4, 9747 AG Groningen, The Netherlands Dutch Polymer Institute (DPI), Eindhoven, The Netherlands

Jorge Emigdio Sánchez Pólito
Institute of Engineering, Coordination of Environmental Engineering, National Autonomous University of Mexico, Post Box 70-472, Coyoacán 04510, D. F. Mexico, Mexico

Ajay Vasudeo Rane
Department of Mechanical Engineering, Durban University of Technology, Durban, South Africa. E-mail: ajayrane2008@gmail.com

Renjanadevi B.
Department of Chemical Engineering, Government Engineering College, Thrissur, India

M. Neftalí Rojas-Valencia
Institute of Engineering, Coordination of Environmental Engineering, National Autonomous University of Mexico, Post Box 70-472, Coyoacán 04510, D. F. Mexico, Mexico. E-mail: nrov@iingen.unam.mx

Dalibor Susa
Institute of Chemical and Environmental Engineering, Faculty of Chemical and Food Technology, Slovak University of Technology, Radlinského 9, 812 37 Bratislava, Slovakia

Sabu Thomas
School of Chemical Sciences, Mahatma Gandhi University, Kerala, India

Abitha V. K.
School of Chemical Sciences, Mahatma Gandhi University, Kerala, India. E-mail: abithavk@gmail.com

A. Y. Zahrim
Faculty of Engineering, University Malaysia Sabah, Kota Kinabalu, Sabah, Malaysia.
E-mail: zahrim@ums.edu.my

LIST OF ABBREVIATIONS

AAS	atomic absorption spectrophotometer
AF	alternative fuel
ANOVA	analysis of variance
AnPOMW	anaerobically treated wastewater
ASR	automobile shredder residue
AT	anaerobic treatment
BFA	bagasse fly ash
BOD	biological oxygen demand
CAC	commercial activated carbon
CCD	central composite design
CCFP	charred citrus fruit peel
CCFPIP	charred citrus fruit peel immobilized papain
COD	high chemical oxygen demand
CPO	crude palm oil
DAP	diammonium phosphate
DCP 40	dicumyl peroxide 40
DCPD	dicyclopentadiene
DEA	diethanolamine
DEG	diethylene glycol
DSC	differential scanning calorimetric
EDS	energy dispersive X-ray spectroscopy
ENB	ethylidene norbornene
EPA	Environmental Protection Agency
EPDM	ethylene propylene diene monomer
EPM	ethylene–propylene rubber
FAO	Food and Agriculture Organization of the United Nations
FBRs	fast breeder reactors
FTIR	Fourier-transform infrared
GDP	gross domestic product
GHGs	green house gases
GTZ	Deutsche Gesellschaft für Technische Zusammenarbeit
HHV	higher heating value

IE	Institute of Engineering
IPN	National Polytechnic Institute
ITESM	Monterrey Institute of Technology and Higher Education
LGPGIR	General Law on the Prevention and Comprehensive Management of Wastes
LMMCC	low molecular mass colored compounds
MA	maleic anhydride
MA-g-EPDM	maleic anhydride-grafted EPDM
MAP	monoammonium phosphate
MDF	medium-density fiberboard
MOX	mixed oxide
NBR	acrylonitrile-butadiene rubber
OCDE	Organization for Economic Co-operation and Development
OMW	olive mill wastewater
PAC	polyaluminum chloride
PAni	polyaniline
PDMS	polydimethylsiloxane
PE	polyethylene
PEEK	polyether ether ketone
PER	pentaerythritol
PET	poly(ethylene terephthalate)
PHA	polyhydroxyalkanoate
POME	palm oil mill effluent
POMW	palm oil mill wastewater
PP	polypropylene
PS	polystyrene
PSOM	pseudo-second-order model
PU	polyurethanes
PUF	polyurethane foams
PVC	polyvinyl chloride
PVPD	poly-(vinyl 1,4-phenylenediamine)
REPDM	reclaimed EPDM
RHC	rubber hydrocarbon
RPP	recycled PP
RSM	response surface methodology
Sc	spreading coefficient

SCCFPIP	spent charred citrus fruit peel
SD	saw dust
SEA	specific enzymatic activity
SEI	secondary electron image
SEM	scanning electron microscope
PDMC	polyacrylamide and poly(2-methacryloyloxyethyl) trimethyl ammonium chloride
TCPP	tris(2-chloropropyl) phosphate
TDA	toluene diamine
TDI	toluene diisocyanate
TEP	triethyl phosphate
TESPT	triethoxysilyl propyl tetrasulfide
TG	thermogravimetric
TMP	trimethyl phosphonate
TSS	total suspended solid
UNAM	National Autonomous University of Mexico
USW	urban solid wastes
VNB	vinyl-norbornene
WA	work of adhesion
WEEE	waste electrical and electronic equipment
WPCs	wood–plastic composites

PREFACE

Nowadays recycling and reuse issues of materials have received considerable attention from scientists, engineers, technologists, and industrialists. This book will be an effective tool for better understanding of the recycling and reuse issues of materials. Recycling is a process to change materials into new products to prevent waste of potentially useful materials, reduce the consumption of fresh raw materials, reduce energy usage, reduce air pollution and water pollution by reducing the need for "conventional" waste disposal, and lower greenhouse gas emissions as compared to new material production. Recycling is a key component of modern waste reduction and is the third component of the "Reduce, Reuse, and Recycle" waste hierarchy.

This book will be a very valuable reference source for university and college faculties, professionals, postdoctoral research fellows, senior graduate students, polymer technologists, and researchers from R&D laboratories working in the area of recycling.

Finally, the editors would like to express their sincere gratitude to the contributors of this book who have extended excellent support for the successful completion of this venture. We are grateful to them for the commitment and the sincerity they have shown toward their contributions in the book. Without their enthusiasm and support, the compilation of this volume could not have been possible. We would like to thank all the reviewers who have taken their valuable time to make critical comments on each chapter. We also thank the publisher, Apple Academic Press, for recognizing the demand for such a book and for realizing the increasing importance of the area of recycling and reusing issues of advanced materials.

CHAPTER 1

RECYCLING OF MATERIALS AND REDUCTION OF WASTE

UDAYAN DE*

West Bengal Academy of Science and Technology, Kolkata, India

**E-mail: ude2006@gmail.com*

CONTENTS

ABSTRACT

Reuse and recycling of different important resources is becoming increasingly essential due to several factors triggered mainly by population explosion and modern lifestyle, increasingly exploiting world's natural resources such as materials, land, water, and energy. Used resources usually turn into waste. Earlier, mother earth had enough space and capacity to accommodate or recycle such wastes, generated by a smaller population with a "lower" lifestyle. Natural recycling is at its best in the water cycle of nature, now pushed to the point of breakdown by unforeseen problems such as global warming. So, we propose to extend existing advice of 3 Rs (Reduce, Reuse, and Recycle) to 4 Rs (Restore, Reduce, Reuse, and Recycle), stressing the need to Restore natural recycling processes. For example, in addition to highly needed artificial recycling or purification of water, serious efforts must be made to preserve and restore nature's water cycle. Some priority steps for these 4 Rs have been explored. However, implementation of the concepts through public initiative and government laws is as important as the development of recycling technologies that are economically viable and safe. Here, we look for a solution to the health hazards of ship breaking and recycling of some electronic items and batteries. Fuels burnt in industry and transport sectors generate pollution and green house gases (GHGs). Increased global warming due to GHGs has endangered water cycle of nature threatening life on earth. Planting more trees, developing green technologies and banking on green energy sources including nuclear power are discussed as best possible ways to combat global warming and pollution. Recycling of spent nuclear fuel and thorium fuel cycle are discussed as long term solution to growing energy need of India and the rest of the world.

1.1 INTRODUCTION

Need for recycling of different materials including water is becoming more acute in last few decades or more due to several factors, triggered mainly by population explosion and modern lifestyle. Modern lifestyle, based on gadgets and electricity, increasingly exploits world's natural resources, land and materials. Materials include fuels (to produce energy) and water. Used resources mostly turn into waste. Even a few hundred or hundred years ago, mother earth had enough space and capacity to accommodate or recycle the wastes, generated by a smaller population with a "lower" lifestyle. It is no longer true, and our solid, liquid and gaseous wastes are affecting life on earth.

To sustain life in our planet properly, world must adopt, what we will call 4 Rs (Restore, Reduce, Reuse, and Recycle). These 4 Rs imply: (1) Restore (natural recycling), (2) Reduce (use of different materials and energy), (3) Reuse (materials and energy), and (4) Recycle (materials). Last 3 Rs (Reduce, Reuse, and Recycle) are well acknowledged,[1,2] and the first R is being proposed by us as the top priority step to further conserve nature and hence the natural resources. Adopting 4 Rs will automatically reduce waste.

Natural fuels (petroleum, natural gas, and coal) meet bulk of the energy need worldwide. These are not being freshly produced in nature, and likely to be exhausted in near future. Moreover, burning of these fuels produces green house gases (GHGs) and other pollutions. So, search for environment friendly energy sources have been intensified in last few decades. Renewable energy sources such as solar power, wind power, and wave power are welcome. But these have limitations in producing power in large scale and require correspondingly large areas. Recovery of waste energy involves the intelligent concept of recycling of energy. But it will generate small amounts of power from situations such as vibrations, walking, etc. So, these sources will only partially meet the growing energy demand of the world. We project nuclear energy with proper nuclear fuel cycle and safety measures[3,4] as the best green energy option. The words "nuclear reactor" usually refers to fission type nuclear reactor. Energy production by controlled nuclear fusion is definitely a superior option. But its discussion can wait, as its development is still at a preliminary stage.

1.2 RESTORING NATURAL RECYCLING PROCESSES

Carbon-, oxygen-, nitrogen-, and water cycle in nature help sustain life on earth. Natural recycling is at its best in the water cycle of nature. It ensures removal of excess salt and purification of seawater by evaporation under sunshine, and its return to land masses and mountain glaciers (such as Gangotri in the Himalayas) as rain and snow. Increasing emission of GHGs, such as carbon dioxide from industry and transport, is trapping infrared emissions from earth to cause extra warming, referred to as global warming.[5] GHG generated in one region warms up all regions due to air flow, making it a global effect. Global warming pushes the snowline up, shrinking the glaciers and reducing their contribution to the rivers.[5] Mouth of the Gangotri Glacier is moving to higher altitude, change in a decade being visible even to the visitors. This continuing shortening of different glaciers on a global scale will drastically reduce the water flow in rivers such as Ganga and lead to a global water crisis. Water crisis is already intense in parts of Australia.

To restore the natural water cycle, the world has to arrest global warming by reducing GHG emission. This has to be done primarily by shifting to green energy and green chemistry[6] technologies in industry and by possible trapping of emitted GHG. Planting and growing more trees will convert emitted carbon dioxide into oxygen and assimilated carbon. These two products, oxygen and wood or green plants are precious gains for mankind and living beings in general.

Buildings, concrete or similar pavements and roads, take rain-water into city-drains and prevent recharging of ground water, further depleted by huge tube wells. Loss of groundwater leads to many problems. So, ground water recharging from rainwater needs to be arranged in these concrete jungles of cities and towns. Man's disturbance to natural recycling of groundwater can thus be partially undone.

Still about 20% of India has forest cover, and it is not too late to start utilization cum conservation of certain forests including waste-lands. A scientifically planned cut and plant policy can effectively recycle wood, as is done in many countries. Sunita Narain[7] feels "We need … to make money on our forest wealth … without destroying the forests." That needs a positive change in Government outlook and policy.[7] Recycled (implying replanted) wood will be renewable and better than metals and plastics in many applications.

1.3 REDUCING USE OF DIFFERENT MATERIALS AND ENERGY

Minerals, stones such as marble and granite, and conventional fuels are not going to last forever. Fresh water is becoming rare day by day. So, to sustain our civilization, logically needed minimum amount of these items should be used without any wastage. This reduces the load on recycling and reuse that are difficult, costly, and often inefficient. Conservation of fuels and power has the additional benefit reducing global warming and environmental hazards. Walking and cycling, for shorter distances, will save fuel and health.

Let us stress that waste of filtered water, either as overflow from overhead tanks or from street taps has to be avoided. Waste of pumped irrigation water must also be stopped. City authorities in advanced countries such as Germany tax not only the water supplied to a house but also the waste water draining out of the house. This checks wastage of water in the house.

1.4 REUSE AND RECYCLING OF MATERIALS

1.4.1 GENERAL

One kind of reuse of materials involves using of reusable items instead of disposable items (e.g., disposable cups, paper napkins, and disposable plastic bags). One should sell or donate used but still usable items (unwanted gadgets, tools, etc.). Reuse will include stitching of "kantha," a soft cotton mattress made in some parts of India, from multiple layers of old cotton clothes ("saris," in particular). Used cotton cloth is usually softer than freshly manufactured cotton cloth. To overcome faded color and poor strength of old saris, the kantha is innovatively stitched with colored threads and intricate designs. So, the kanthas are colorful, soft and absorbent—ideal for childcare. A "Bottle Boat" reuses hundreds of 1 or 2 L plastic water bottles in empty but capped condition. Refilled with ordinary water, such bottles have been reused also to construct temporary huts, benefitting from the high specific heat of water. People recover phosphorous, due to its low world stock (to last 50–100 years at the current rate of use), and costly elements such as gold and lead from wastes.

Recycling recovers from waste materials, by chemical and/or physical processes, the original useful material or useful new products. Many materials can be recycled—reducing the recycling cost being most important. Being similar in effect, composting or other reuse of biodegradable wastes—such as food or kitchen waste is considered recycling. A few selected cases of reuse and recycling will be highlighted in Section 1.4.3 without any attempt to be exhaustive.

1.4.2 RECYCLING OF NUCLEAR FUELS

As of May 2014, 435 nuclear reactors in 30 countries are generating electricity, and 72 new nuclear plants are under construction in 15 countries (vide: http://www.nei.org/Knowledge-Center/Nuclear-Statistics/World-Statistics). About 14.2% of global electricity is produced from these nuclear power reactors.[8] The website shows a higher concentration of nuclear reactors in the United States, Europe, and Japan, countries trying to restrict development of nuclear energy in developing countries. France generates 75% of its electricity from nuclear power in 59 reactors. France is world's largest net exporter of electricity due to its very low cost of generation and earns € 3 billion per year from this export. It has safely recycled nuclear fuel for decades.

Natural uranium (U) is a mixture mainly of two isotopes: ~99.3% of U-238 and ~0.7% of U-235. The isotope U-235 can, under certain conditions, be split by a slow neutron yielding a lot of energy and few neutrons. It is "nuclear fission" of fissile U-235, as shown in one of the three diagrams that follow. The neutrons, under certain conditions, can split more U-235 atoms and sustain a fission chain reaction. It can either be a controlled-chain reaction (as in a nuclear reactor with control rods) or an uncontrolled-chain reaction (as in a nuclear bomb). Fission energy of 200 MeV per splitting implies that 1 kg of natural uranium (with 0.7% U-235) gives 50,000 kWh energy, which is very high compared to ~3 kWh energy, obtained by burning 1 kg of coal. Let us go to some details.

Fission chain reaction:

Heavy fission product (A = 139, etc.)

n + U-235 ⇒ + ↗ n + U-235 chain reaction

Light fission product (A = 095, etc.) + 200 MeV + few neutrons

↘ Absorption or loss of n

U-238, present in spent fuel, falling under nuclear waste, can be "recycled" by proper neutron irradiation into U-239, undergoing beta decay first (1/2 life ~ 23.5 min) into Np-239 and then (1/2 life ~ 2.36 day) into Pu-239, a nuclear fuel. This and production of fissile U-233 by reactor irradiation of Th-232 (present in monazite) are shown below.

Production of fissile Pu-239 from U-238:

$$n + \text{U-238} \longrightarrow \text{U-239} \longrightarrow \beta + \text{Np-239} \longrightarrow \beta + \text{Pu-239}$$

Production of fissile U-233 from Th-232:

$$n + \text{Th-232} \longrightarrow \text{Th-233} \longrightarrow \beta + \text{Pa-233} \longrightarrow \beta + \text{U-233}$$

Huge stock of monazite sand in Kerala coast of India makes above-mentioned production of U-233, a nuclear fuel, by reactor irradiation of nonfissile Th-232, very important for India. It can overcome the shortage of uranium ore in India. Fuel Assembly is removed from a reactor after it has reached the end of its useful life (~18 to ~36 months depending on the design), and is called spent fuel. It can be (1) either reprocessed to recover and recycle (Fig. 1.1) the usable portion of it before final storage and safe disposal or (2) subjected to long-term storage and safe disposal without reprocessing. Recycling under option (1) reduces radioactivity level in the final waste, and this option generates more power from the fuel.

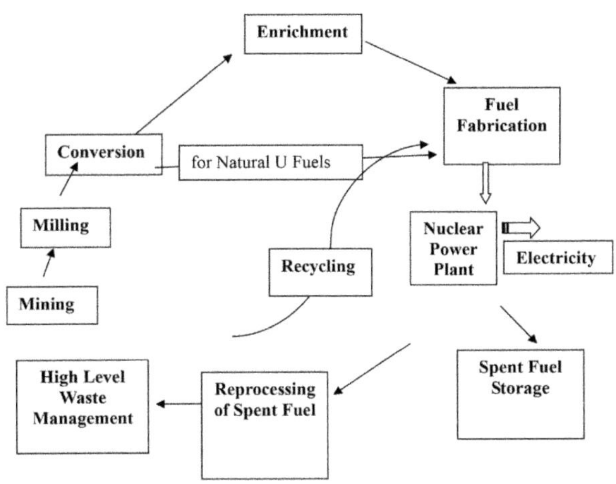

FIGURE 1.1 For the highly radioactive spent nuclear fuel from the "nuclear power plant" there are two options: (1) "reprocessing" to "recycle" remaining useful components to "fuel fabrication" plant and send the rest to HLW management and (2) safe storage in "spent fuel storage." The reprocessing route closes the nuclear fuel cycle with advantages discussed in the text.

Reprocessing and recycling are backbones of India's long-term nuclear power program that also involves pioneering utilization of her thorium reserve through fast breeder reactors (FBRs). This program is a closed nuclear fuel cycle consisting of three stages:

I: Natural uranium fuelled pressurized heavy water reactors, generating electrical power and Pu-239 in the spent fuel; reprocessing of spent fuel.
II: FBRs utilizing mixed oxide (MOX) fuel based on above-recovered Pu-239 (29%) and Th-232 of monazite sand, for example. FBRs produce or breed more fuel than it consumes so that more reactors can be set up.
III: Advanced nuclear power systems for optimal utilization of thorium.

1.4.3 RECYCLING CIVIC WASTES—SCOPE AND ETHICS

Compared to the use and throw practice of the upbeat Western world of first half of last century, countries such as India has a long tradition of repair, reuse, and recycling, one example (kantha) being already cited. However, these traditional practices have not been significantly upgraded in recent decades in India, while the West has moved forward along 3 Rs.

Separation of components from mixed garbage is costly and difficult. So, separation of waste components at source is implemented in modern cities and towns—with separate collection for waste paper, plastic wrappings, and other garbage. Different types of glass are collected in different bins.

As much as 50% of waste in dustbins in United Kingdom could be composted, and this figure must be higher in India. Dry leave component of street garbage in some Indian cities is disposed by burning instead of composting. This is a bad practice on two counts—loss of the compost fertilizer and creation of GHGs and smoke pollution. Up to 60% of the rubbish in a dustbin is recyclable in United Kingdom. Some European Union countries actually recycle over 50% of their waste. In fact, recycling is very profitable in many cases. One recycled tin can saves enough energy to power a TV for 3 h. Recycling has to be implemented by public initiative and government laws.

A few common things[2] that common people should know about recycling right from childhood are not well advertised in less organized countries such as India. UK websites[2] teach:

> "Paper fibers cannot be recycled if they are contaminated with food. Here are a few tips:
> Put greasy wrappers into your compost/main rubbish.
> Tear out contaminated portions (e.g., a cheesy pizza box lid), and recycle the clean remainder.
> Use tissues as compost, as their dense fibers make them unsuitable for paper recycling."

Moreover, we must "remove paper clips, staples, and plastic envelope windows from paper"[2] before giving it for recycling to make recycled paper. "Removing the cap before putting any plastic container into a recycling bin is important, as the cap and container are often made from different polymers, and therefore, have different melting points. Too many caps will contaminate the load during recycling." is also advised.[2] Mixing glass with other wastes makes it a hazard for ever. Everyone must note that glass is 100% recyclable and can be used again and again. So, glasses must go to recycle bins, and there should be different bins for glasses of different colors.

Present e-waste includes discarded computer equipment (monitors, printers, hard drives, and circuit boards). It contains toxic substances—up to 2 kg of lead in a personal computer. Such items should never be disposed as household rubbish. But many countries do not have any arrangement and rule for their disposal. So, "e-waste often ends up in the developing world," and the UN's Environment Program is alarmed by "the amount of electronic goods which is improperly disposed off overseas" by the richer nations.

1.4.4 ARTIFICIAL RECYCLING OF WATER FROM THE SEA

Over 97.5% of the earth's water is saline. There will be increasing scarcity of fresh water in different parts of the world in near future in view of global warming and other factors. One large-scale solution is electrically operated desalination cum purification plants to artificially recycle seawater in coastal areas and brackish water in semi-desert areas. As of now, more than 16 billion gal of desalinated water are produced daily, mostly in the Persian Gulf countries. Since desalination, by either multistage flash distillation or reverse osmosis or a hybrid of the two, requires huge electrical power,[9] it is often coupled to a small nuclear power plant[8] to reduce the cost. Cost reduction is to something between one-third and half for a project sponsored by Tunisia.

1.4.5 NO WASTE DEVICES

Fuel cells, producing electricity and water by an electrochemical combination of hydrogen (H_2) and oxygen (O_2) gases, involve green technology. If the fuels (H_2 and O_2) are produced by a green technology such as electrolysis of water using green electricity like that from nuclear energy, this combination is a no waste device to produce a portable energy source. Introduction of fuel cell cars by Toyota and others led us to point out that here the exhaust or output is pure water, not a waste, and the fuels can be produced from water. This is also a recycling of water, in fact into a purer form. For higher efficiency energy management, Toyota is combining hybrid technology (powered by petrol and battery) and fuel cell technology to power the vehicle. At a steady cruising speed, the motor is run by fuel cell energy. When more power is needed, for example, during sudden acceleration, the battery supplements the fuel cell's output. But at low speeds when less power is required, the vehicle is operated by the battery. During slowing and breaking, the motor acts as an electric generator to capture braking energy, which is stored in the battery. It is recycling of energy that is otherwise lost as heat and sound.

1.5 SALIENT POINTS AND CONCLUSION

Different nodal authorities such as United States Environmental Protection Agency recommends "Reduce, Reuse, and Recycle," often called "3 Rs," for saving money, energy, environment and, above all, natural resources. Here, one assumes the huge recycling power of nature to back up our efforts as before. Unfortunately, that recycling power of nature is being seriously weakened by modern civilization through factors such as global warming. So, we add another R that stands for "Restore" implying "Restore Natural Recycling

Processes" as the top priority step, and advocate "4 Rs" (Restore, Reduce, Reuse, and Recycle).These four Rs, discussed in this work, can be outlined here. Regarding the first R, worldwide reduction of per capita contribution to global warming and of all kinds of (unintentional or intentional) foul play with nature is obviously the surest step to "restore natural recycling processes." Rainwater recycling restores groundwater. One must also "Reduce" all nonessential and wasteful consumption of materials or energy. "Reuse" of used objects or materials conserves materials and labor. If reuse is not possible, useful items can be "Recycled," from the used materials. Recycling of seawater by novel techniques such as desalination plants running on green electricity from seaside nuclear reactors must be more seriously explored.

Implementation of reuse and recycling concepts is as important as the development of recycling technologies that are economically viable and more efficient. We further stress that reuse and recycling have to be implemented by public initiative and government laws. However, several recycling industries (ship breaking, recycling of batteries, and electronic goods) are hazardous. So, research on safer recycling processes, clear laws, and implementation of the laws are needed. Pollution in Africa, Asia, and South America, due to hazardous chemicals and heavy metals, mainly from disposal of computers and other electronic goods by the advanced countries, has been reported.[2] A complex mixture of the toxic materials with other materials makes recycling of computers very difficult.

Electrical power is mostly produced by burning a conventional fuel resulting in a large "carbon footprint" and global warming. This global warming weakens natural water recycling, one of the finest and largest recycling processes. So, burning such a fuel for electricity drastically reduces two important materials—the fuel and water, while electricity from nuclear reactors avoids both flaws. As much energy as can be economically and safely produced from other green sources is always welcome. Rest coming from safely managed nuclear power reactors has to be encouraged.

Growing energy demand due to growing industry and population, shortage of conventional and uranium fuels within India, and difficulty of importing these freely put India in a unique position to go for reprocessing-based closed cycle nuclear power, and utilization of her thorium reserve with help of breeder reactors. So, these technologies are being mastered by India now to solve the energy problem, unique to India, and to be faced by other nations at a later stage. India has been recycling spent fuel from its reactors almost from the beginning of her nuclear program, while some other nations stored the highly radioactive spent fuel without any reprocessing. These nations later started work on reprocessing, to close the nuclear fuel cycle.[10]

Recent upgradation of hydrogen–oxygen fuel cells from demonstration model status to powerhouse for real cars generates the hope that this device will find wider and higher power applications. But commercial success will depend on whether electricity to produce its fuels (hydrogen and oxygen) can be cheap enough. Depending on locality, nuclear energy can be the answer.

It is hoped that by practicing 4 Rs, that is, Restore, Reduce, Reuse, and Recycle, life in this beautiful planet will be sustained with undiminished glory.

KEYWORDS

- **recycling**
- **reuse**
- **reduction of waste**
- **restore**

REFERENCES

1. US Environmental Protection Agency. http://www2.epa.gov/recycle/reducing-and-reusing-basics
2. Recycling Guide. http://www.recycling-guide.org.uk/
3. De, U. In *Safety in Fast Breeder Reactors Through Ultrasonic Monitoring—Materials Characterization and Development, Invited Talk, Book of Ext. Abstracts, All India* Seminar on Safety & Disaster Management of Process Industries, WB State Centre, Inst of Engineers, Kolkata 700020, India, March 15–16, 2013.
4. Raj, B. Fast Breeder Programme: An Inevitable Option for Energy Security, March 2, 2013. http://dae.nic.in/
5. De, U. Environmental and Socio-economic Impacts of Climate Change in the Sundarban Delta and the Need for Green Management. In *Knowledge Systems of*

Societies for Adaptation and Mitigation of Impacts of Climate Change, Environmental Science and Engineering; Nautiyal et al., Eds., Springer-Verlag: Berlin/Heidelberg, 2013; pp 601–633. DOI: 10.1007/978-3-642-36143-2_34.

6. Anastas, P.; Warner, J. C. *Green Chemistry: Theory and Practice*; Oxford Science Publications: Oxford, 1998.
7. Narain, S. When Planting Trees Is a Curse. *CSE's Fortnightly News Bulletin*, March 4, 2014.
8. Nuclear Energy Institute, 1201 F *St., NW, Suite 1100, Washington*, DC 20004-1218. *Water Desalination*. http://www.nei.org/Knowledge-Center/Other-Nuclear-Energy-Applications/Water-Desalination
9. NRC-WSTB. 2008. *Desalination: A National Perspective*. Committee on Advancing Desalination Technology, Water Science and Technology Board, Division on Earth and Life Studies, National Research Council of the National Academics. The National Academies Press: Washington, D.C; 2010.
10. U.S. Dept. of Energy (Office of Environmental Management). *Closing the Circle on Splitting of the Atom*; U.S. Dept. of Energy: Washington, D.C., U.S.A.; 1995.

CHAPTER 2

CHEMICAL RECYCLING OF POLYURETHANE FOAMS: A REVIEW

AJAY VASUDEO RANE[1*], ABITHA V. K.[2], K. KANNY[1], and SABU THOMAS[2]

[1]*Department of Mechanical Engineering, Durban University of Technology, Durban, South Africa*

[2]*School of Chemical Sciences, Mahatma Gandhi University, Kerala, India*

[*]*Corresponding author. E-mail: ajayrane2008@gmail.com*

CONTENTS

ABSTRACT

The fate of polymers in nature seems to present an inconsistent problem at times; most polymers are designed and manufactured to resist environmental degradation (photodegradation, hydrolysis, oxidation, biodegradation, etc.) but are used for protective and/or structural purposes. Increasing awareness of solid waste management problems has led to the insist for polymers that do not have an unsafe collision on the environment during any part of their life cycles. The corresponding recycling of carbon, hydrogen, and nitrogen elements as well as of energy through reprocessing and biological and chemical conversion should be taken advantage of in polymer design. Recycling is a crucial area of research in Green Polymer Chemistry. Various developments in recycling are driven by environmental concerns, interest in sustainability and desire to decrease the dependence on nonrenewable petroleum-based materials. Polyurethane foams (PUF) are widely used due to their light weight and superior heat insulation as well as good mechanical properties. As per survey carried out by Polyurethane Foam Association, 12 metric tonnes of PUF are discharged during manufacturing and/or processing and hence recycling of PUF is necessary for better economics and ecological reasons. Hence we propose a write up describing various methods to recycle PUF.

2.1 INTRODUCTION

2.1.1 GREEN TECHNOLOGY

> *"The good solve today's problems, the great design today's world to avoid tomorrow's issues."*

Green Chemistry is not a novel branch of science; it is a new-fangled philosophical approach, all the way through application and extension of the principles of Green Chemistry is capable of to contribute to sustainable developments. Chemistry was once viewed as a ground of innovation resulting in break troughs and modern convenience but now is viewed by many as a fouling the planet earth.[1,2] Green Chemistry encourages environmentally conscious behavior and reduces and prevents pollution and destruction of planet; it involves process of recycling and use of renewable resources for energy. Green Chemistry is the design of chemical products and processes that reduce or eliminate the use and/or generation of hazardous substances, it creates new ways to make desired materials via different feedstock's, different pathways, identifies desired performance characteristics and create new materials. Green Chemistry provides a technical solution to many environmental problems. Green Chemistry is effective due to design stage, an effort starting at the molecular level lets you design out the hazardous properties and design environmentally appropriate features. It is upto us to live up to our greatness in order to dream, design, and realize a truly sustainable world. The modern understanding of sustainability began with the United Nations World Commission of Environment and Development's report *"Our Common Future"* also known as the *"Brundtland Report."* The Brundtland Commission described Sustainable Development as *development that meets the needs of the present without comprising the ability of the future generations to meet their own needs.*[14] The definition does not give us many clues or supply much practical guidance as how to implement sustainable development or move toward sustainable development activities, but it does provide us with a powerful aspiration. It has been upto society collectively and upto us as individuals to develop guidance and tools that will help us to design systems and processes that have the potential to achieve the type of development described in the definition of sustainable development. By using the term *phase triple bottom line* (Fig. 2.1) Elkington tried to highlight the need to consider the intricate interrelationship among *environmental, social, and economic aspects* of human society and the world. In a way, sustainable development can be seen as a very delicate balancing act among the three factors and not always with a strong one to one relationship.[6,8]

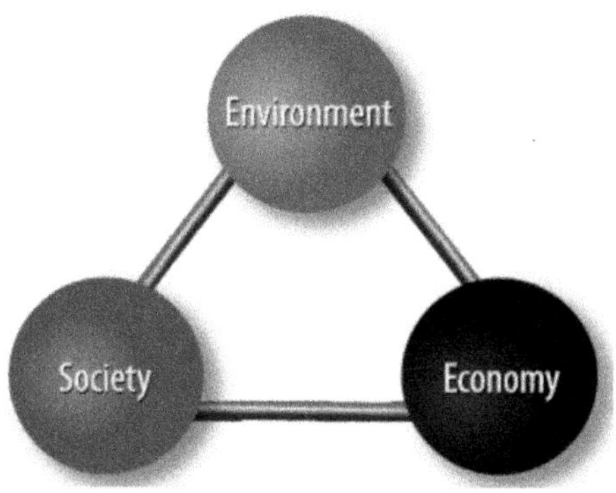

FIGURE 2.1 Phase triple bottom line.

Coupling technological innovations with policies and strategies is not sufficient to bring about the kind of change that is needed to lead society toward more sustainable practices. To influence people, to make continued changes in the face of pervasive, persistent, and prolonged resistance, we need to use all the spheres of influence at our disposal. Influencing is an essential part of driving sustainability; as we have seen that the achievement of greater sustainability would not happen through people working in isolation and requires the dealings and partnership of many years (Fig. 2.2).[14]

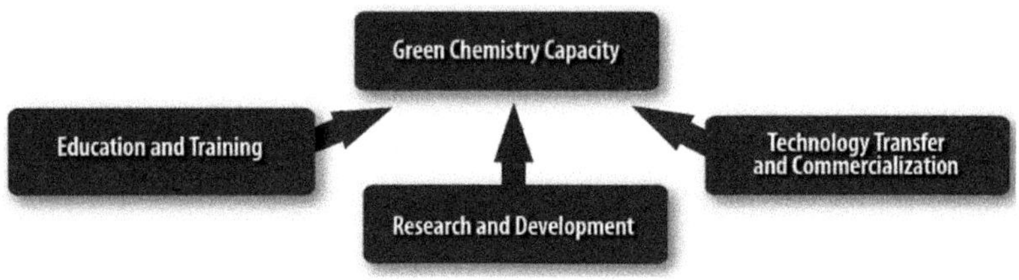

FIGURE 2.2 Development of green chemistry.

2.1.2 THE 12 PRINCIPLES OF GREEN CHEMISTRY

The most important aims of Green Chemistry were defined in 12 principles. The number 12 is highly significant and symbolic like the 12 months of the year as the complete sum of the most important things that we have to do to accomplish a multiple task. Green Chemistry has to cover a broad spectrum of chemical and technological aspects in order to offer its alternative vision for sustainable development.[16] Green Chemistry had to implement fundamentals ways to reduce or to eliminate environmental pollution through dedicated, sustainable prevention programs. Green Chemistry must focus on alternative, environmentally friendly chemicals in synthetic routes but also to increase reaction rates and lower reaction temperature to save energy. Green Chemistry looks very carefully on reaction efficiency, use of less toxic solvents, minimizing the hazards of feed stocks and products, and reduction of wastes. Anastas, an organic Chemist working in the office of Pollution Prevention and Toxins at EPA, and John C Warner developed the 12 principles of Green Chemistry in 1991. These principles can be grouped into "*Reducing risk*" *and* "*Minimizing the Environmental Footprint*."[39]

- **Principle no. 1**—*Prevention*: It is better to prevent than to clean or to treat afterwards (waste or pollution). This is a fundamental principle. The preventative action can change dramatically many attitudes among the scientists developed in the last decades. Most of the chemical processes and synthetic routes produce waste and secondary toxic substances. Green Chemistry can prevent waste and toxic by-products by designing the feed stock's and the chemical processes in advance and with innovative changes.
- **Principle no. 2**—*Maximize synthetic methods, Atom Economy*: All synthetic methods until now were wasteful and their yields were between 70% and 90%. Green Chemistry supports that synthetic methods can be designed in advance to maximize the incorporation of all reagents used in chemical processes into the final product. The concept of Atom Economy was developed by Barry Trost of Stanford University. It is a method of expressing how efficiently, how a particular reaction makes use of the reactant atoms.
- **Principle no. 3**—*Less hazardous chemical use*: Synthetic methods should be designed to use and generate substances that possess little or no toxicity to the environment and public at large.
- **Principle no. 4**—*Design for safer chemicals*: Chemical products should be designed so that they not only perform their designed function but are also less toxic in the short and long terms.
- **Principle no. 5**—*Safer solvents and auxiliaries*: The use of auxiliary substances such as solvents or separation agents should not be used whenever possible. If their use cannot be avoided, they should be used as mildly or innocuously as possible.
- **Principle no. 6**—*Design for energy efficiency*: Energy requirements of chemical processes should be recognized for their environmental and economic impacts and should be minimized. If possible, all reactions should be conducted at mild temperature and pressure.
- **Principle no. 7**—*Use of renewable feedstock*: A raw material or feedstock should be renewable rather than depleting whenever technically and economically practicable. For example, oil, gas, and coal are dwindling resources that cannot be replenished.
- **Principle no. 8**—*Reduction of derivatives*: Use of blocking groups, protection/deprotection, and temporary modification of physical/chemical processes is known as *derivatization*, which is normally practiced during chemical synthesis. Unnecessary derivatization

should be minimized or avoided. Such steps require additional reagents and energy and can generate waste.

- **Principle no. 9**—*Catalysis*: Catalytic reagents are superior to stoichiometric reagents. The use of heterogeneous catalysts has several advantages over the use of homogeneous or liquid catalysts. Use of oxidation catalysts and air is better than using stoichiometric quantities of oxidizing agents.
- **Principle no. 10**—*Design for degradation*: Chemical products should be designed so that at the end of their function they break down into innocuous degradation products and do not persist in the environment. A life cycle analysis (beginning to end) will help in understanding its persistence in nature.
- **Principle no. 11**—*Real-time analysis for pollution prevention*: Analytical methodologies need to be improved to allow for real-time, in-process monitoring and control prior to the formation of hazardous substances.
- **Principle no. 12**—*Inherently safer chemistry for accident prevention*: Substances and the form of a substance used in a chemical process should be chosen to minimize the potential for chemical accidents, including releases, storage of toxic chemicals, explosions, and fires.

- *Five main foci emerge from these 12 principles, namely,*[15]

  1. less,
  2. safe,
  3. process-oriented,
  4. waste-reducing, and
  5. sustainable.

- *All five key words could be grouped and written as*[15]*:*

  1. Uses fewer chemicals, solvents, and energy.
  2. Have safe raw materials, processes, and solvents.
  3. Process should be efficient, without waste, without derivatization, and should use catalysts.
  4. Waste generated should be monitored in real time and should degrade.
  5. All chemicals, raw materials, solvents, and energy should be renewable or sustainable.

The history of polyurethanes (PU) started in the 1930s in Germany when Otto Bayer proposed using diisocyanate and diol for preparation of macromolecules. The first commercial PU, based on hexamethylene diisocyanate and butanediol, had similar properties to polyamides and is still used to make fibers for brushes. However, fast growth of the production and expanded application range started in the 1950s with the building of toluene diisocyanate (TDI) and polyester polyol plants for flexible foams. However, the real jump in applications came with the introduction of polyether polyols in foam formulations. Today, PU are about the sixth largest polymer by consumption, right behind high-volume thermoplastics, with about 6% of the market. The largest part of the urethane application is in the field of flexible foams (about 44%), rigid foams (about 28%), while 28% are coatings, adhesives, sealants, and elastomers applications. PU are a broad class of very different polymers, which have only one thing in common—the presence of the urethane group (Fig. 2.3).

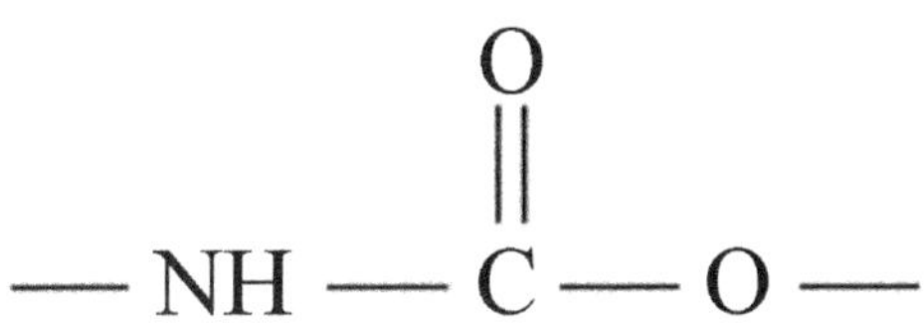

FIGURE 2.3 Urethane group.

The number of these groups in a polymer can be relatively small compared with other groups in the chain (e.g., ester or ether groups in elastomers), but the polymer will still belong to the PU group. Varying the structure of PU, one can vary the properties in a wide range. PU are formed by reaction of polyisocyanates with hydroxyl-containing compounds, most frequently during processing. By selecting the type of isocyanate and polyols, or combination of isocyanates and combination of polyols, one can tailor the structure to obtain desired properties. For this, however, it is necessary to know the relationship between the structure and properties. The flexibility to tailor the structure during processing is one of the main advantages of PU over other types of polymers. Urethane groups form strong hydrogen bonds among themselves and with different substrates. Strong intermolecular bonds make them useful for diverse applications in adhesives and coatings, but also in elastomers and foams.[18] One of the great advantages of PU arises from the high reactivity of isocyanates, which can react with a number of substances having different functional groups. This allows polymerization at relatively low temperatures and in short times (several minutes). One group of polymers, which is conditionally treated as urethanes, is polyurea, because urea is often formed during urethane production. Urea is formed in the reaction between isocyanates and amines. The urea group is similar to the urethane group, except that it has two –NH– groups (Fig. 2.4), and can form more hydrogen bonds than the urethane group.

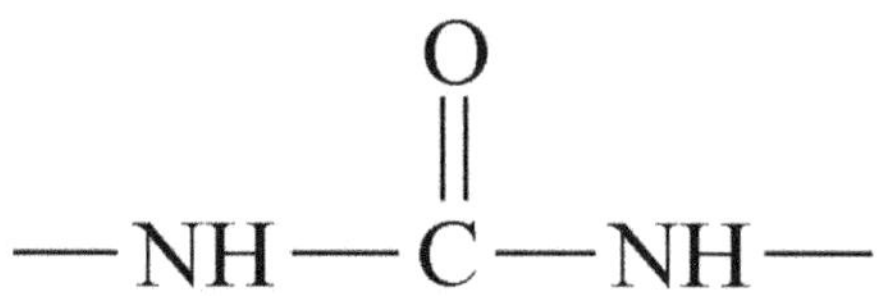

FIGURE 2.4 Urea group.

2.1.3 SYNTHESIS OF PU

The basis of PU chemistry is the high reactivity of isocyanates. They react under mild conditions with all compounds that contain "active" hydrogen atoms. These are mainly alcohols (OH group) *but also* amines. When the isocyanate group (NCO) reacts with alcohols, amines, carboxylic acids, and water, urethane, urea, and amide linkages are formed (Fig. 2.5). PU are usually classified as engineering polymers characterized by the presence of the carbamate group (–O–CO–NH–). The major route for the preparation of PU is the reaction between a diisocyanate and a hydroxyl-rich compound with at least two hydroxyl groups, according to following reaction.

FIGURE 2.5 Polyurethane synthesis by reaction of diisocyanate and diol.

In Figure 2.5, R can be an aliphatic, cycloaliphatic, or polycyclic group. Of the different diisocyanate used, the most common are methylene-4,4'-diphenyldiisocyanate, toluene 2,4-diisocyanate, and hexamethylene diisocyanate. Likewise, low-molecular weight hydroxyl-terminated polyesters or polyether's are typically employed as dihydroxyl compounds.[18]

2.1.4 RECYCLING METHODS

Recycling is a crucial area of research in Green Polymer Chemistry and developments in recycling are driven by environmental concerns, interest in sustainability, and desire to decrease dependence on petroleum-based materials. Sustainability refers to the development that meets the needs of present without comprising the ability of future generations. Polyurethane foams (PUF) are widely used due to their light weight and superior heat insulation as well as good mechanical properties. However, large quantities of PUF are discharged during manufacturing/processing as a waste and hence recycling of PUF is necessary. Recycling allows the wastes to be reintroduced into the consumption cycle, generally in secondary applications because in many cases, the recycled products are of lower quality than the virgin ones. Recycling must be applied only when the amount of energy consumed in the recycling process is lower than the energy required for the production of new materials. Plastics can be recycled using two different approaches: mechanical and feedstock recycling. In the first case, the plastics are recycled as polymers, whereas in the second, plastic wastes are transformed into chemicals or fuels.[13]

PU recycling methods can be categorized into four groups, namely, primary, secondary, tertiary, and quaternary recycling. There is also a so called "zero-order" recycling technique, which involves the direct reuse of a PU waste material. There are many other terminologies used for these recycling categories.[25]

- **Primary recycling**: Primary recycling, also known as reextrusion is the oldest way of recycling PU. It refers to the "in-plant" recycling of the scrap materials that have similar features to the original products. This process ensures simplicity and low cost, but requires uncontaminated scrap, and only deals with single-type waste, making it an unpopular choice for recyclers.
- **Secondary recycling**: Secondary recycling, also known as mechanical recycling, was commercialized in the 1970s. It involves separation of the polymer from its contaminants and reprocessing it to

granules via mechanical means. Mechanical recycling steps include sorting and separation of wastes, removal of contaminants, and reduction of size by crushing and grinding, extrusion by heat, and reforming. The more complex and contaminated the waste is, the more difficult it is to recycle mechanically. Among the main issues of secondary recycling are the heterogeneity of the solid waste and the degradation of the product properties each time it is recycled. Since the reactions in polymerization are all reversible in theory, the employment of heat results to photooxidation and mechanical stresses, causing deterioration of the product's properties. Another problem is the undesirable gray color resulting from the wastes that have the same type of resin, but of different color.

- **Tertiary recycling**: Tertiary recycling, more commonly known as chemical recycling, involves the transformation of the PU polymer chain. Usually by means of solvolytic chain cleavage, this process can either be a total depolymerization back to its monomers or a partial depolymerization to its oligomers and other industrial chemicals. Since PU is a polymer with functional ester, ether groups, it can be cleaved by some reagents such as water, alcohols, acids, glycols, and amines. Also, PU is formed through a reversible polycondensation reaction, so it can be transformed back to its monomer or oligomeric units by pushing the reaction to the opposite direction through the addition of a condensation product. These low-molecular products can then be purified and reused as raw materials to produce high-quality chemical products. *Among the recycling methods, chemical recycling is the most established and the only one acceptable according to the principles of "sustainable development," defined as development that meets the needs of present generation without compromising the ability of future generations to meet their needs (World Commission on Environment and Development, 1987), because it leads to the formation of the raw materials (monomers) from which the polymer is originally made. In this way, the environment is not surcharged and there is no need for extra resources for the production of PU.*[14]

Chemolysis is true depolymerization applicable to the recycling of PU and other polyaddition materials as well as to condensation polymers such as polyesters (e.g., PET) and polyamides (e.g., nylon). In this type of treatment, the molecules are broken down into smaller building blocks, which may then be reassembled into polymers suitable for use in quality applications similar to those for which the original components were employed. Because it delivers high-grade products that largely retain their original properties and functionality, chemolysis offers an attractive alternative to mechanical recycling and the recovery of petrochemical feedstock's or energy.

For chemolysis of PU (Fig. 2.6), it is preferable to process feedstock of known composition in order to obtain consistent and predictable regenerated products. Water (hydrolysis), glycols (glycolysis), and amines (aminolysis) typically serve as reagents to break the urethane bonds. The resulting liquid can be used as such, or the individual components separated. Several options exist for further reprocessing. These may involve purification and chemical processing before use in PU applications.[13]

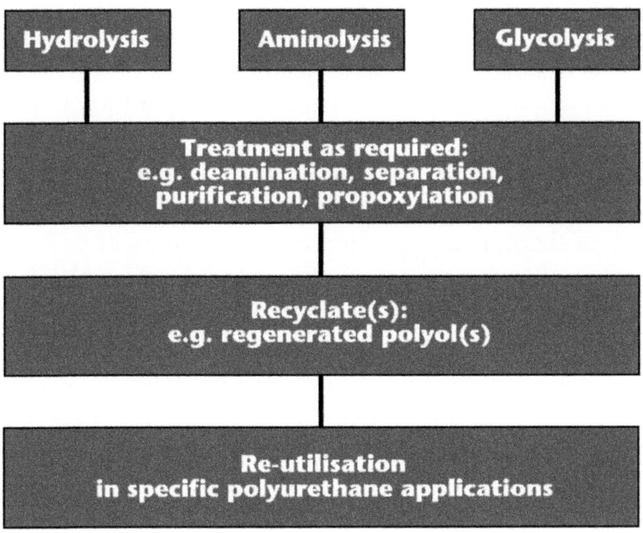

FIGURE 2.6 Chemolysis of polyurethanes.

- **Hydrolysis, Aminolysis,** and **Glycolysis** is a process whereby the PUF is reacted with water under pressure at elevated temperature. Hydrolysis produces the original polyether polyols together with diamines, which are the hydrolysis products of the original diisocyanate. The various components are then separated in order to permit their reprocessing and reuse.

- **Aminolysis:** In aminolysis, the PUF is reacted with amines such as dibutylamine, ethanolamine, diethanolamine (DEA), lactam, or lactam adduct under pressure at elevated temperatures. Aminolysis is still at the research stage and not much has been carried out.
- **Glycolysis** is a process wherein the PUF is reacted with diols at elevated temperature (200°C) with cleavage of covalent bonds. The high-molecular weight, cross-linked, solid PU are broken down to lower molecular weight liquid products.

Single-phase glycolysis has been optimized by ISOPA members and independent researchers (e.g., catalyst selection, posttreatment for minimization of the aromatic amine content).

Split-phase glycolysis has been developed up to pilot scale for MDI flexible foams. The glycolysis product separates into two phases:

- **Quaternary recycling:** Quaternary recycling represents the recovery of energy content from the plastic waste by incineration. When the collection, sorting, and separation of plastics waste are difficult or economically not viable, or the waste is toxic and hazardous to handle, the best waste management option is incineration to recover the chemical energy stored in plastics waste in the form of thermal energy. However, it is thought to be ecologically unacceptable due to potential health risks from the airborne toxic substances.[25]
- the top layer is a flexible foam polyol which after purification can be used alone to make the same flexible foam again
- the bottom layer, after posttreatment with propylene oxide, can be converted into a high-quality rigid foam polyol.

2.2 RECYCLING OF PUF

2.2.1 HIGHLIGHTS

The usage of PU materials in daily life as well as industrially has continuously increased over the last 30 years. The amount of PU consumed per sector in the industrialized countries has increased over this period, while the generation of foam wastes has grown at a similar rate. Recycling of PUF is now an important field in the polymer industry, not just an activity born under environmental pressure. The recycling processes include industrial operations in which secondary materials are reprocessed and/or monomers recovered for further polymerization; such processes are termed secondary and tertiary recycling. At present, there are three main alternatives for the management of polymeric wastes in addition to land filling: (1) mechanical recycling by melting and regranulation of the used polymers, (2) feedstock recycling, and (3) energy recovery. Consequently, feedstock recycling appears as a potentially interesting approach, based on the conversion of polymeric wastes into valuable chemicals useful as fuels or as raw materials for the chemical industry. The cleavage and degradation of the polymer chains may be promoted by temperature, chemical agents, catalysts, etc. The purpose of this work is to describe and review the different alternatives developed for the chemical recycling of PUF wastes, with emphasis on both the scientific and technical aspects. Solid polymeric materials undergo both physical and chemical changes when heat is applied usually resulting in undesirable changes in the properties of the material. A clear distinction needs to be made between thermal decomposition and thermal degradation. The American Society for Testing and Materials defines thermal decomposition as "A process of extensive change in chemical species caused by heat" and thermal degradation is "A process whereby the action of heat or elevated temperature on a material, product or assembly causes a loss of physical, mechanical or electrical properties."[9]

The alternative methods of polymer recycling consist of the breakdown of the polymer by reaction with certain chemical agents, leading back to the starting monomers. These monomers are identical to those used in the preparation of virgin polymers, hence the polymer prepared from both depolymerization and fresh monomers are expected to have similar properties and quality. According to this approach, polymeric wastes are reintroduced to the market as polymers, as happens in the case of material recycling, but without the loss of resin properties typically associated with the mechanical process. Polymer recycling by chemical depolymerization is the most established method of polymer chemical recycling. Different chemolysis processes have been applied on an industrial scale for several years. The major disadvantage of chemical depolymerization is that it is almost completely restricted to the recycling of condensation polymers and is of no use for the decomposition of most addition polymers, which are the main components of the plastic waste stream. Condensation polymers are obtained by the random reaction of two molecules, which may be monomers, oligomers, or higher molecular weight intermediates, which proceeds with the liberation of a small molecule as the chain bonds are formed. Chemical depolymerization takes place by promoting the reverse reaction of the polymer formation, usually through the reaction of those small molecules with the polymeric chains.[13]

With regard to the chemical recycling of PU, two aspects must be highlighted. PU are used to manufacture durable goods, which mean that it takes several years for these items to be disposed off into the solid waste stream. Moreover, PU contain around 4 wt% of N, which may hinder their recycling by oxidative treatments such as incineration or gasification due to the potential release of significant amounts of NO, in the gaseous effluents.[13] The most important chemolysis methods so far developed to reverse the PU polymerization reaction are glycolysis and hydrolysis. These processes are reviewed next, together with other less widely investigated treatments like aminolysis.

2.2.2 GLYCOLYSIS

A variety of processes for PU degradation by reaction with different glycols has been described in the literature. PU glycolysis is usually carried out with an excess of glycols at temperatures around 200°C and in many cases, working at atmospheric pressure. After several hours of reaction, the PU is completely liquefied and depolymerized, with or without catalysts. The chemistry of the glycolytic reaction has been described by Ulrich et al.[37,38] It involves the transesterification of the carbamate group by addition and reaction with the glycol. These authors verified this scheme by reacting benzyl phenyl carbamate and ethylene glycol at 195°C, a total conversion of the former being observed after 3 h according to the reaction shown in Figure 2.7. Because water-blown PUF contain diarylurea linkages, due to the reaction of diisocyanates with water, they also investigated the glycolysis of N, N-diphenylurea as a model compound. The results obtained clearly showed that the urea linkages are also glycolyzed in these reaction conditions, indicating that the polyols derived from water-blown PUF will present amino end groups.

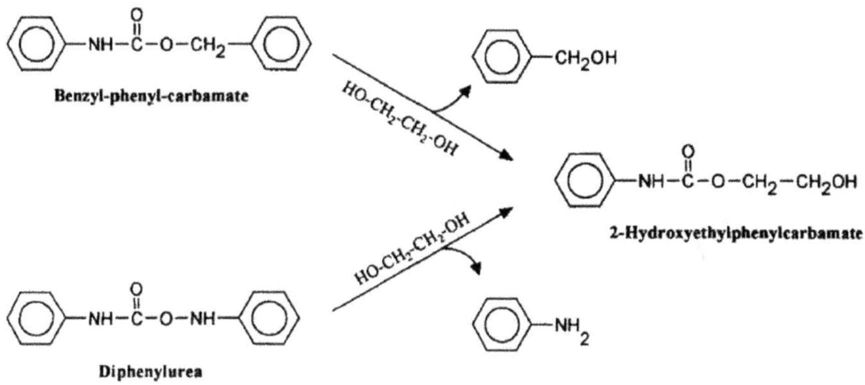

FIGURE 2.7 Glycolysis of benzyl phenyl carbamate and diphenyl urea.

Glycolysis has been successfully applied to rigid and flexible PUF, as well as to polyisocyanurate foam. The polyols obtained by degradation are blended with polyols in a proportion of around 50%. These mixtures can be used in the formulation of new rigid and semirigid foams, which exhibit properties similar to those of completely virgin foams. However, the use of 100% PU degradation polyols is limited by their lower reactivity and higher viscosity compared to virgin polyols.[13] In a recent patent, glycolysis has been reported to be a feasible chemical recycling method for the degradation of PU/polyurea or polyurea wastes. Conventional alcoholysis of these polymeric wastes leads to products with a high content of urea groups and low-molecular weight primary aromatic amines, which limits their application in the isocyanate polymerization reaction because the amine groups are much more reactive toward isocyanates than the alcohol groups of the polyols. These problems have been solved by a two-stage process: reaction of the polymeric wastes at 200°C with a diol or polyol [ethylene glycol, diethylene glycol (DEG), hexanediol, glycerol, etc.] followed by treatment of the alcoholysis products with urea or a carbamic acid ester. It has been found that this second reaction greatly reduces the amine content, so that the final product is suitable for reuse in the isocyanate addition polymerization. Today, several industrial plants based on glycolytic treatment are in operation for the chemical recycling of PU wastes, mainly those generated from the insulation and automotive sectors. Glycolysis process with ethylene glycol, when applied to PU with a high content of urea, that is, of flexible water-blown foams, forms a polyphasic system whose main components are, together, with the triol, the ethylene glycol solution of the products coming from the interaction between the aromatic part of the polymer (carbamates and ureas) with ethylene glycol and a solid phase with urea bonds. In the final product, the free isomeric toluene diamine (TDA) is about 1.52% w/w. The glycolysis process carried out in the presence of hexamethylenetetramine avoids the formation of a solid phase in the product that presents a low free amine content, lower than 100 ppm. This product is safe and easy handling, mainly presents the components required by a material for general use, without separation or purification of the phases, for new PU.[26]

In another study, the degradation of PU wastes was carried out with ethylene glycol, 1, 2-propylene glycol, triethylene glycol, polyethylene glycol, and DEA and the results of the degradation experiments with various EG–DEA mixtures was found to be the fastest at a 1:1 EG–DEA ratio, and it was acceptable even if the ratio of the reactants was 1:6. Glycolysis in this way results in a two-phase mixture at the end of the reaction. To identify the upper liquid phase, industrial polyol standards were used and it was found that this upper liquid phase was the starting polyol in the case of both the flexible foam and the elastomer.[10]

Among the suitable processes, glycolysis, especially in two phases, allows better quality products. In this study, glycolysis reactions of flexible PUF were conducted in "split-phase" with different catalysts, in order to study their activity. DEA, titanium n-butoxide as well as potassium and alcium octoate salts, which are novel compounds for this application, showed suitable catalytic activity. All the catalyzed processes showed appropriate activity compared with the noncatalyzed process, allowing the complete recovery of polyols from the PU matrix. Potassium and calcium octoates, specially the first, have been found advantageous for the recovery process. Potassium octoate leads to the complete degradation of the polymeric chain at low reaction time and the recovery of the polyol in high concentration. These reaction time and concentration are comparable to those obtained with DEA, the most active catalyst studied. Times to reach complete conversion, chemical properties of the polyol phase, and its purity depend on the catalyst employed. The novel catalysts developed have been proved to be a worthy and economic alternative to traditional catalysts.[30,31] In another experiment, glycolysis of flexible PUF in "split phase" was conducted with different glycols, in order to study their activity and select a system to obtain the highest quality recovered polyol. Times required to reach complete conversion, chemical properties of the polyol phase, and its purity depended on the glycol employed. DEG proved to be the most suitable glycol to obtain a high purity in the polyol phase.[29] Glycolysis of integral skin PUF products with DEG/diethanol amine and NaOH have two liquid layers—upper and lower phases, which the upper phase was homogeneous polyol that can be reused in the manufacture of integral skin PUF. Results calculated from gel permeation chromatography (GPC) analysis show that upper phase of product has a structure similar to virgin polyol. Results collected from quality control of recycled polyol maintained that it is applicable in a polyol blend for foam formulation.[32]

In the test of ability, potassium octoate showed a proper performance as catalyst in the glycolysis of PU wastes. An increase in the reaction temperature and catalyst concentration enhances the degradation rate; however, this also negatively affects the process by promotion of secondary reactions and contamination of the polyol phase. As the catalyst concentration is raised, the more polluted is the polyol obtained. The catalyst of 2.2% seems to be the best choice. Temperatures lower than 170°C provide too slow degradation rates, whereas values higher than 200°C also promote secondary decarboxylation reactions. In this interval, an increase in the reaction temperature speeds the degradation up markedly. Related to the glycolysis agent amount, the minimum quantity required to split the phases has been determined, as well as the optimal ratio. Increasing this ratio does not provide any additional improvement of the process and affects economy of the process negatively. As an increment of the foam/glycol ratio larger than 1:1.5 does not produce a significant improvement of the process that proportion can be selected as the best value to carry out the process.[28]

Researchers proposed a system where glycerin was used as a destroying solvent and sodium hydroxide as the catalyst, respectively. In order to study the ability of glycerin as a glycolyzing agent and for selecting a system to obtain high-quality recovered polyol, the effects of various reaction times were investigated and the characterization and comparison of upper and lower phases were performed. Investigation of the obtained results from Fourier-transform infrared (FTIR) spectroscopy indicates that chemical structure of recycled polyol after 1 h and 3 h are too similar to virgin polyol. Because the vibrations in the spectral regions >3300, 2869, 1455, 1373, and 1109 cm^{-1} in virgin polyol are repeated similarly in the recycled polyol. Except the vibrations at 1617 and 1516 cm^{-1} characteristic in the recycled polyol which are corresponding to the bending vibrations of amine N–H bands originated from the starting isocyanates as contaminant which slightly has been dissolved in the upper phase.

Also proton nuclear magnetic resonance (^{1}HNMR) and Carbon-13 (C13) nuclear magnetic resonance (^{13}CNMR) spectra of virgin and recycled polyols regardless the peaks about 6.6–7.2 ppm in ^{1}HNMR and the peaks in the region 110–140 ppm in ^{13}CNMR that are relative to the aromatic by-products, are quite equal. Also GPC results showed the similarity of recycled and virgin polyols. Another observation in our study was the higher hydroxyl value of recovered polyols in comparison with virgin one due to partial solubility of glycerin in the recycled polyol. According to the results, glycerin could be used as glycolysis agent to recovering of high-quality polyol, which can be reused for production of new flexible PUF (Alavi et al., 2007). It is of course possible to convert the waste cold cure flexible PUF to a double-phased product which is completely applicable in new foam formulation as a portion of polyol blend by mixture of solvents containing 95% DEG, 5% DEA, and 1% NaOH as the catalyst is a low amine content mixture that is appropriate for PUF recycling. Not being necessary to remove the contamination from waste foams and providing the systems free of chlorofluorocarbons (CFCs) were some of other advantages of this research. Large-scaled usage of PUF production would make huge amount of wastes. Thus, large-scaled glycol treating would initially run on production waste and is an essential step toward encouraging the development of post consumer waste logistics for flexible foams. A scaled-up version of the process would be combined with a batch reaction as an industrial plant for cold cure flexible PUF recycling, producing a commercially valuable recyclate. By the way, it is necessary to transfer the result of recycling R&D group works to the PUF producers sponsored by environmental protecting organization. This implies that successful recycling via split-phase glycolysis will only be feasible with cooperation or joint development between scrap suppliers, palletizes, and process operators, all supported by an organization (Alavi et al., 2007). Researchers attempted to develop a process where, flexible PUF can be advantageously treated by two-phase glycolysis in order to recover their constituent polyols with an improved quality respect to the single-phase processes. It has been demonstrated that the entire family of commercial metal octoates shows a certain ability catalyzing glycolysis process.

The octoates have showed different catalytic activities according to their hardness and coordination ability. Hardness concept of cations is related to their response level in an electrical field, concretely in relationship to the interactions with other atoms and ions. Therefore, it is connected to their polarizability. It represents how easily a metal ion can be deformed in the presence of an electric field; the softer the cation, the more polarizable it is. The activity of alkaline an alkaline-earth metal octoates is basically related with their hardness as cation that determines their potential for the formation of a metal alkoxylate. In the case of transition metals, the mechanism involves several steps, including the formation of a metal alkoxylate, coordination insertion of the alkoxide into the urethane group, and transfer from recovered polyol to glycol. The presence of signals in the IR spectrum of the recovered polyols corresponding to transesterification carbamates, DEG, and hydrolysis products such as aromatic amines characteristic of certain paths of the coordination mechanism proposed demonstrated it veracity. Among the octoates studied, lithium and stannous octoates showed a remarkable catalytic activity. They yielded the greatest quality for the recovered polyol as well as the highest decomposition rates.[27]

A study was conducted where, PUF was dissolved in a mixture of DEG and pentaerythritol (PER). PER is a useful choice in recycling processes because of its OH functional groups and its structural similarity to polyols. The solvent system contained DEG/PER as 9/1 ratio. Sodium hydroxide was used as the catalyst and the optimum reaction time was 4 h. The regenerated polyol provided by PER-assisted glycolysis of flexible PUF wastes is capable of high-quality foam formulation production. By optimizing the recycling approach, it is possible to replace about 40% of the virgin polyol by glycolysis product. Although the economic aspects of the proposed method are of high interest, the environmental pollution reduction is the main advantage of the studied polyol regeneration approach.[5]

It is useful to know that stannous octoate shows a proper performance as catalyst in the glycolysis of PU wastes, yielding a great quality for the recovered polyol in the shortest reaction time. Furthermore stannous octoate also presents the advantage that the catalyst does not need to be removed from the polyol for using in further foaming. An increase in the reaction temperature and catalyst concentration enhances the degradation rate; however, this also negatively affects the process by extent of side reactions and contamination of the polyol phase. High mass ratios of glycolysis agent to PUF displace the equilibrium to the glycol substitution, promote the phase splitting which allows the obtaining of a polyol-rich phase, and avoid problems related to agitation due to undissolved PU portions in the glycol. However, a too large excess of glycol would imply huge volume equipment requirements and larger amounts of bottom phase that has to be recycled by distillation. Taking all into consideration, equilibrium between reaction rates, polyol content in the upper phase and recovered polyol properties must be achieved.[33,34]

Further in a process, three phases are obtained: an upper phase which contains the polyol, a bottom phase which has the subproducts of the reaction and the excess of glycol, and a third phase which is in the middle and it is formed by the unreacted PU. Several factors that affect the reaction have been studied to optimize the yield of obtained polyol. These factors are the temperature of reaction, the time of reaction, the catalyst (DEA) to solvent (DEG) mass ratio, and the catalyst + solvent to PU mass ratio. The catalyst reduces the reaction time and allows the complete breakdown of the PU chain against the process with a lower quantity of DEA. Also, a high catalyst concentration affects negatively the process because the secondary reactions and the contamination of the polyol phase occur. Increasing the relative amount of glycol does not provide any additional improvement in the yield of polyol and affects negatively to the economy of the process[1].

2.2.2.1 MECHANISM OF DEPOLYMERIZATION USING METAL SALT AS CATALYST

For transition metal salt catalysts, the activity observed cannot be justified only on the basis of the different electronegativities or on the cationic–ionic ratio. Some of the cations are located in the border of strong and soft acids and also present some coordination abilities. As mentioned before, the transesterification takes place by means of a nucleophilic attack of the glycol on the carbon of the carbonyl group. For this reason, and taking into account several theories applied to polyester transesterification, it can be postulated that the glycolysis of PU in the presence of transition metal carboxylates is not only a result of a simple addition of an alkoxide but there could be also an intermediate formed by a coordination complex of the metallic species with the carbonyl group of the urethane, specially favorable in the case of the stannous salt. In this way, the proposed mechanism for catalysts studied would be composed by a three step cycle.

2.2.2.1.1 Step No. 1: Initiator Alkoxide Formation

The first step corresponds to the reaction of glycol (e.g., DEG) with the metal octoate, producing a metal alkoxide species and free 2-ethyl hexanoic acid shown in Figure 2.8. This was the starting acid used to form the octoate salt. In this step, there is equilibrium between alkoxide and carboxylate; the stabilities of both species being different, which in addition to the great excess of glycol, move the equilibrium to the alkoxide formation.

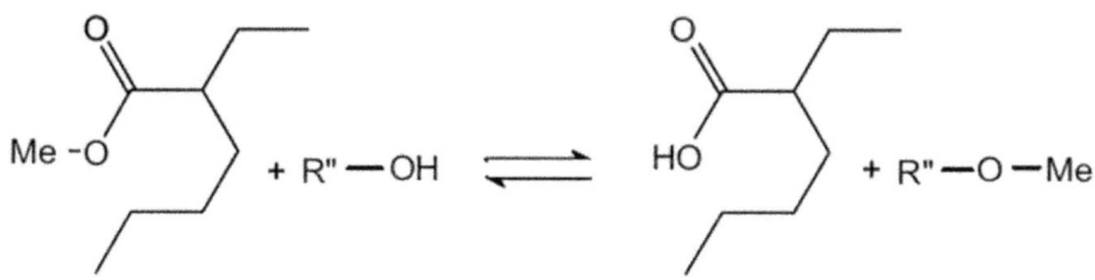

FIGURE 2.8 Reaction of DEG with the octoate, producing a metal alkoxide species and free 2-ethyl hexanoic acid.

2.2.2.1.2 Step No. 2: Coordination/Insertion and Exchange

After the alkoxide formation, in the case of cations with coordination ability, a coordination of the cation and the carbonyl oxygen atom would be produced. This interaction decreases electron density in the carbonyl group, which enhances the nucleophilic insertion of the alkoxide moieties as in Figure 2.9. After the exchange, the polyol is released to the reaction medium in an irreversible way, since the glycol is added in a large excess and polyol molecules are so big that statistics does not allow substitution. This step in the mechanism implies that metal cations able to produce coordination would improve the process, as does tin.

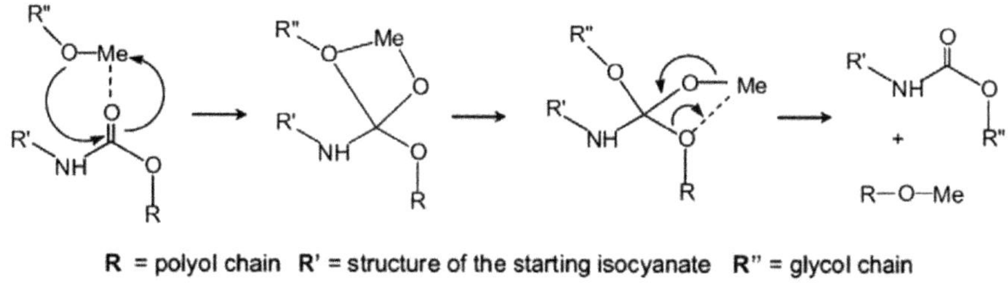

FIGURE 2.9 Coordination of the cation and the carbonyl oxygen atom/nucleophilic insertation of the alkoxide moieties.

2.2.2.1.3 *Step No. 3: Active Species Regeneration*

The third step requires the active species regeneration in order to continue the transesterification. The regeneration undergoes by a fast intermolecular exchange of the metal alkoxide moiety in the polyol for a proton from hydroxyl groups of glycol shown in Figure 2.10. As in the first step, the interaction between species, concretely cation–polyol and cation–glycol, is responsible for the equilibrium shift. In the case of using a glycol and a divalent cation, the substitution yields a more stabile structure if a ring is formed described in Figure 2.11; this means that polyol regeneration would be enhanced. This fact has also been reported for stannous octoate in ester exchange reactions with glycol. Depending on the catalyst type (alkaline, alkaline earth, or transition metal salt), the first step, namely, alkoxide formation, or the coordination for the insertion would be promoted. As it has been observed experimentally, lithium and stannous octoates are the most active catalysts in good agreement with their greater hardness and coordination ability, respectively.

$$\text{R–O–Me} + \text{R''–OH} \rightleftharpoons \text{R–OH} + \text{R''–O–Me}$$

FIGURE 2.10 Intermolecular exchange of the metal alkoxide moiety in the polyol for a proton from hydroxyl groups of glycol.

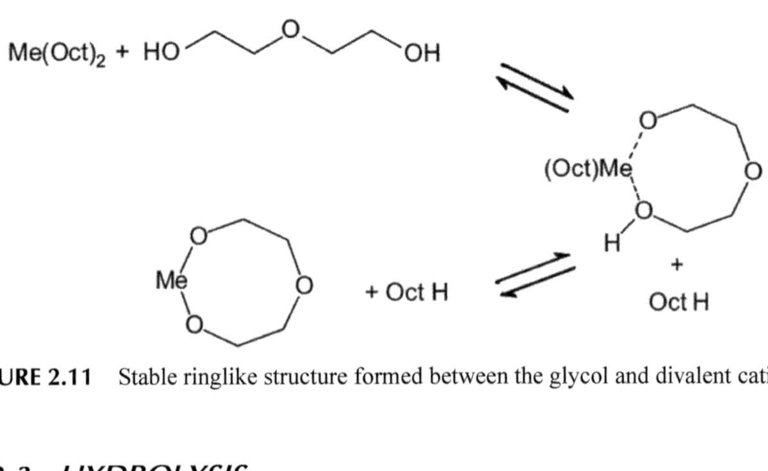

FIGURE 2.11 Stable ringlike structure formed between the glycol and divalent cation.

2.2.3 *HYDROLYSIS*

Hydrolysis is the second most important method of chemical recycling of PU. Various studies have been published dealing with PU degradation by reaction with liquid water (150–200°C) or steam (200–320°C). The hydrolytic reaction proceeds as shown in Figure 2.12.[13]

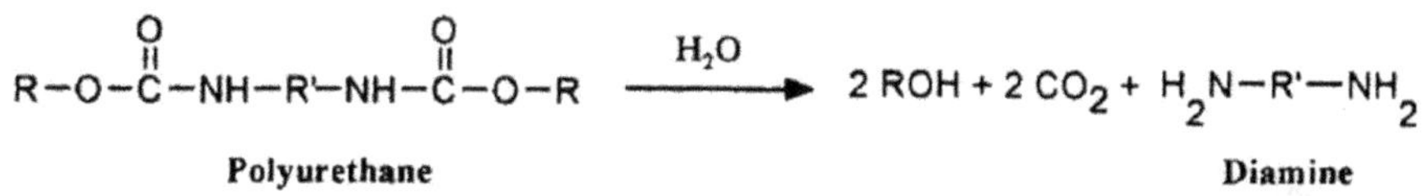

FIGURE 2.12 Polyurethane hydrolysis.

Polyols, diamines, and carbon dioxide are the final products formed by PU hydrolysis. The reaction between the diamine and phosgene allows the corresponding isocyanate to be formed, whereas the subsequent polymerization of this isocyanate and the polyols yields the starting PU again. Mahoney et al.[24] have described the reaction of PUF and superheated water at 200°C for 15 min, which leads to TDA and polypropylene oxide. Hydrolysis of PU and rubber mixtures has been used as a method not only of recovering valuable chemicals from the PU fraction but also to separate the polymers because rubber is inert to hydrolysis. The degradation takes place by contact with saturated steam at 200°C for 12 h. This process may find particular applications in the treatment of rubber/PU laminations. The mechanism and kinetics of the reaction of PUF with dry steam have been investigated by Gerlock et al.[17] using a polyether-based TDI as starting material. Reaction with steam at temperatures between 190°C and 250°C caused the destruction of all urea and urethane linkages to yield a high-quality polyol product, although a significant difference in reactivity between these two types of linkages was observed. The authors propose that the urethane bonds are rapidly broken by hydrolysis according to the reaction shown in Figure 2.13, whereas the urea linkages undergo a slow thermal dissociation to form the corresponding isocyanate and amine. Finally, the formed isocyanate is also hydrolyzed, increasing the yield of TDA. The effect of basic catalysts on these reactions was also investigated, a sharp acceleration being observed with the addition of sodium hydroxide at levels of about 2.9 mg per 100 mg of foam. The polyols obtained by this treatment were mixed with virgin polyol in a ratio of 20/80, yielding a high-quality flexible foam.[13]

A study carried out in which neutral hydrolytic depolymerization of the PU waste was done using 0.5 L high-pressure autoclave at temperatures of 150°C, 180°C, 200°C, and 240°C, the autogenious pressures of

75, 160, 220, and 480 psi, and time intervals of 30, 45, 60, and 90 min. The obtained product was characterized by measuring its amine value. The optimum amount of catalysts such as zinc acetate and lead acetate was found to be 1 g. Zinc acetate was more effective catalyst than lead acetate for the depolymerization of PUF reaction. On the basis of amine value and residual weight of the depolymerized product, the velocity constant was obtained and found to be in order of 10^{-3} min^{-1}, and the reaction was found to be first-order (Nemade et al., 2009).

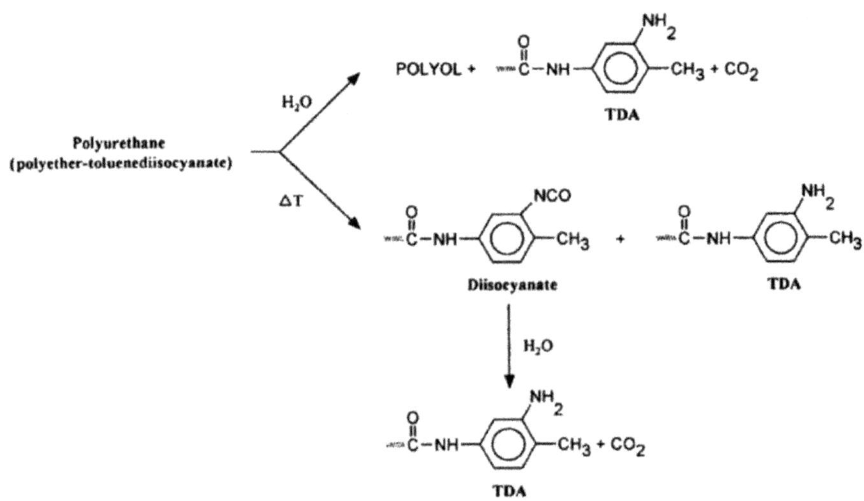

FIGURE 2.13 Mechanism for hydrolysis of polyether-based toluene diisocyanate polyurethane.

A successful attempt wherein post-use PUF was degraded in super-heated water at 423 to 623 K. The yield of TDA, one of the products, reached near 90%. The hydrolysis conditions—time and temperature were important and the perfect liquid products could be obtained under the economic conditions at 523 K for 30 min. TDA was added to the reaction to investigate the effect on the decomposition of PUF waste.[15]

2.2.4 AMMONOLYSIS AND AMINOLYSIS

PU chemolysis by reaction with ammonia or amines has been described by Sheratte et al. (1978); he had proposed a process based on PU decomposition by various agents. Several examples were provided for PU degradation with ethanolamine (120°C), ammonia and ammonium hydroxide (180°C), diethylene triamine (200°C) and other basic reagents. In all cases the process involves, simultaneously or subsequently, reaction with propylene oxide, which allows the different amines obtained to be quantitatively converted into polyols according to the reaction shown in Figure 2.14. The polyols derived from this process were used in the reformulation of new PU by polymerization with the corresponding isocyanate and were suitable for application in rigid foams.

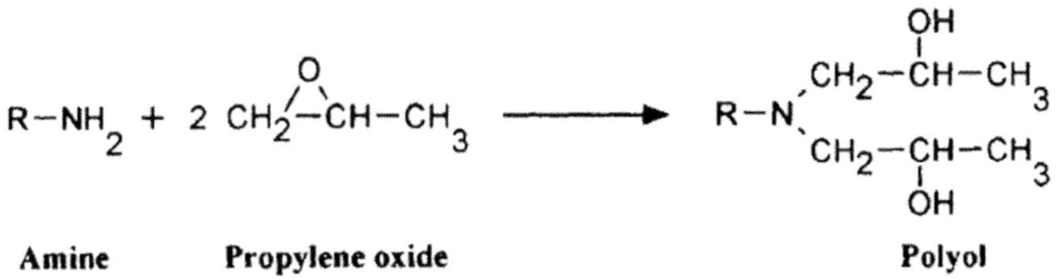

FIGURE 2.14 Conversion of primary amines into polyols by reaction with propylene oxide.

An interesting ammonolysis process has recently been developed based on the reaction of PU with ammonia under supercritical conditions, which favors both the degradation reactions and separation of the polyols produced. The flow diagram of this process is shown in Figure 2.15. Two different PU were used as starting materials: a solid elastomer based on a trifunctional polyethertriol, 1,4-butanediol, and methylene bis(phenyl isocyanate) and a flexible foam where the diol was replaced by water. The ammonolysis reactions were carried out at 139°C and 140 atm for 120 min, and with a PU/ammonia weight ratio of 1. Under these conditions the PU conversion was practically total. The ammonolysis reaction transforms the CO group into urea and the ester groups and derivatives of carboxylic acids into amides, whereas ether and hydroxyl groups are inert toward ammonia.[13] Figure 2.16 illustrates the stoichiometry proposed by the authors for the ammonolysis of the polyether urethane. The diamines and the diol can be separated by distillation or precipitation. The phosgenation of the amine leads to the corresponding diisocyanate, which together with the polyol and the diol may be used in the recovery of the raw PU.

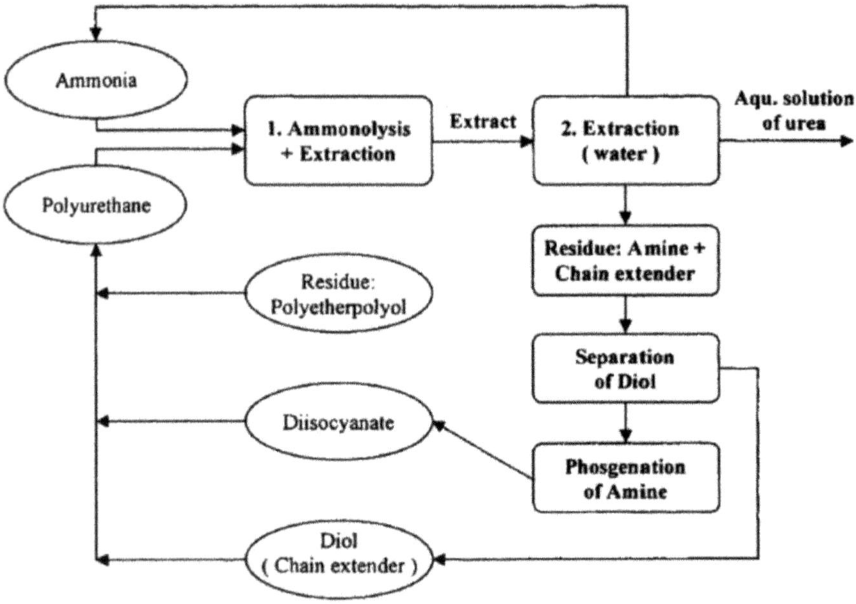

FIGURE 2.15 Ammonolysis of PU by treatment with supercritical ammonia.

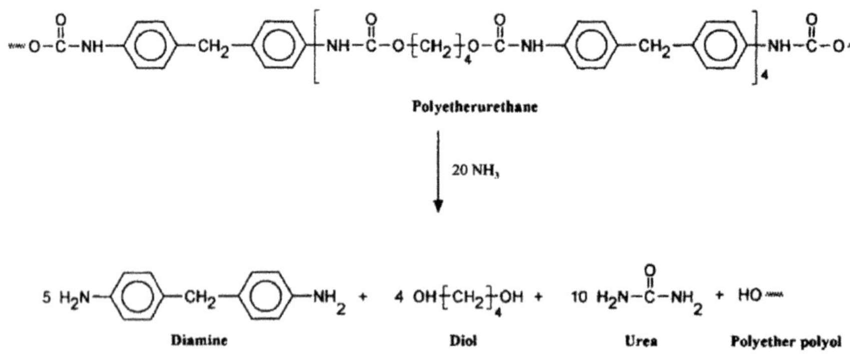

FIGURE 2.16 Stoichiometry of ammonolysis of polyether polyurethane.

After the reaction, urea is separated by extraction with water, whereas the polyol remains as a residue in the reactor. Therefore, under supercritical conditions, the polyether polyols are separated from the mixture at the same time that the ammonolysis reaction progresses. Watando et al.[40] has proposed a process based on PUF decomposition using DEA in an extruder and the resulting decomposed product could be used as an alternative virgin polyol in reclaiming PU elastomers. In an experiment by Fukaya, a reheating process for the products in the chemical recycling of rigid PUF by an extruder with DEA as a decomposing agent was effective for improving the product stability. In another experiment by Chuayjuljit, rigid PUF scrap was depolymerized by aminolysis using diethylene triamine as a degrading agent and sodium hydroxide as both a reactant and catalyst, resulting in 4,4′-methylenedianiline.[11] Kanaya proposed a system wherein flexible foams were decomposed by alkanolamines without a catalyst at 150°C and decomposed products were completely separated into two layers, the upper liquid layer being a polyol and the lower liquid layer of methylene diphenyl amine.[23] Lee et al. carried out aminolysis of rigid PU using diethylene triamine and studied the application of aminolyzed products as hardeners of epoxy resins.[20,21,41]

2.2.5 COMBINED CHEMOLYSIS

Several promising alternatives have been described for PU chemical recycling through a combination of treatments. The Ford hydroglycolysis process is a good example of these combined alternatives because it couples hydrolytic and glycolytic reactions to degrade the PU chains. This process was developed while trying to solve some of the problems present in conventional PU glycolysis, in which a complex mixture of products is obtained, made up of diamines, aminocarbamates, urea linked aminocarbamates oligomers, glycols, and polyols. The composition is difficult to predict, hence the properties of the PU produced using this recycled mixture directly are difficult to control. An alternative is the separation and purification of the produced polyol, but this is not an easy task. On the contrary, the product obtained in the hydroglycolysis process is made up of significantly fewer products, which makes purification of the polyols more feasible. Figure 2.17 is a flow diagram of the Ford hydroglycolysis process. In the reactor, PUF is degraded in the presence of water, DEG and alkali metal hydroxides at 200°C. When NaOH is added to the reaction mixture as a catalyst, a cleaner polyol is obtained due to the absence of carbamates and ureas in the product—they are transformed into amines and alcohols by hydrolysis. After 4 h of reaction, all the PU is decomposed, yielding polyols, amines, isocyanates, and carbon

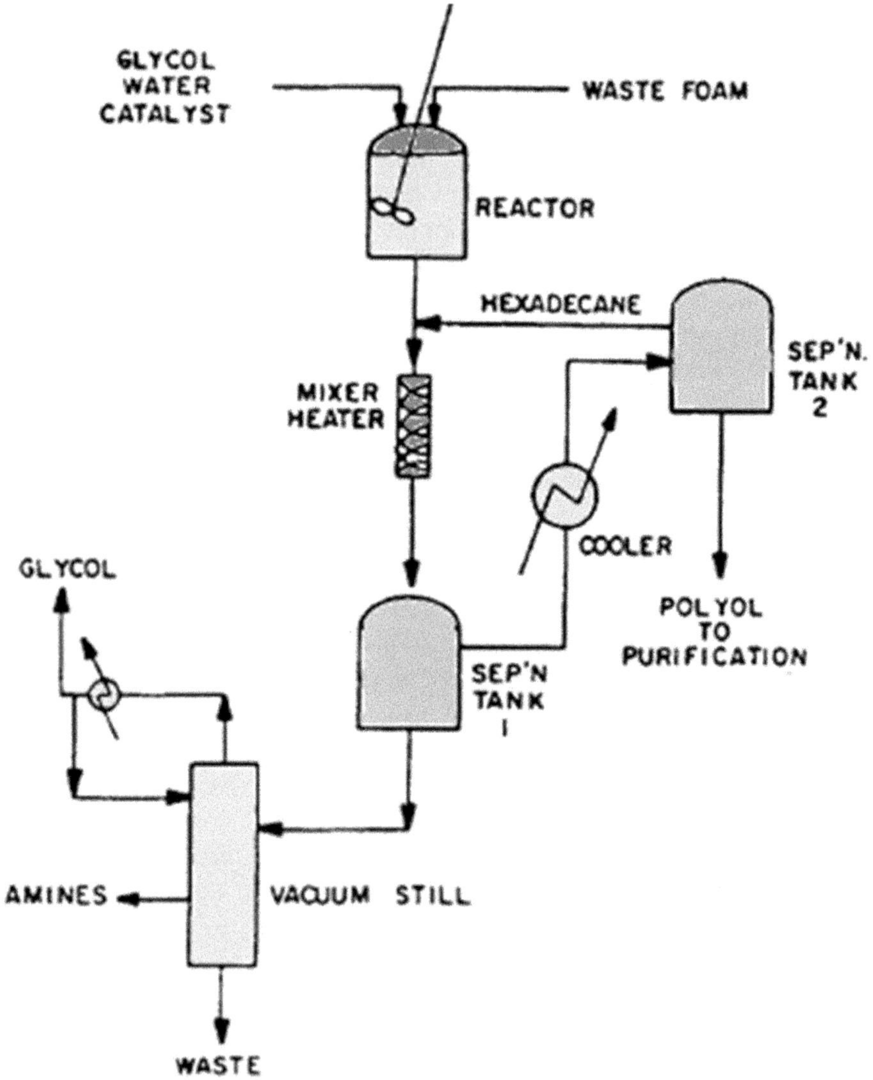

FIGURE 2.17 Ford hydroglycolysis of polyurethane foam.

dioxide. Separation of the products takes place by extraction with hydrocarbons such as hexadecanes (separation Tank l), which dissolve most of the polyols. Moreover, phase separation occurs at about 160°C and the polyols are expelled from the hydrocarbons (separation Tank 2), being purified by vacuum distillation. Finally, the glycol-rich phase, coming out of the first settling tank, is filtered to remove impurities and also vacuum distilled to separate the amines and DEG, the latter being recycled back to the reactor. Variations of this process have been patented by isolation of the glycolytic and hydrolytic treatments in different steps and/or the use of steam as hydrolyzing agent. The quality of the polyols produced in this hydroglycolysis process was evaluated by using them in the formulation of new PU by partial replacement of commercial polyols. The foam so produced did not have significantly different physical properties than the all-virgin foam, even with replacement levels of 50%.[13] In a novel study, the combination of glycolysis and aminolysis has also been described for the chemical recycling of PU. Kondo et al.[22] have reported the degradation of both PU and polyisocyanurate foams by reaction with a mixture of a glycol (DEG, dipropylene glycol, 1,4-butanediol, or 1,5-pentanediol) and monoethanolamine at around 200°C for 30 min. The liquid product obtained was used in the production of rigid PUF that, although slightly brown in color, exhibited properties suitable for application as a heat insulating material. Similarly, a recent work combines aminolysis and hydrolysis reactions for achieving PU decomposition. Thus, scrap PU is reacted with a mixture of DEA and aqueous sodium hydroxide. The simultaneous attack of these agents on the polymeric chains allows the reaction time to be appreciably shortened. The reaction product, obtained as an emulsion, is subjected to a second treatment with propylene oxide in order to transform the amines and ureas present in the mixture into polyols, giving a final product which is substantially free of any hydrogen-containing nitrogen atoms. The polyols produced have been found to be particularly suitable for the preparation of fresh PU polymer which can be used as an elastomer or flexible foam.

2.2.6 MICROWAVE DEPOLYMERIZATION

In this study, microwave irradiation has been used as an energy source for glycolyzing flexible PUF wastes. In order to study the susceptibility of the foam and select a system to obtain the highest quality recovered polyol, the reactions were performed in the presence of various basic catalysts and microwave powers. The completion and dissolution of the foam was measured, the obtained products separated in two split phases, and the recovered polyol in the upper phase was characterized and compared by virgin polyol by using FTIR, ^{1}HNMR, and ^{13}CNMR spectroscopic methods. All the assayed catalysts showed appropriate activity compared with the noncatalyzed process, allowing the recovery of polyols from the PU matrix. According to obtained information, potassium hydroxide, especially, has been found advantageous for the recovery process due to the reaction performance completely at the shortest reaction time, whereas the reactions performed with zinc acetate in comparison with the other basic catalysts has longer reaction time, potassium hydroxide and sodium hydroxide lead to the complete degradation of the polymeric chain at low reaction time and the recovery of the polyol in high quality.[2]

Researchers attempt to carry out glycolysis reaction of cold cure flexible PUF in a microwave oven at atmospheric pressure. In order to obtain high-quality recovered polyols via "split-phase" condition and short-reaction times, glycerin and sodium or potassium hydroxides were used as solvent and catalysts, respectively. Reactions were carried out at various temperatures, namely, 160°C, 180°C, 200°C, and 220°C. Decreasing of reaction time was observed by the increasing in reaction temperature. On the order hand, microwave irradiation accelerated the conversion reaction 20–30 times faster than conventional heating methods. Hence, microwave can be used as an energy source in glycolysis reactions and their performance was monitored at short time with simple controlled conditions[3].

In another experiment, the chemical recycling of flexible PUF wastes are performed through hydroglycolysis process under microwave irradiation at atmospheric pressure and 160°C. Mixtures of glycerin, water, and sodium hydroxide are used as hydroglycolysis agent. Water is added at different ratios containing 5%, 10%, 15%, 20%, 25%, 30%, 35%, and 40% of solvent system. The effect of water in the foam dissolution rate as well as water content of final recovered polyol is investigated. Although addition of water as the second reagent in the hydroglycolysis agent slows foam dissolution (such that only 20% of the foam dissolves after 15 min), it is a potential low-cost replacement for up to 40% of the glycerin, since reasonable dissolution times are still achievable (compared with 10 h for conventional heating).[7] Hydroglycolysis of flexible polyurethane rubber (PUR) foam has been studied under the microwave radiation with glycerol/sorbitol/water mixtures, differing in compositions, and NaOH were used as glycolysis agents and the catalyst, respectively. An increase in sorbitol part in the mixture being glycolysis agent prolongs the glycolysis process.[4]

2.2.7 DEGRADATION

Researchers propose a system in which dimethyl phosphonate is used as a degrading agent for PU elastomer based on diphenylmethane diisocyanate and polyester polyol. The results obtained reveal that the chemical degradation proceeds at 142°C without any catalysts. The degradation products are phosphorus containing oligomers with phosphonate end groups. These products, directly or after chemical treatment, can be used for the synthesis of PU with reduced flammability or as additives to polymers improving their fire resistance. It is likely that other diesters (alkyl and aryl) of phosphonic acid, esters of alkyl(aryl)phosphonate acid, as well as alkyl(aryl)ester of phosphoric acid, could also participate in the abovementioned reactions and could also degrade polyesters, polycarbonates, and phosphorylated polyamides.[34] Flexible PUF based on polyester or polyether polyol and TDI can be converted quantitatively into liquid form by treatment with triethyl phosphate (TEP) by reaction between the urethane group and ethoxy groups of phosphoric acid triethyl ester. The degraded products are phosphorus containing oligourethanes which can be used as a flame retardant.[35]

In another experiment, diethyl phosphonate and tris (1-methyl-2-chloroethyl) phosphate were used as degrading agents for microporous PU elastomer. Degraded products clearly confirm that the liquefaction of the microporous PU elastomer results from the proceeding of exchange reaction between ethoxy groups of diethyl phosphonate or 1-methyl-2-chloroethoxy groups of tris (1-methyl-2-chloroethyl) phosphate and urethane groups. The rate of the exchange reaction depends on the type of the α-carbon atom of the degrading agents. Degraded products are phosphorus or phosphorus and chlorine containing oligomers.[36]

In another study, chemical degradation of used PU was intentionally made by the addition of flame retardants such as tris(2-chloropropyl) phosphate (TCPP), TEP, and trimethyl phosphonate (TMP). Final product obtained after the degradation reaction turned out to be phosphorous containing oligourethanes. Rigid PUF was produced using the degraded products as flame retardants. The flammability and thermal stability of recycled rigid PU was investigated. The mechanical properties such as compressive strength and tensile strength of recycled PU were also studied. The recycled PU shows reduced flammability and higher thermal stability over virgin PU. Mechanical strength of recycled PU also shows as high as that of virgin PU. In order to evaluate flame retardant properties of the recycled PUF with various amounts of depolymerized product, heat release rate of the foam was measured by cone calorimeter. Scanning electron micrograph of recycled PU shows uniform cell morphology as virgin PU.[12]

Polyester PUR had attracted attention because of its biodegradability. There are many reports on the degradation of polyester PUR by microorganisms, especially by fungi. Microbial degradation of polyester PUR is thought to be mainly due to the hydrolysis of ester bonds by esterases. Recently, polyester PUR degrading enzymes have been purified and their characteristics have been reported. Among them, a solid-polyester-PUR-degrading enzyme (PUR esterase) derived from *Comamonas acidovorans* TB-35 had unique characteristics. This enzyme has a hydrophobic PUR surface binding domain and a catalytic domain, and the surface-binding domain was considered as being essential for PUR degradation. This hydrophobic surface-binding domain is also observed in other solid-polyester-degrading enzymes such as polyhydroxyalkanoate (PHA) depolymerases. There was no significant homology between the amino acid sequence of PUR esterase and that of PHA depolymerases, except in the hydrophobic surface-binding region. Thus, PUR esterase and PHA depolymerases are probably different in terms of their evolutionary origin and it is possible that PUR esterase come to be classified as a new solid-polyester-degrading enzyme family.[19]

2.3 CONCLUSION

The purpose of chemical recycling is the recovery of polyols, which can be reused in the production of PUF, by the depolymerization of PUF scrap. Numerous processes applicable to chemical recovery of PUF have been proposed including hydrolysis, aminolysis, glycolysis, alcoholysis, and combinations thereof such as hydroglycolysis. These processes vary according to the quality of the PU feed that can be used, the quality of the end products, and the number of purification and washing steps involved. In hydrolysis, PU are broken down into their original precursors; the base polyol and an amine, by the application of superheated steam In principle, it is possible to separate the amine obtained from the hydrolysis process and, after purification, use it again as a raw material for the isocyanate process. In aminolysis, the PU scrap is chemically cleaved by amines such as dibutylamine; ethanolamine or lactam are added to the PUF scrap. The decomposition of PUF with amines leads to quite a different product mix than with glycolysis and hydrolysis. Aminolysis converts the urethane linkages to polyol and disubstituted urea which in turn, breaks down to yield oligomeric ureas and amines. Aminolysis is not much explored route for recycling, hence we have proposed aminolysis as a tool to recycle PUF in chapter ahead. Glycolysis of PU is by far the most promising chemical recycling route for PUF.

KEYWORDS

- **recycle**
- **polyurethane**
- **foams**
- **degradation**
- **green chemistry**

REFERENCES

1. Aguado, A.; Martínez, L.; Moral, A.; Fermoso, J.; Irusta, R. Chemical Recycling of Polyurethane Foam Waste via Glycolysis. *Chem. Eng. Trans.* **2011,** *24*, 1069–1074.
2. Alavi Nikje, M. M.; Nikrah, M. Microwave-assisted Glycolysis of Polyurethane Cold Cure Foam Wastes from Automotive Seats in "Split-Phase" Condition. *Polym. Plast. Technol. Eng.* **2007,** *46*, 409–415.
3. Alavi Nikje, M. M. "Split-phase" Glycolysis of Flexible PUF Wastes and Application of Recovered Phases in Rigid and Flexible Foams Production. *Polym. Plast. Technol. Eng.* **2007,** *46*, 265–271.
4. Alavi Nikje, M. M.; Mohammadi, F. H. A. Sorbitol/Glycerin/Water Ternary System as a Novel Glycolysis Agent for Flexible Polyurethane Foam in the Chemical Recycling using Microwave Radiation. *Polimery* **2009,** *54*, 541–546.
5. Alavi Nikje, M. M.; Tavassoli, K. M. Chemical Recycling of Semi-rigid Polyurethane Foams by using an Eco-friendly and Green Method. *Curr. Chem. Lett.* **2012,** *1*, 175–180.
6. Alavi Nikje, M. M.; Nikrah, M. Glycerin as a New Glycolysing Agent for Chemical Recycling of Cold Cure Polyurethane Foam Wastes in "Split-phase" Condition. *Polym. Bull.* **2007,** *58*, 411–423.

7. Alavi Nikje, M. M.; Nikrah, M.; Mohammadi, F. H A. Microwave-assisted Polyurethane Bond Cleavage via Hydroglycolysis Process at Atmospheric Pressure. *J. Cell. Plast.* **2008,** *44*, 367–380.
8. Alavi Nikje, M. M.; Nikrah, M.; Haghshenas, M. Microwave Assisted "Split-phase" Glycolysis of Polyurethane Flexible Foam Wastes. *Polym. Bull.* **2007,** *59*, 91–104.
9. Beyler, C. L.; Marcelo, M H. Thermal Decomposition of Polymers. In *SFPE Handbook of Fire Protection Engineering.* **2002**, *2*, 111–131.
10. Borda, J.; Pasztor, G.; Zsuga, M. S. Glycolysis of Polyurethane Foams and Elastomers. *Polym. Degrad. Stab.* **2000,** *68*, 419–422.
11. Chuayjuljit, S.; Norakankorn, C.; Pimpan, V. Chemical Recycling of Rigid Polyurethane Foam Scrap via Base Catalyzed Aminolysis. *J. Met. Mater. Miner.* **2002,** *12*(1), 19–22.
12. Chung, Y.; Kim, Y.; Kim, S. Flame Retardant Properties of Polyurethane Produced by the Addition of Phosphorous Containing Polyurethane Oligomers (II). *J. Ind. Eng. Chem.* **2009,** *15*, 888–893.
13. Clark, J. *Feedstock Recycling of Plastic Wastes,* 1st Ed., Royal Society of Chemistry: UK, 1999; pp 31–56.
14. Clark. J.; Macquarrie, D. *Handbook of Green Chemistry and Technology*, 1st Ed., Blackwell Publishing: New Jersey, 2002; pp 1–9, 56–60.
15. Dai, Z.; Hatano, B.; Kadokawa, J.; Tagaya H. Effect of Diaminotoluene on the Decomposition of Polyurethane Foam Waste in Superheated Water. *Polym. Degrad. Stab.* **2002**, *76*, 179–184.
16. Doble, M.; Kruthiventi, A. K. *Green Chemistry and Engineering*; Academic Press, Elsevier Inc.: London, 2010.
17. Gerlock, J. L.; Braslaw, J.; Mahoney, L R.; Ferris, F. C. Hydrolysis of Polyurethane Foam Waste. *J. Appl. Polymer Sci.* **1980,** *18*, 541.
18. Hans, R. K.; Oskar, N.; Graham, S. *Handbook of Polymer Synthesis*, 2nd Ed., Marcel Dekker: New York, 2004; pp 511–517.
19. Kambe, Y.; Nakajima, T.; Akutsu, S.; Nomura, N.; Onuma, F.; Nakahara, T. Microbial Degradation of Polyurethane, Polyester Polyurethanes and Polyether Polyurethanes. *Appl. Microbiol. Biotechnol.* **1999,** *51*, 134–140.
20. Keesuwan, S.; Chen, J. Recycling of Polyurethane Foam as a Hardener for Epoxy Resin. *J. Sci. Res.* **1998,** *23*(2), 155–165.
21. Kim, B. C.; Lee, D. S.; Hyun, S. W. Curing Behavior of Epoxy Resin Using Aminolysis Products of Waste Polyurethane as Hardners. *J. Ind. Eng. Chem.* **2001,** *7*(6), 449–453.
22. Kondo, O.; Hashimoto, T.; Hasegawa, H. US Patent 4 014 809, 1977.
23. Kanaya, K.; Takahashi, S. Decomposition of Polyurethane Foams by Alkanolamines. *J. Appl. Polym. Sci.* **1994,** *51*, 675–682.
24. Mahoney, L. R.; Weiner, S. A.; Ferris, F. C. Hydrolysis of Polyurethane Foam Waste. *Environ. Sci. Technol.* **1974,** *8*, 135.
25. Mantia, F. P. *Recycling of Plastic Materials,* 1st Ed. ChemTec Publishing: Canada, 1993; pp 1–14.
26. Modesti, M.; Simioni, I.; Munari, R.; Baldoin, N. Recycling of Flexible Polyurethane Foams with Low Aromatic Amine Content. *React. Funct. Polym.* **1995,** *26*, 157–165.

27. Molero, C.; de Lucas, A.; Rodriguez, J. F. Activities of Octoate Salts as Novel Catalysts for the Transesterification of Flexible Polyurethane Foams with Diethylene Glycol. *Polym. Degrad. Stab.* **2009,** *94,* 533–539.
28. Molero, C.; de Lucas, A.; Rodriguez, J. F. Recovery of Polyols from Flexible Polyurethane Foam by "Split-phase" Glycolysis: Study on the Influence of Reaction Parameters. *Polym. Degrad. Stab.* **2008,** *93,* 353–361.
29. Molero, C.; de Lucas, A.; Rodriguez, J. F. Recovery of Polyols from Flexible Polyurethane Foam by "Split-phase" Glycolysis: Glycol Influence. *Polym. Degrad. Stab.* **2006,** *91,* 221–228.
30. Molero, C.; de Lucas, A.; Rodriguez, J. F. Recovery of Polyols from Flexible Polyurethane Foam by "Split-phase" Glycolysis with New Catalysts. *Polym. Degrad. Stab.* **2005,** *91,* 894–901.
31. Nemade, A. M.; Mishra, S.; Zope, V. S. Kinetics and Thermodynamics of Neutral Hydrolytic Depolymerization of Polyurethane Foam Waste Using Different Catalysts at Higher Temperature and Autogenious Pressures. *Polym. Plast. Technol. Eng.* **2010,** *49,* 83–89.
32. Nikje Alavi, M. M.; Haghshenas, M.; Garmarudi, A. B. Preparation and Application of Glycolysed Polyurethane Integral Skin Foams Recyclate from Automotive Wastes. *Polym. Bull.* **2006,** *56,* 257–265.
33. Simon, D.; Garcia, M. T.; de Lucas, A.; Borreguero, A. M.; Rodriguez, J. F. Glycolysis of Flexible Polyurethane Wastes Using Stannous Octoate as the Catalyst: Study on the Influence of Reaction Parameters. *Polym. Degrad. Stab.* **2013,** *98,* 144–149.
34. Troev, K.; Grancharov, G.; Tsevi, R.; Tsekova, A. A Novel Approach to Recycling of Polyurethanes: Chemical Degradation of Flexible Polyurethane Foams by Triethyl Phosphate. *Polymer* **2000,** *41,* 7017–7022.
35. Troev, K.; Grancharov, G.; Tsevi, R. Chemical Degradation of Polyurethanes 3. Degradation of Microporous Polyurethane Elastomer by Phosphonate and Tris(1-methyl-2-chloroethyl) Phosphate. *Polym. Degrad. Stab.* **2000,** *70,* 43–48.
36. Troev, K.; Atanasov, I.; Tsevi, R.; Grancharov, G.; Tsekova, A. Chemical Degradation of Polyurethanes. Degradation of Microporous Polyurethane Elastomer by Dimethyl Phosphonate. *Polym. Degrad. Stab.* **2000,** *67,* 159–165
37. Ulrich, H.; Odinak, A.; Tucker, B.; Sayigh, A. R. R. Recycling of Polyurethane and Polyisocyanurate Foam. *Polym. Eng. Sci.* **1978,** *18*(11), 844–848.
38. Ulrich H. Recycling of Polyurethane and Isocyanurate Foam. *Adv. Urethane Sci. Technol.* **1978,** *5,* 49.
39. Valavanidis, A.; Vlachogianni, T. *Green Chemistry and Green Engineering,* 1st Ed. Synchrona Themata, Non Profit Publishing House: Athens, 2012; pp 17–44, 45–60.
40. Watando, H.; Saya, S.; Fukaya, T.; Fujieda, S.; Yamamoto, M. Improving Chemical Recycling Rate by Reclaiming Polyurethane Elastomer from Polyurethane Foam. *Polym. Degrad. Stab.* **2006,** *91,* 3354–3359.
41. Xue, S.; Omoto, M.; Hidai, T.; Imai, Y. Preparation of Epoxy Hardeners from Waste Rigid Polyurethane Foam and Their Applications. *J. Appl. Polym. Sci.* **1995,** *56,* 127–134.

CHAPTER 3

IMPORTANCE OF REUSING AND RECYCLING ELECTRONIC WASTES

M. NEFTALÍ ROJAS-VALENCIA* and LUZ QUINTERO LÓPEZ

Institute of Engineering, Coordination of Environmental Engineering, National Autonomous University of Mexico, Post Box 70-472, Coyoacán 04510, Mexico, D. F. Mexico

Corresponding author. E-mail: nrov@iingen.unam.mx

CONTENTS

ABSTRACT

A large number of academic and administrative staff works at universities and research centers. In their daily activities, they use computers, printers, and multifunctional devices. At the end of their service life or because of obsolescence, some of them become waste electrical and electronic equipment (WEEE). It they are not appropriately disposed off, they can release pollutants, such as heavy metals and organic and inorganic compounds, pollute the environment, and affect public health. Thus, the aim of this study is to propose techniques for reusing and recycling them to help prevent waste generation and reduce environmental pollution. In order to fulfill the proposed objective, the methodology was divided into four stages: in the first stage, the state of the art of the WEEE was analyzed; in the second phase, the study area was delimited, while sampling and a generation study were performed in the third stage. Finally, a SWOT analysis and a prospective analysis were conducted in the fourth stage. The results showed that, globally, the two institutions (the Institute of Engineering, IE and the College of Engineering of UNAM) generated 1829 devices in 6 months (the IE generating the main part), mostly CPUs and keyboards and to a lesser extent, projectors. Video cameras, videoconference equipment, sound equipment, photocopiers, regulators, scanners, printers, laptops, monitors, and mice were also among the WEEE generated at those institutions. It is worth mentioning that most WEEE could be reused. It is urgent to disclose these results to raise awareness in the university community and establish plans for preventing the generation of theses wastes.

3.1 INTRODUCTION

In 2012, worldwide, 48.9 million metric tons of electronic wastes were generated, equivalent to 7 kg per person. If nothing is done, by 2017, the annual volume of electronic wastes will increase by 33% compared to 2012, to a total of 65.4 million tons. Mexico is not immune to this situation. In 2012, approximately 1.03 million tons of electronic wastes were discarded, equivalent to 8.9 kg per person.[1–4]

A large number of academic and administrative staff works at universities and research centers. In their daily activities, they use computers, printers, multifunctional devices. Moreover, they also use other apparatus and devices such as intercommunication radios, TV sets, video cameras, videoconference equipment, sound equipment, photocopiers, electrical and electronic typewriters, desk and pocket calculators, fax, telephones, cellular phones, automated answering machines, audio equipment, modems, routers, hard disks, power supplies and regulators, scanners, electronic cards, cables, chargers, stereos, faxes, recorders, mice and keyboards, lighting apparatus, such as fixtures for fluorescent lamps, compact fluorescent lamps, and straight fluorescent tubes, and high intensity lamps, including pressure sodium lamps and metal halide lamps.[5–7] At the end of their service life or because of obsolescence, some of them become waste electrical and electronic equipment (WEEE). It they are not appropriately disposed of, they can release pollutants, such as heavy metals (copper, arsenic, selenium, antimony, or mercury), polybrominated biphenyls and polychlorinated biphenyls that are toxic, bioaccumulative, and persistent. Since we do not have a management scheme for their reduction, treatment and/or reuse, they contaminate the environment and affect public health.[8,9]

The Institute of Engineering (IE) and the College of Engineering (CE) of National Autonomous University of Mexico (UNAM) are not exempt from this environmental problem. It is estimated that they decommission approximately 470 devices (5%) annually, generating globally over 10 tons/year of electronic wastes, belonging thus to the group of large generators of electronic wastes.

3.1.1 BACKGROUND

In Mexico, no figures or studies are available as regard to the generation of electronic wastes in universities. Despite the fact that the General Law on the Prevention and Comprehensive Management of Wastes (LGPGIR) classifies electronic wastes as requiring a special handling, there are very few initiatives about this matter.

The generation of electronic wastes in educational institutions has increased in the past years and there are few studies regarding this matter. Currently, no data are available as regard to their contribution to this growing problem, and there are no figures about the quantities that are generated. It is estimated that governmental institutions, including large universities, are responsible for 20% of WEEE.

It is a new topic, essentially identified with minor sectors of society, regarding which little information is available. The generation and accumulation of electronic wastes in educational areas has led to a search for actions in various universities.

3.1.2 RELEVANCE OF UNIVERSITIES IN WEEE GENERATION

Despite the shortage of information about Universities involved in this effort in Mexico, contributions made by some institutions can be mentioned.

The National Polytechnic Institute (IPN), together with the municipality, carried out the program "Atizapán recycles with social responsibility" in an attempt to reuse and recycle electronic wastes within a socially accountable scheme. The materials obtained from computer parts consisted of ferrous, nonferrous and precious metals (32%), plastics (55%), and glass from discarded monitors (13%).

Monterrey Institute of Technology and Higher Education (ITESM) conducted various campaigns to collect obsolete or damaged electrical and electronic devices so that some of them could be reused by students of Engineering. Besides, they have entered into an agreement with Nokia to use technological waste in experiments for the manufacture of new cell phones.

Facing the need to reduce the contamination generated by electronic waste such as computers, cell phones, TV sets, and other household appliances, a group of students from the Juarez University of the State of Durango designed the e-Waste Recycling Center, a plant dedicated to classify and reintegrate said apparatus as raw materials for manufacturing new products.

As can be seen, very few universities have started implementing actions with regard to this problem and thus the objective of this study was to analyze the generation of electronic wastes at UNAM's IE and CE through a generation study and prospective analysis setting the benchmark as regard to the current situation and future scenario for taking necessary actions for the appropriate handling of WEEE, emphasizing the importance of reuse and recycling.

3.2 METHODOLOGY

The methodology is divided into three stages:

3.2.1 DESK WORK

This stage consists of a brief bibliographical review of the state of the art of the WEEE, including part of the problem caused by the increase of these electronic apparatus. Moreover, some of the initiatives developed that some universities to prevent environmental damages were analyzed.

3.2.2 DELIMITATION OF THE STUDY AREA

In order to perform their research and teaching activities, the IE and the Faculty of Engineering (see Figs. 3.1 and 3.2) engage a large number of academic and administrative personnel. Two hundred faculty members, 140 permanent administrative staff, 180 contract employees, and 600 trainees work for the IE. One thousand eight hundred and thirty-eight full time professors are employed by the CE. They are also supported by the various administrative departments. In their daily activities, all of them use computers, printers, multifunctional devices, and, in some cases, laboratory equipment. Altogether, 9422 electronic devices are used.

FIGURE 3.1 Building 5 of UNAM Institute of Engineering.

FIGURE 3.2 Faculty of Engineering.

3.2.3 FIELD STUDY

In this stage, the current generation of WEEE at the IE and the CE of the UNAM is identified and described.

3.2.4 PROJECTION STUDY

Based on the current circumstances, a prospective analysis was conducted to permit better future planning.

This phase consisted of the following activities:

1. Conduction of an inventory of the volumes of electronic wastes generated.
2. Classification of the main electronic wastes generated.
3. Prospective analysis
    a. Current perspective
    b. Future perspective
    c. SWOT analysis
4. Importance of the reuse and recycling, function of the universities in the generation of electronic wastes.

3.3 RESULTS AND DISCUSSION

3.3.1 DESK WORK

In this case, the definition and classification of electronic wastes according to the Directive 2002/96/CE of the European Parliament and of the Council on wastes electrical and electronic equipment (WEEE), were adopted and thus it was considered that any electronic apparatus becoming waste shall include all its components. Figure 3.3 shows the WEEE classification according to the European Union.

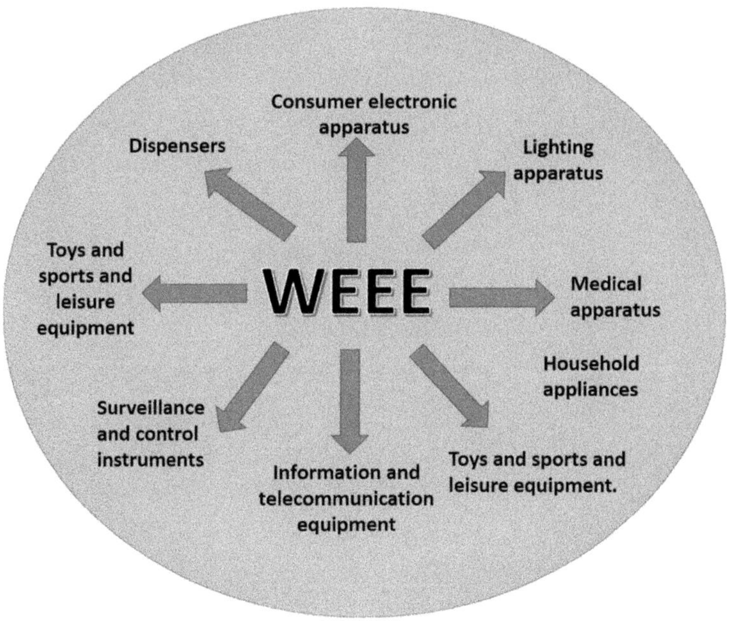

FIGURE 3.3 WEEE classification according to the European Union.

3.3.2 PROBLEM RELATED TO THE INCREASE OF ELECTRONIC WASTES

Short-term waste proliferation is an ever-growing worldwide problem. In Latin America, approximately 120 tons of electronic wastes are generated yearly. Without people being aware of it, contamination generation starts since the manufacturing process of the electronic apparatus, because the CO_2 released to the atmosphere, upon processing, part of the metals used to obtain an electronic product, pollutes the atmosphere.

As can be seen in Figure 3.4, Mexico is a country that stands out as WEEE generator in Latin America.

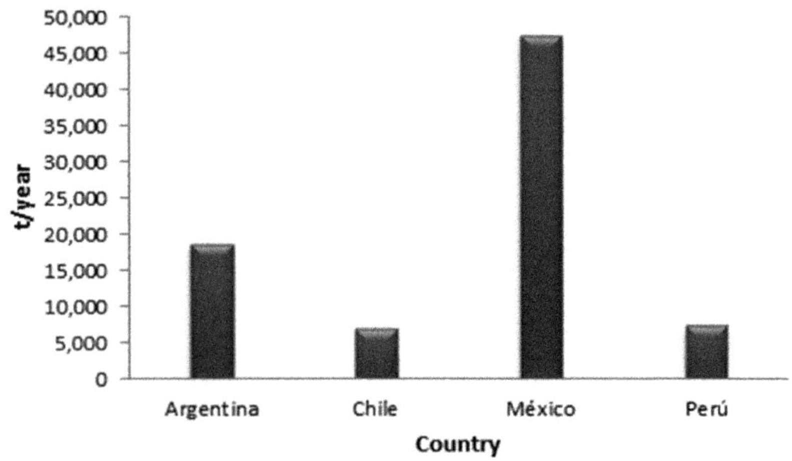

FIGURE 3.4 Generation of electronic wastes in Latin America (tons/year).

Some studies have shown that contaminants were present in electronic wastes because they contain hazardous components such as lead in cathode ray tubes and welding, arsenic in the oldest cathode ray tubes, antimony trioxide, and flame retardants. These substances have nefarious effects on the health of the people upon being dispersed in the air, water, and soil, causing a deterioration of air quality, and originating numerous health, economic, and environmental problems.

3.3.3 FIELD STUDY

3.3.3.1 RESULTS OF THE INVENTORY OF THE VOLUMES OF ELECTRONIC WASTES

The inventory carried out shows the quantities of electronic wastes that are generated at the CE and IE of the UNAM, as shown in Table 3.1. For its elaboration, data about electronic equipment definitively decommissioned

TABLE 3.1 Inventory for Estimating the WEEE Volumes Generated at UNAM's CE and IE.

Number	Quantity	Product description	Control number	Trademark	Model	Serial number	Reason for the decommissioning	Destination of the product
100	1	Laptop	2276810	Toshiba	PLL10U-017RL3	X8243262Q		Definitive decommis-sioning
101	2	Laptop	2286655	Toshiba	PSLC8U-03FRL1	29052518Q	Obsolescence	
102	1	Laptop	2188144	Dell	PP12L	J5896A02		
103	1	Laptop	2205873	Dell	PP20L	UF230A01		
104	1	Laptop	2268158	HP	8710W	CND81035QN		
105	1	Laptop	2269233	Apple	Macbook air	W88110GDY51		
106	1	Video camera	1955928	Sony	DCR-TVR110-NTSC	72389		
107	1	Printer with scanner	2208754	HP	LASERJET2840	CNFC64903G		
108	1	Video Camera	1910860	Canon	ES8100VA	621798		
109	1	Camera	2193369	Canon	PC1114	528409055		X
110	1	Equipment	2243388	Lacie	INNSO4-42200	1721011230		X
111	1	Flat screen monitor	2224102	HP	HPVS15	BF62601489		X
112	1	Printer with scanner	1943231	HP	LASERJET8500	JPDB011973		X
113	1	CPU	1877377	BTC	ESXKBS301	4407	Not working	X
114	1	Keyboard	1922454	Dell	RT7D20	43227459	Not used	X
115	1	Keyboard	2117421	HP	KB9970	TH007N12437	Not working	X
116…	1	Monitor	1868797	Dell	ESS1	19404396B	Not used	X

Source: Based on the inventories of electronic devices decommissioned in 2013.

were collected, and warehouses where wastes are temporarily kept were visited. Altogether, UNAM's CE and IE generate over 10 tons/year and are thus among the large electronic waste generators.

In Table 3.2, the results of WEEE generated at UNAM's IE and CE during a period of 6 months are summarized.

TABLE 3.2 Quantities of WEEE Generated at the IE and CE During a Period of 6 Months.

Month	Institute of Engineering	Faculty of Engineering
1	297	39
2	125	35
3	215	25
4	270	34
5	336	52
6	324	77
	1567	262
Total		1829

3.3.2 CLASSIFICATION OF THE MAIN ELECTRONIC WASTES

Based on checklists and with the support of the inventory, statistical data were obtained that permitted to evaluate the main types of electronic waste generated. Figure 3.5 shows the main devices or the ones appearing continuously in the area of study.

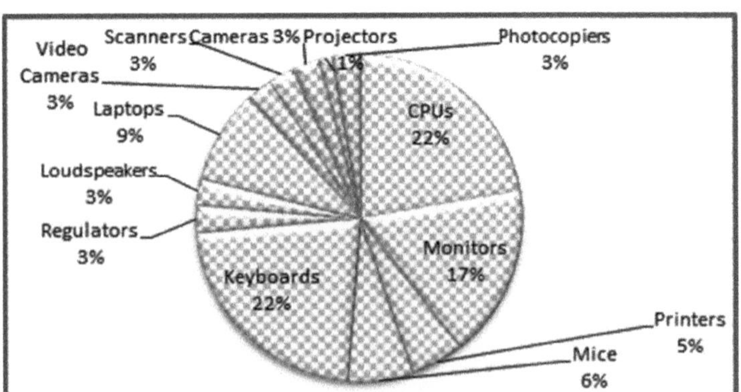

FIGURE 3.5 Main electronic wastes generated at UNAM's College of Engineering and Institute of Engineering.

From Table 3.3 and Figure 3.6, it is derived that the most generated WEEE were CPUs and keyboards and, to a lesser extent, projectors.

TABLE 3.3 Most Generated Electronic Wastes at UNAM's IE and CE According to the EU Classification.

Electrical waste (information and telecommunication equipment)	Generated quantities	% generation
CPU	282	22
Speakers	38	3
Video camera	38	3
Photocopier	38	3
Camera	38	3
Printer	64	5
Laptop	115	9
Monitor	218	17
Mouse	77	6
Regulator	38	3
Projector	13	1
Scanner	38	3
Keyboard	282	22
Total	1279	100

The most representative data samples were taken and a general exhibition was performed, regarding the most commonly discarded devices. These statistical results are shown in Figure 3.6.

From the results obtained in Table 3.4, it can be seen that the information and telecommunication equipments make out the greatest percentage of wastes, representing 70% of the electronic wastes generated.

TABLE 3.4 Main Electronic Wastes Generated at UNAM's IE and CE, According to the EU Classification.

Classification according to the EU	Generated quantities	Percentage (%)
Information and telecommunication equipments	1280	70
Other products	549	30
Total	1829	

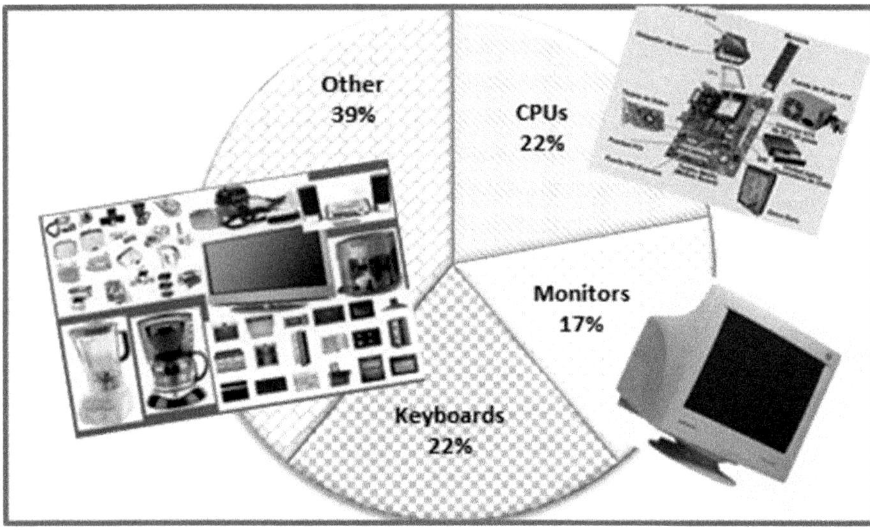

FIGURE 3.6 Generation of waste computer and peripheral equipment.

One of the main limitations of the study of waste electronic devices at UNAM's IE and CE is that, since there are many participants, it is not always possible to obtain complete information. Moreover, no previous studies of the situation are available, and only a record of the electronic wastes being generated is kept.

3.3.3 PROSPECTIVE ANALYSIS

3.3.3.1 CURRENT PERSPECTIVE

With the statistical analysis, supported and based on the WEEE inventory and classification, it was possible to appraise the current situation of the evolution of electronic wastes production for both institutions (IE and CE) in a period of 6 months. It was determined that, apparently, changes are not significant and a reduction was observed during the second and third months, as shown in Figure 3.7.

However, the IE and CE follow the pattern of the law of electronic waste growth. Figure 3.8 shows a comparison of what is occurring in both institutions.

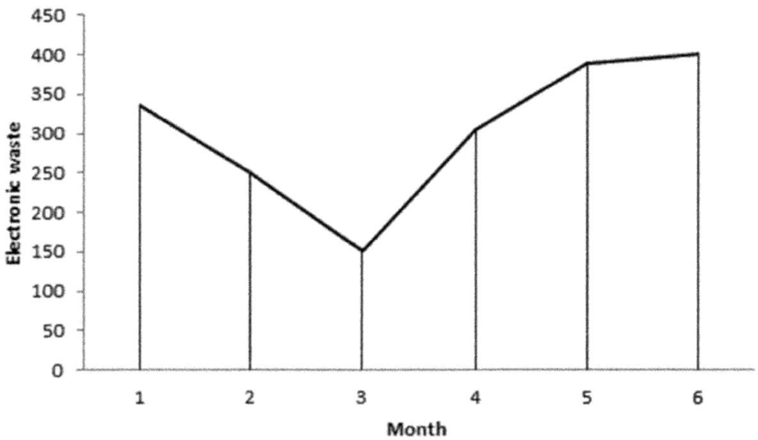

FIGURE 3.7 Current WEEE generation growth trend at the IE and CE.

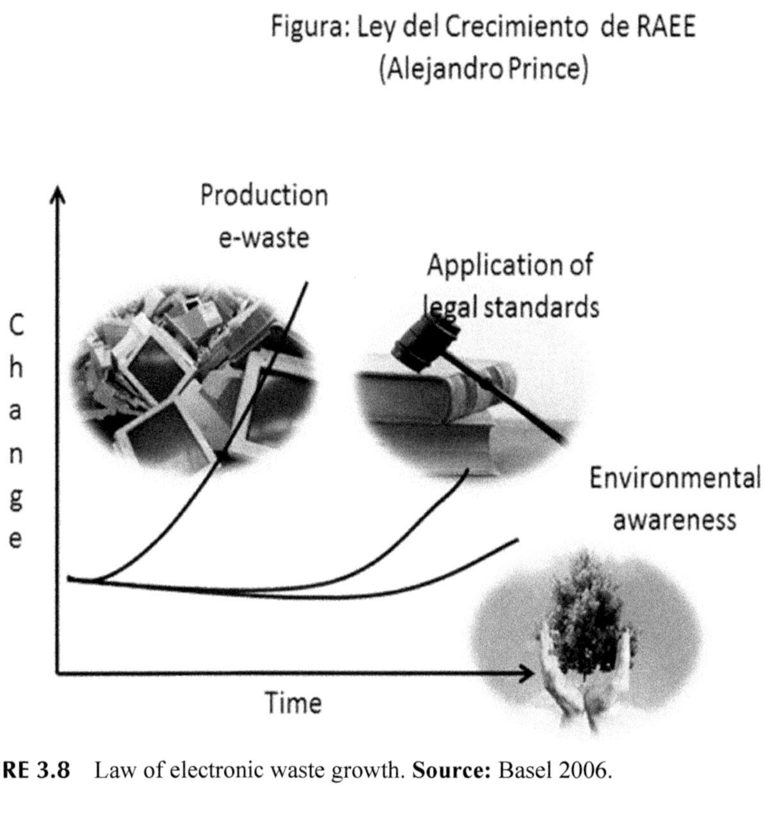

FIGURE 3.8 Law of electronic waste growth. **Source:** Basel 2006.

Strong technological renewal, among students, researchers, and workers, leads to a great accumulation of equipments, and the decommissioned electronic apparatus and their components (parts and materials) grow at a faster pace than the adoption of new equipments and much more quickly than the environmental awareness of the users.

3.3.3.2 FUTURE PERSPECTIVE

The projection of a future perspective evidences possible scenarios that could become reality if actions are not taken to modify the current trend.

It is considered that the national consumption of electronic apparatus in the coming years will follow the consumption and growth prospect illustrated in Figure 3.9.

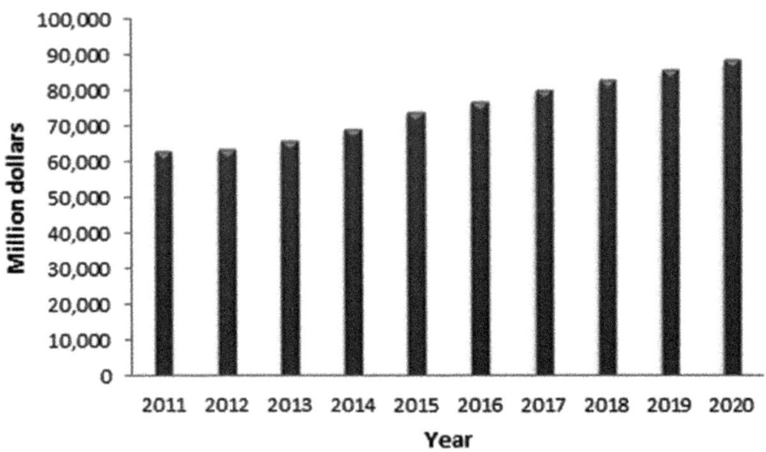

FIGURE 3.9 National consumption prospect of electronic apparatus and its growth during the years 2011–2020 in thousands million dollars (tmd).

Source: ProMexico, INEGI, Global Trade Atlas and Global Insight 2008.

A future projection of WEEE growth trend in IE and CE, Figure 3.7, shows and supports effectively an exponential growth in these institutions in the coming months. The results registered in 6 months, as already mentioned, showed a slow growth (see Figure 3.7), however, absent awareness raising, an exponential growth could occur such that the one shown in Figure 3.10.

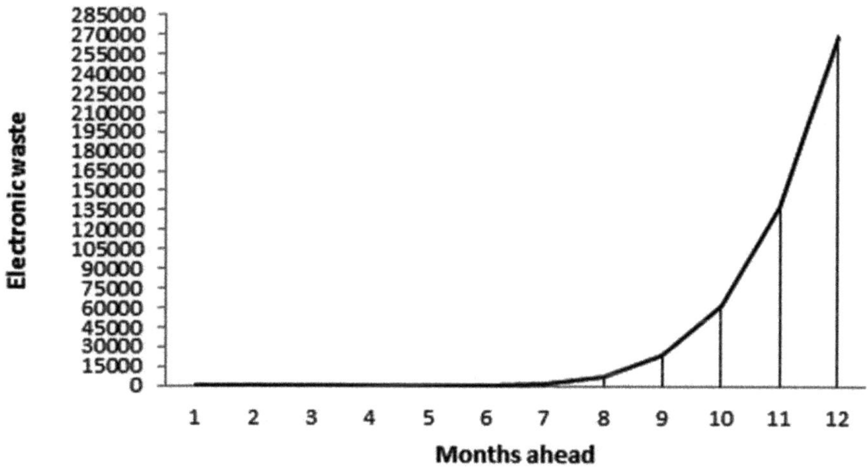

FIGURE 3.10 Future WEEE generation growth trend at UNAM's IE and CE during the coming 12 months.

Prospective analysis as a methodological tool facilitates the conduction of a reflection as regard to the topic in question and gives the opportunity of building images and future scenarios in order to use them as basis and thus currently face changes in a real environment.

3.3.3.3 SWOT ANALYSIS

Based on the above mentioned results, the weaknesses and threats as well as the strengths and opportunities in these Institutions were identified.

3.3.3.3.1 Strengths

- Trained personnel are available to optimally fulfill the functions.
- The documents regarding the management of electronic wastes are being computerized, and thus it will be possible to rely on updated data as regard to their generation and destination in updated time.
- Because of their composition and through an effective segregation at the origin, these electronic wastes are attracting the attention of the recovery and recycling industry.

- There is a powerful manufacturing sector and this should permit to start various actions for the sustainable management of electronic wastes in both institutions.

3.3.3.3.2 Opportunities

Creation of a data base and a part bank in order to help reuse the components of the equipments the useful life of which is not over.

- Formation and awareness of involved sectors, researchers, students, trainees, dissertationists, etc.
- Incorporation of responsible electronic apparatus consumption in the students and workers community.
- Possibility of reuse, valorization, recycling, and use as secondary, auxiliary materials or by-products.
- The inclusion of recycling and reuse of "electronic scrap" would be extremely important to help protect the environment.
- Being the first to implement a recycling management of "electronic waste" at the Campus that could incorporate other dependencies.

3.3.3.3.3 Weaknesses

- Existence of deficiencies and initiative regarding the management of WEEE generation flows.
- High WEEE elimination rates in short periods of time.
- Lack of records following a line for correctly quantifying and reproducing the situation.
- Lack of current precise inventories and generation rates because of the inaccuracy of the above point.

3.3.3.3.4 Threats

- If the current WEEE generation growth is not stopped, a long-term management problem could occur.
- This evidences a scarce social perception of the problem and its origin.

- Since those are research and educational institutions, where research lines are developed and new contributions are made, they should participate in the handling of said situations.
- Trend to become a major electronic wastes generator.

3.3.4 IMPORTANCE OF REUSE AND RECYCLING, PART PLAYED BY THE UNIVERSITIES IN THE GENERATION OF ELECTRONIC WASTES

Studies conducted in the European Union state that, on average, electrical and electronic apparatus are made of 25% of reusable components, 72% of recyclable materials, and 3% of elements that are potentially hazardous such as lead, mercury, cadmium, arsenic, and asbestos, among others.[10–13]

The recycling and reuse of electrical and electronic apparatus generates positive impacts:

- It permits to recover metals and materials that are everyday scarcer and whose obtainment, through mining and industry, generates a high environmental and social impact.
- It reduces the impact generated by these residues on the environment upon being degraded in dumps or landfills, polluting water, soil, and air.

It is important to take into account that the development and continuous updating of the database regarding the generation of electronic wastes make it possible to visualize the current reality, and this compilation and integration of the information should be useful to offer a more complete vision of the WEEE generation situation and the environmental resources to the general public. We also trust that this information will be used as raw material for academicians and people interested in environmental matters at UNAM's IE and CE who, after analyzing it, could contribute new ideas and proposals to help make better decisions with regard to the environmental care and the handling of our natural resources. The results show the need to start working on strategies for the management of electronic wastes, considering their reuse and creating infrastructure for appropriate recycling since most residues generated at UNAM's IE and CE correspond to the information and telecommunication classification, stressing the point that, due to their composition, computers, and part of their components are among the main contaminants. In view of the above, it is necessary to stimulate and reinforce the reuse and recycling of electronic wastes that are clearly evident in the economy of the consumers, offer tangible benefits reflected in savings with regard to the consumption of new raw materials, generate a reduction of the management costs, and reduce the environmental impact caused by their toxic components.

Some of the universities mentioned above offer a totally different situation. Aware of the importance of reusing and recycling electronic wastes, they have endeavored to control all waste related activities. They have sought to capitalize on the devices whenever they are still in working conditions, emphasizing reuse and recycling. It is noteworthy to mention that the Secretariat of Environment of the Government of the Federal District organized a WEEE collection program at UNAM in October 2013 during which 32,762.5 kg were collected.

3.4 CONCLUSIONS AND RECOMMENDATIONS

The information obtained from UNAM IE and CE establishes the basis for decision making. These institutions generate in excess of 10 tons/year of waste. These results show the need to start working on electronic waste management strategies, with a view to reusing them, and to develop an understanding of the most convenient recycling method, looking for the most appropriate strategies offering the best results.

It is necessary to stimulate and reinforce the work already initiated by some universities in this field where there exist areas of opportunities to reduce the impact caused by the release of organic compounds or toxic substances from electronic waste.

The importance of reusing and recycling electronic waste is clearly evident for the economy of the consumers, offering tangible benefits through raw materials saving, administrative cost reduction, and environmental impact minimization.

WEEEs are currently the object of research because of the accelerated technological growth and the lack of development of alternatives for their treatment, and universities play a relevant part in the formulation of strategies.

It is of the utmost importance to provide a follow-up to this project because WEEE represents a time bomb for the University because although it is true that there is a patrimonial department, it is also true that a recycling and reuse culture is lacking.

The definition of a management plan will offer strategies for the exploitation and/or appropriate treatment of WEEE, and will permit to estimate waste generation, to have a more precise overview with regard to the magnitude of the problem at the university in order to look for sustainable measures, the conservation of resources and environmental protection.

The SWOT analysis allows to identify part of the current problem and is helpful to record useful ideas and proposals.

Finally, we hope that this project will contribute to raise environmental awareness in the university community.

KEYWORDS

- **reuse**
- **recycling**
- **WEEE**
- **universities**
- **electronic waste**

REFERENCES

1. Aguilar, A. 2014. El valor del Reciclaje (Recycling Value). http://ganar-ganar.mx/pdf/r57/62.pdf (accessed Feb 9, 2014).
2. Armijo de Vega, C. 2011. Aplicación de un plan de manejo de residuos electrónicos en Baja California" (Application of a Plan for the Management of Electronic Wastes in Baja California). Ensenada, Baja California Internet: http://www.inecc.gob.mx/descargas/sqre/2011_plan_manejo_res_elec_ensenada.pdf (accessed March 20, 2014).
3. Ávila González, E. 2014. Análisis de ciclo de vida de residuos electrónicos y propuesta de manejo en ciudades fronterizas de Coahuila, Nuevo León y Tamaulipas (Analysis of Life Cycle of Electronic Wastes and Management Proposal in Border Cities of Coahuila, Nuevo León and Tamaulipas). http://www.inecc.gob.mx/descargas/sqre/2011_taller_ree_pres_mgonzalez.pdf (accessed Feb 9, 2014).
4. BBC Mundo (2010). Rodríguez Margarita. Se ahoga México en basura electrónica? [Is Mexico Drowning in Electronic Scrap?] http://www.bbc.co.uk/

mundo/ciencia_tecnologia/2010/06/100603_basura_electronica_mexico_mr.shtml (accessed July 30, 2014).

5. Betancourt, D. I. 2009. Propuesta financiera para el desarrollo de una planta recicladora de e-waste en la ciudad de Bogotá (Financial Proposal for the Development of an E-waste Recycling Plant in the City of Bogota). Thesis Universidad de la Sabana, Bogotá Colombia.
6. Barbosa Espinosa, C. M. 2014. Evaluación de los residuos tecnológicos generados por las Universidades de Cali y sus estudiantes (Evaluation of Technologial Wastes Generated by the Cali Universities and Their Students). http://biblioteca.universia.net/html_bura/ficha/params/id/57286845.html (accessed Feb 8, 2014).
7. Desechos electrónicos 2011. Consecuencias que trae la basura electrónica (Electronic Wastes, 2011. Consequences of Electronic Scrap). http://desechoelectronicos.blogspot.mx/ (accessed July 29, 2014).
8. Godet, M. How to Be Rigorous with Scenario Planning. *Foresight* **2000,** *2*(1), 5.
9. Investigación y Desarrollo ID (Research and Development ID). 2014. http://www.invdes.com.mx/medio-ambiente/2764-disenan-universitarios-duranguenses-sistema-para-reciclar-basura-tecnologica (accessed Feb 9, 2014).
10. López Sardi, M. 2014. La tecnología y su lado oscuro "E-scrap: el impacto de la tecnología sobre el medio ambiente" (The Technology and Its Dark Side "E-scrap: the Impact of Technology on the Environment"). http://www.palermo.edu/ingenieria/downloads/pdfwebc&T8/8CyT04.pdf (accessed Feb 8, 2014).
11. Nnorom, Innocent C.; Oladele, Osibanjo. Overview of Electronic Waste (E-Waste) Management Practices and Legislations, and Their Poor Applications in the Developing Countries. *Resour. Conserv. Recycling* **2008,** *52*(6), 843–858.
12. Tsydenova, O.; Bengtsson, M. Chemical Hazards Associated with Treatment of Waste Electrical and Electronic Equipment. *Waste Manag.* **2011,** *31*(1), 45–58.
13. Suarez Ezquievel, M. 2014. Reciclatrón, sitio para los residuos eléctricos y electrónicos (Reciclatron, Site for Electrical and Electronic Wastes). http://www.jornada.unam.mx/2013/06/22/capital/030n2cap (accessed Feb 9, 2014).

CHAPTER 4

PRODUCTS FROM CLOTHING WASTE

SUMAN D. MUNDKUR[1*] and ELA M. DEDHIA[2*]

[1]*Department of Textiles and Apparel Designing, SVT College of Home Science (Autonomous), S.N.D.T Women's University, Juhu Road, Mumbai 400049, India*

[2]*Textiles and Fashion Technology, College of Home Science, Nirmala Niketan (Affiliated to Mumbai University), 49, New Marine Lines, Mumbai 400020, India*

[*]*Corresponding author. E-mail: suman_mundkur@yahoo.com; elamanojdedhia@yahoo.com*

CONTENTS

ABSTRACT

The growth in the apparel retail industry there are a large number of clothing that is discarded from households much earlier than its life. Indian traditions promoted the use of secondhand clothes from older members in the family to younger ones. Used clothing is reused when given in charity or sold as secondhand clothes for which there is a great demand owing to the large population in both rural and urban India. Whether used clothes are passed on for reuse, given in charity, or exchanged for economic gains, the key word is recycling. The process of reuse and redistribution of used clothes in India is underresearched. Earlier studies indicated that the homemakers, nowadays, lack time and motivation to develop products themselves from their discarded family clothing. With the aim of promoting recycling, products can be made using used clothing. An experimental design is used to explore the possibilities of getting the products made for the homemakers by women of self-help groups in Mumbai, thus providing opportunities for income generation. The purpose of this study is to understand the consumer acceptability of the products. The women trained who had received skill training in basic stitching and bag making were selected by purposive sampling to develop sample products. Additional training in sorting, color coordination, mix and match, and better finishing was given. Pricing and costing and inculcating a spirit of competitiveness were areas that needed to be developed. Additional inputs in sourcing for lining, interlining, canvas, fasteners, and trims were provided. Further support was required in developing of 14 utility products. The products were divided into two categories—miscellaneous products and bags. The products have been tested for consumer acceptability in terms of design details, price, and functionality. Six products in all received favorable responses. There is a need to diversify and customize products. Creating awareness of the recycling facility will be important more for the environmental concern than economic benefits.

4.1 INTRODUCTION

Although the majority of textile waste originates from household sources, waste textiles also arise during yarn and fabric manufacture, garment-making processes, and from the retail industry. As reported in the *Indian textile Journal* (2009), recycling in the textile and clothing sector can take several forms.[1] The best known method involves the manufacture of a textile or clothing product from recycled consumer waste—such as plastic bottles, waste polyester yarns, or fabrics.[2] Other forms involve the reuse of waste textile and clothing products in a way which avoids throwing the items away, such as: shredding the products into fibers (e.g., for sound insulation); redistributing the items in the form of secondhand clothing via charity shops or textile merchants (also known as rag collectors); and reusing fabrics for "eco-fashion".[3] Gulich,[4] in his intensive studies on reclaimed fibers, stressed that the raw materials and waste disposal are becoming more and more expensive and looking for suitable raw materials to make reclaimed fibers, household textile waste as well as industrial waste should receive more attention. The growth of the retail industry and increased affordability have given rise to overconsumption of clothing. This can lead to early disposal and generation of voluminous waste. The problem of disposal can be severe with the rising population. India's population according to 2011 census is 1.21 billion, 17.5% of the world's population. Although used clothing is being reused though charity and sale as secondhand clothing; there is a need for recognizing the value of reuse and recycling. It is most often that the homemaker decides on how to deal with disposal of clothing. There is a need not only on creating awareness among homemakers as consumers on facilities available but also opportunities in participating in recycling.

4.1.1 RECYCLING USED CLOTHES IN INDIA

Norris[10] observed that clothing is too replete with social meaning to be wasted until it is literally falling apart. Used clothing is a valuable resource; it can always be strategically gifted, used up, or exchanged for some more desirable, textiles are rarely found among garbage dumps for domestic waste; unwanted clothing follows a different route out of the house altogether. Unlike plastics, metals, and glass recovered by rag-pickers, textiles are rarely found among garbage dumps for domestic waste; unwanted clothing follows a different route out of the house altogether.[11–13] According to an article, the journey of a garment does not end as a landfill all the time. They may be recycled in two ways; either it will be sold again as used clothing or will be shredded and mechanically recycled into raw material for the manufacture of other recycled apparel products. Clothes are being utilized in many ways; efficiently. Some of them even undergo a fundamental treatment and are being transformed into completely a different clothing; altogether. Threads from a woolen sweater may be wound up into balls which are used for embroidering on costly materials while manufacturing women's slips. Tailors buy small pieces of clothes, sew them together, dye them and sell it in the market as bed linen (www.fibre2fashion.com/a-second-life-for-fashion-used-clothing1.asp). Nakano's[9] survey studies on perceptions toward clothes with recycled content show that there is a contradiction between the public reaction toward products with recycled content and their awareness toward environmental issues. Developing end markets for recycled materials is necessary for recycling to be successful, and the achievements of those markets depends on consumer demand for products.

4.1.2 TRADITION IN RECYCLING

Cloth and clothing is never just thrown away in India. Indian traditions promoted the use of secondhand clothes from older members in the family to younger ones. A newborn was dressed and wrapped in used clothing instead of new. There was a time when women took pride in inheriting grandma's nine yard sari or lehengas in brocade with pure golden or silk threads woven. Expensive clothes were well preserved and stored carefully to prolong the life of garments even though care labels did not exist in those times. For decades, Bengali women have recycled and reused old clothes. The run-stitch Kantha used to create patterns on quilts two to four layers of mostly old cotton clothes with rural designs are now being used on dress material and saris. (www.andhranews.net/.../17-Kantha-embroidery-makes-95360.asp) In ancient India, patchwork or appliqué consisting of a cut or pieced design of one fabric is applied to the surface of another. Among the traditional textiles of India, the art of appliqué work occupies a distinguished place. According to the Webster's dictionary, appliqué is a "decoration or trimming made of one material attached by sewing, gluing, etc. to another." In India, appliqué art is widely prevalent in the western states, especially in Gujarat and Rajasthan, and in the eastern coast of Orissa (http://www.indiasite.com/arts/appliquework.html).

4.1.3 SECONDHAND CLOTHING

Secondhand clothing has a history, as old as manufacturing of clothes itself. A profitable market for used clothing is conspicuous both in the urban and village areas. The current era of internet boosts up the resale of clothing.[14] Many people sell their old clothing directly to other individuals through sites such as eBay. Consignment and thrift shops are also becoming increasingly popular, growing at a rate of 5% every year providing options for the sale of used clothing. (www.fibre2fashion). There is a demand for secondhand clothing. The entire supply chain of secondhand clothing market is thus managed by thousands who live on daily earnings from the sale of used clothes.[15] Hiller Connell[14] also suggested that there is a need to educate consumers about eco-conscious nature of secondhand clothing consumption more for the environmental benefits than economic ones. A study is under publication by Ekstrom and Salomonson[3] on finding solutions on the problem of sustainable reuse and recycling clothing in a macromarketing perspective involving different actors in the society.

4.1.4 OVERCONSUMPTION

According to evidence from a survey of consumer behavior conducted by WRAP in 2013, the average UK household owns around 30% of clothing in wardrobes has not been worn for at least a year; the cost of this unused clothing is around £30 billion; extending the average life of clothes by just 3 months of active use would lead to a 5–10% reduction in each of the carbon, water, and waste footprints; an estimated 350,000 ton of used clothing goes to landfill in the United Kingdom every year.

In the Indian context, Norris[10] observed that "…middle class Indian families … have more disposable income, give and receive more clothes, purchase more fashionable items, and have fewer avenues for disposal of them. Women's wardrobes overflowing with clothing, … often with more clothing coming into the home than can be properly dealt with by former strategies of domestic reuse and recycling. Often clothing are not fully worn out before it must be got rid of … and more fashionable forms generated by the increasingly rapid changes in fashion."

The apparel retail industry has grown in the last decade. There has been a greater purchasing power among Indian consumers. Homemakers as consumers of clothes have therefore a surplus of clothes to be discarded.

There are a large number of branded and unbranded clothing that is discarded from households much earlier than its life. The clothes may not be torn, faded, or discolored; out of boredom, consumers may want to discard them. Families with children need to discard more often as they outgrow the clothes. Discarding clothes makes more room and space in the wardrobe, thus gives scope for new purchases. Consumers from the middle and upper class are well aware of different ways of remodeling old clothes. Primary consumers are aware of old clothes being used for other utility purposes. In practice, this is seldom done.[7] Lang et al[6] opined that clothing products remain important in today's consumer culture, but the sustainability of that consumption is questionable as it often leads to excess waste.

4.1.5 PROBLEM OF DISPOSAL

Norris identified three reasons why it has probably always been difficult to redistribute excess clothing effectively; people seem to feel that the excess is increasing. There are three reasons for this:

- First, it is harder to hand clothes on to extended kin networks and residents of natal villages, their traditional recipients.
- Second, commercial products to furnish the home and use as cleaning cloths are increasingly available and somewhat superseded the reuse of cloth within the home.
- Third, the increasing speed of changes in clothing fashion may have a determining role—many women are acquiring more clothes to keep up with the changes.

Women begin to feel that they cannot keep outdated clothes that they would no longer wear but that are not worn out, a feeling that conflicts with the practice of thrift within domestic and personal special networks.

4.1.6 ROLE OF HOMEMAKERS AS CONSUMERS IN RECYCLING

The women have to decide how to balance the moral expectations surrounding them: whether to keep valuable or auspicious clothes circulating within the family, hand serviceable ones down to servants, or thriftily exchange them for something more desirable. Their decision is affected by both family's circumstances and the meanings inherent in every piece of cloth. The surplus clothes unsuitable for family use are sacrificed, and in the process, the "remains" acquire exchange value; barter is the most appropriate form of commodity exchange by which to get rid of them, and in north India, pots are particularly suitable to be exchanged for old clothes. Treasured pieces can be preserved for favorite younger relatives, suitable serviceable clothes gifted to a maid and rags reused in the house. Used clothing is made to work, to produce value for the home.[10]

4.1.7 PRODUCTS FROM USED CLOTHING

The demand for recycled sari clothes in the Western style dresses, such as halter-neck tops, capri pants, and hot pants, restyled from old wedding saris in the Western countries and among tourists to Delhi and Rajasthan have been reported.[10] The findings of a survey on recycling of Indian clothes by homemakers from middle and upper middle class are well aware of different ways of remodeling old clothes to prolong the usage. Expensive clothes, saris, or their borders are stitched into decorative cushion covers, quilts, and carry bags. Faded clothes are re-dying in a darker or a different color. Dupattas (stoles) are used to stitch kurtis (tops). Remodeling takes away the aesthetic value (cut/fit) of the dress, it is time-consuming and a futile effort. Reuse of clothes for other utility purposes such as bags, quilts, laundry bags, pillow covers, tea-cozy covers, tray cloth, kitchen napkins, and table mats is carried out.[7] At the household level, many utility products can be made such as pouches, bags covers for equipment, table cloth, table mats, cover for storing silver cutlery, as a substitute for gloves, soft furnishing including, bed covers, cushion covers, bolster covers, quilts called sujni, comforters, duvet cover. Miscellaneous articles such as photo frame, display board for family photographs, hand puppets, rag dolls, shaped cushions, and stuffing soft toys, small-sized quilted mats for prayers, yoga, and meditation, floor rugs, and shaped door mats are also prepared[15]. A study on materialism and clothing post purchase behaviors[5] regarding apparel hoarding, disposing, and participation in recycling suggests that marketing media should address benefits and ways to recycle and educate consumers in sustainable consumption behaviors.

Although there is an awareness of what kind of products can be made from used clothes, there remains a very little understanding of how to increase the number of consumers and who will recycle their clothing into products. There is a need for research that focuses on homemakers who do not currently recycle even a portion of their unused clothes. The problem is that the homemakers are not in a position to recycle their clothes into products for reuse. This is due to lack of time and motivation. This is the problem that the present study attempts to address.

Hiller Connell,[14] studied the degree of acquiring and the perceived barriers in acquiring apparel from secondhand sources among American University students. The findings suggested the attitudinal and contextual barriers as obstacles in purchasing and using clothing from secondhand stores. Such stores are not popular in India, although a large population uses secondhand clothing from street vendors and weekly markets[8]. It was felt that these barriers would not exist if the family clothing was recycled for their own household needs. The consumers would not mind getting products made from their own clothing instead of those worn and belonged to someone else. In this study, therefore, sample products were designed and displayed to the consumers so that they could place an order for similar products to be made from their own clothing waste. This study would help understand the willingness to accept more attractive and useful products made from their discarded clothing.

4.2 OBJECTIVES

- To develop household utility products from clothing waste.
- To use the entrepreneurial skills of women from self-help groups (SHGs) in product development and provide income generation opportunities.
- To understand the acceptability of "made to order" sample products developed from clothing waste.

4.3 METHODOLOGY

An experimental design was used for this study. The study was conducted in Mumbai, India over a period of 7 months in three stages.

In Stage 1 of the study, members of three women's SHGs who had undergone skill oriented training in bag making and basic stitching (provided by a nationalized bank as part of their Community Rating System activity) were purposively selected and guided in developing sample products from clothing waste. Most products required the use of foam or canvas and lining cloth to maintain the stability and shape. The objective was to extend support and motivate them to move toward self reliance. Guiding them step by step in the product development stage gave an opportunity to understand their problems. It was necessary to regularly interact with them individually and in groups and solve their issues, and ensure that action plan to sustain and enhance the growth of the SHGs known as Bachat Gats in India.

In Stage 2, further refining of products with color coordination, better finishing, and mix and match design ideas were provided by undergraduate students. The necessary surface ornamentation was done for attractiveness. Additional inputs in sourcing for lining, interlining, canvas, fasteners, and trims were provided. In all, 14 products developed as samples specifically designed for household uses. The sample products were identified by a name and code number. The cost of making the product including material and labor was labeled on the product.

In Stage 3, the semistructured opinionaire as a research tool was prepared using a 4-point rating scale. These products were displayed for inspection to 49 homemakers by convenience sampling method on the basis of their willingness to participate in the study. The respondents were informed that the products were not for sale and they could place an order for making similar products from their own clothing. Eleven responses were rejected due to incomplete responses. In all, 38 responses were statically analyzed and the results are discussed.

4.4 RESULTS AND DISCUSSION

4.4.1 OVERALL APPEARANCE, FUNCTIONALITY/UTILITY, AND DURABILITY

Consumer perception through visual examination of the products displayed, a majority of 97% (n = 37, N = 38) found the overall appearance of the products to be good and the remaining 3% found the overall appearance and finishing of the products satisfactory as seen in Figure 4.1.

There was no respondent who was undecided on this opinion. Functional utility of the products was perceived by the respondents easily. Alternative uses of some bags were suggested by a few respondents.

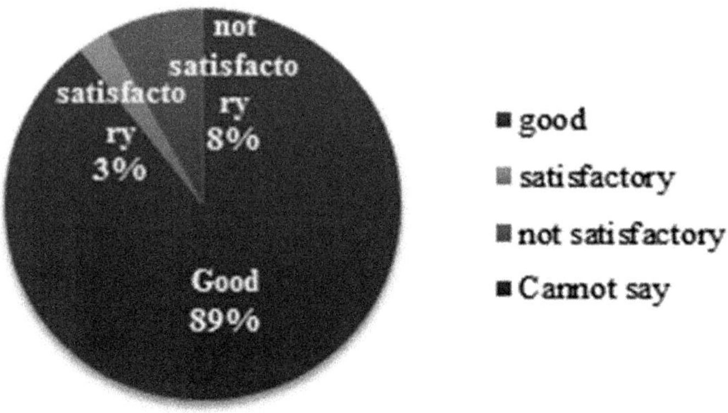

FIGURE 4.1 Perception of functionality/utility of the product.

A majority of 89% ($n = 34$, $N = 38$) opined that the products were good and 3% said they were satisfactory. Contrary to this, 8% of respondents were not satisfied. Further, the perceived durability of products developed from clothing waste fabrics along with suitable lining of the same fabric with foam and fasteners, was judged by the appearance.

As in Figure 4.2, 68% of the respondents said that products may be good but 32% were not sure, whereas 92% were open to the idea of upcycling their clothes.

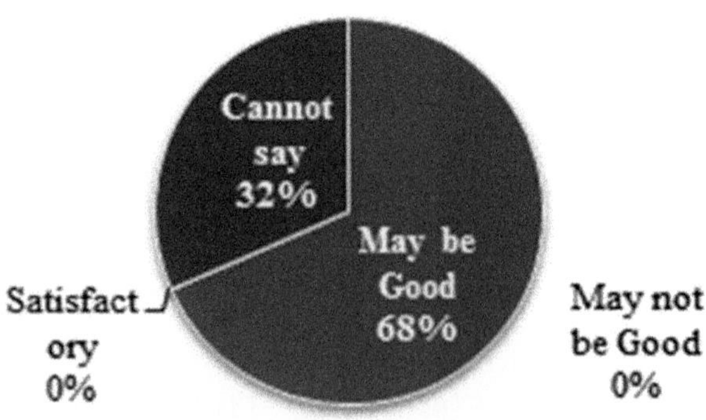

FIGURE 4.2 Perception of durability of the product from the appearance.

4.4.2 CATEGORY OF PRODUCTS AND RATING

For ease in analysis of opinions of respondents, all the 14 products were classified into two categories:

Category I: Miscellaneous products including table linen and bed linen and

Category II: Bags for various purposes.

The details of the same are given below in Table 4.1. In Category I, the large cushion covers was ranked highest and the most preferred product. In Category II, the most preferred bag was Product No. 14, the pouch followed by Product No. 6, the box bag.

TABLE 4.1 Details of Products and Rating of Design and Value for Money.

S. No	Category	Product number	Description of the product	Ranking—product design	Ranking—value for money
1	**I**	Product 1	Quilted bedcover	3.236	2.421
2	**Table and bed linen**	Product 2	Sujni quilt	3.342	2.921
3		Product 3	Table mats set	3.315	2.921
4		Product 4	Cushion covers small	3.315	3.0
5		Product 5	Cushion covers large	3.526	3.0
6	**II**	Product 6	Box bag	3.5	2.868
7	**Bags**	Product 7	Tote bag	3.421	2.894
8		Product 8	Damru bag	3.421	3.0
9		Product 9	Sari cover	3.0	2.842
10		Product 10	Jhola	3.157	2.763
11		Product 11	Shiba bag	3.263	2.763
12		Product 12	Lunch bag	3.263	3.0
13		Product 13	Travel pouch	3.421	3.0
14		Product 14	Pouch	**3.684**	3.0

4.4.3 ACCEPTABILITY OF DESIGN OF PRODUCT

Each of the respondents inspected each of the products and recorded their responses on a 4-point rating scale.

In total, 58% (n = 22, N = 38) respondents found the design of large size cushion covers 15″ × 15″ as very good, whereas 37% found it good as seen in Figure 4.3. The design of table mats was reported to be very good by 53% followed by sujni quilt. Both the cushion covers and table mats were made from sari. As seen in Figure 4.4, the design of Damru bag, travel pouch, and small pouch were found most acceptable.

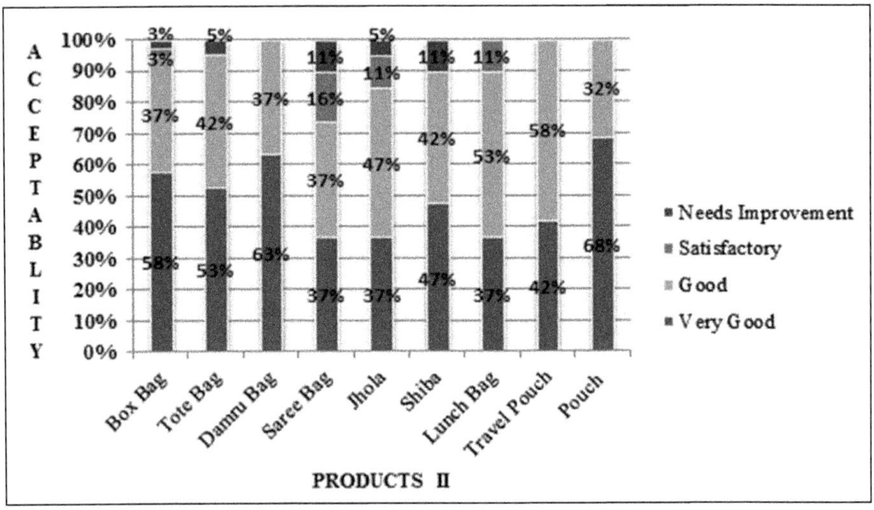

FIGURE 4.3 Acceptability of product design—Category I.

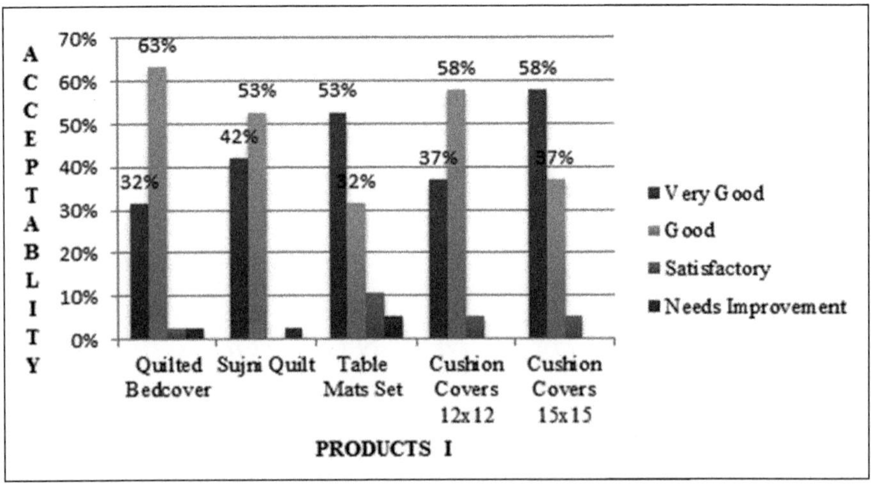

FIGURE 4.4 Acceptability of product design—Category II.

Rating scale: Very good—4, Good—3, Satisfactory—2, Needs improvement—1.

There were less than 16% respondents who found the other products satisfactory or needed some improvement in design.

4.4.4 COMPARATIVE RATINGS OF THE PRODUCTS ON VALUE FOR MONEY SPENT ON RECYCLING

Cushion covers, Damru bag, lunch bag, and travel pouch got the maximum overall rating for value for money spent on upcycling. As seen in Table 4.1 from Category I, table mats and sujni quilt were rated over the quilted bedcover. Respondents found more value for money spent on developing the quilted bed cover and sari bag. These were rated higher the bags made from denim jeans. Products in the category II of bags, the Damru bag, lunch bag, travel pouch, and small pouch were rated higher over other bags.[16,17]

4.4.5 ADDITIONAL FEEDBACK ON THE PRODUCTS

Some of the respondents through informal oral responses suggested that the lunch bag could also be used for multiple uses as vanity bag, pooja kit (prayer items), etc. Sujni quilt, they said, seemed narrow and needed more than the sari width of 45–48 in. Size of the quilted bedcover needed to cater to lager beds of king and queen size. In addition, it may be necessary to customize according to the width of the mattress. A couple of respondents opined that not all are very well designed—only four to five products are exceptionally good. Damru bag must be made bigger in size for overnight travel. Handle of lunch bag needs to be broader, stiffer, and longer. Queries included the number of saris required to make quilts, the number of bags that can be made from one sari, and whether an assortment of bags be made out of one sari. They were answered in the affirmative.[18,19]

4.4.6 INCOME GENERATION THROUGH PRODUCT DEVELOPMENT

This facility of made to order recycled products from clothing waste would help the homemakers recycle their clothes for a nominal cost. This would certainly help the women from SHGs earn an income. Pricing and costing and inculcating a spirit of competitiveness were areas that needed to be developed. During the course of administering the opinionnaire an order was placed for 10 travel pouches. Leaflets of the products and other services were distributed to spread awareness of the recycling facility. This facility has been available in Mumbai city at present. Few orders have been tricking in and the concept of made to order products is yet to catch up. Measures need to be taken in promoting environmental conscious behavior by encouraging Indian consumers to recycle their clothes.

4.5 CONCLUSIONS

India faces a duel problem of increased affordability, overconsumption, and problem of disposal on one hand and poverty and dependence on secondhand clothing on the other. The study would help in creating awareness in the homemakers of the facility of getting their unused waste clothes recycled to utility products for a nominal cost. Acceptability of the products developed and exhibited was measured through the opinionnaire. The products such as cushion covers, Damru bag, and travel pouches were most preferred products followed by table mats and sujni quilts for recycling from clothing waste. Cushion covers, Damru bag, travel pouch, and small pouch were the most accepted designs and value for money spent on making them. Further inputs are required of improvement of design, customizing products, and training the women of SHGs. Recycling helps in extending the life of the product instead of discarding. There is scope for further research in identifying barriers and reducing them in acquiring recycled products from their own clothing waste.

ACKNOWLEDGMENTS

The authors acknowledge the contribution of the undergraduate students of the Department of Textiles and Apparel Designing, SVT College of Home Science, Mumbai for their mix and match design ideas; Mrs. Armaiti Shukla, HOD, SVT and Mrs. Anju Tulshyan for their valuable inputs; Mrs. Vandana Joshi, retired Officer In-Charge of Cell for Entrepreneurial Development for women and Mr. S. W Joshi, Manager, Microfinance Branch Dharavi, Canara Bank for their cooperation in identifying the women SHGs.

KEYWORDS

- **clothing waste**
- **recycle**
- **tradition in recycling**
- **reuse**
- **secondhand clothes**

REFERENCES

1. Anon. Boost to T&C Recycling in EU. *Indian Text. J.* **2009,** *5*, 8–10.
2. Hiller Connell, K. Y. Internal and External Barriers to Eco-conscious Apparel Acquisition. *Int. J. Consum. Stud.* **2010,** *34*, 279–286.
3. Ekstrom, K. M.; Salomonson, N. Reuse and Recycling of Clothing and Textiles—A Network Approach. *J. Macromarketing*. DOI: 10.1177/0276146714529658. http://jmk.sagepub.com/content/early/2014/04/01/0276146714529658.abstract (accessed April 2, 2014).
4. Gulich, B. Development of Products Made from Reclaimed Fibres. In *Recycling in Textiles*; Wang, Y., Ed.; Woodhead Publication Ltd.: Cambridge, England, 2010; pp 117–135.
5. Hyun-Mee, J. Materialism and Clothing Post-purchase Behaviors. *J. Consum. Market.* **2013,** *30*(6), 530–537.
6. Lang, C., Armstrong, C. M., Brannon, L. A. Drivers of Clothing Disposal in the US: An Exploration of the Role of Personal Attributes and Behaviours in Frequent Disposal. *Int. J. Consum. Stud.* **2013,** *37*(6), 706–714. DOI: 10.1111/ijcs.12060 http://onlinelibrary.wiley.com/doi/10.1111/ijcs.12060/abstract (accessed August 29, 2013).
7. Mundkur, S. Recycling of Used Clothing in India. *J. Asian Reg. Assoc. Home Econ. (ARAHE)* **2010,** *17*(1), 26–30.
8. Mundkur, S.; Dedhia, E. Waste Management and Supply Chain for Second-hand Clothes in Mumbai. *Text. Value Chain* **2012,** *1*(3), 57–58.
9. Nakano, Y. *Perceptions Towards Clothes with Recycled Content and Environmental Awareness: The Development of End Markets in Eco-textiles—The Way Forward for Sustainable Development in Textiles*; Miraftab, M.; Horrocks, A. R., Eds.; Woodhead Publishing Ltd.: Cambridge, England, 2007; p 3.
10. Norris, L. *Recycling Indian Clothing—Global Contexts of Reuse and Value*; Indiana University Press: Indiana, U.S.A, 2010. p 9, 37,130,136,151.
11. India at a Glance: Census 2011 Provisional Population Totals. http://censusindia.gov.in/2011-prov-results/indiaatglance.html (accessed July 27, 2014).
12. Valuing Our Clothes; WRAP Report, UK; 2013 http://www.wrap.org.uk/content/valuing-our-clothes (accessed June 26, 2014).

13. Groot, H.; Luiken, A. In *Research Areas for Upgrading Textile Recycling*. Proceedings of Conference; Ecotextiles '98—Sustainable Development; pp 159–164; R. Horrocks Ed.; The Bolton Moat House: Woodhead Publishing Ltd., 1998.
14. Hiller Connel, K.Y. *Book of ITAA Proceedings, International Textile and Apparel Association;* Inc. #66 Bellevue, Washington, USA, 2009. www.itaaonline.org (accessed June 27, 2014).
15. Mundkur, S.; Dedhia E. In *Book of Abstracts*, XXII Word Congress International Federation of Home Economics (IFHE) at Melbourne, Australia; July 16–21, 2012. p 51.
16. Sinha, P.; Dissanayake, D. G. K., Hussey, C.; Barltett, C. In *Recycled Fashion: Taking up the Global Challenge*, Proceedings of 15th Annual International Sustainable Development Research Conference; Utrecht University, the Netherlands; July 5–9, 2009. http://globalchallenge2009.geo.uu.nl/ 22 Apr 2013 16:26 (accessed June 18, 2014).
17. Menon, A. Kantha Embroidery Makes Bengali Rural Women Self-reliant. www.andhranews.net/.../17-Kantha-embroidery-makes-95360.asp (accessed June 25, 2014).
18. Anon. Applique Work. Retrieved from http://www.indiasite.com/arts/appliquework.html (accessed June 25, 2014).
19. Anon. A Second Life for Fashion: Used Clothing, Fibre2Fashion. http://www.fibre2fashion.com/industry-article/15/1452/a-second-life-for-fashion-used-clothing1.asp (accessed June 25, 2014).

CHAPTER 5

RECYCLING SOLID WASTE BY COPROCESSING

M. NEFTALÍ ROJAS-VALENCIA*, DAVID MORILLÓN GÁLVEZ, SERGIO MARÍN MALDONADO, and JORGE EMIGDIO SÁNCHEZ PÓLITO

Institute of Engineering, Coordination of Environmental Engineering, National Autonomous University of Mexico, Post Box 70-472, Coyoacán 04510, D. F. Mexico, Mexico

**Corresponding author. E-mail: nrov@pumas.iingen.unam.mx*

CONTENTS

ABSTRACT

The object of this study was to analyze the current situation, both at national and international levels, as regard to the recycling of solid wastes through the thermal treatment currently known as coprocessing. An exhaustive study of the companies in Mexico that have introduced the coprocessing approach was conducted focusing on the materials that are integrated in the process, the calorific capacity of the solid wastes used, and the savings generated by this strategy in terms of nonrenewable fossil fuels. For this purpose, articles, textbooks, journals, and congress proceedings were reviewed. The results showed that worldwide, 10 million tons of wastes and residues are coprocessed every year. Countries such as Japan, the United States of America, Canada, and members of the European Union have used coprocessing as an environmentally sustainable and economically viable alternative in their production processes. The cement industry in Germany, Belgium, Austria, and Switzerland currently substitutes over 40% of its fuel needs by wastes especially conditioned for said purpose, reaching over 80% in Holland. In Latin America, in countries such as Argentina, Brazil, Chile, Costa Rica, Colombia, Guatemala, Mexico, and the Dominican Republic, among others, substitution ranges from 7% to 18%. Based on the review of various studies, it can be concluded that if appropriately handled, coprocessing should not have an impact on the health of the people or the environment. For waste sustainable management, it is necessary to responsibly appraise the best disposal alternative.

5.1 INTRODUCTION

In Mexico, as well as in other developing countries, waste management is not as efficient as in developed countries. Wastes are dumped in streets where they end up choking sewage systems, inappropriately incinerated in vacant lots or in generator installations, buried, unlawfully dumped in inappropriate places, or taken to final disposal sites that were not designed for urban solid wastes (USW), let alone industrial ones. All the above activities negatively impact the environment, causing soil, water, and air contamination that lead to the deterioration of living and health conditions of the neighboring settlements. They also have a negative impact at social and economic levels.

Statistics for the year 2011 show that in Mexico, about 41 million tons of USW were generated. According to the Secretaría de Medio Ambiente y Recursos Naturales (SEMARNAT) [Secretariat of Environment and Natural Resources], only 4.8% of USW is recycled, and thus about 10 million tons of textiles, plastics, paper, cardboard, and paper products remain in the various sites of final disposal. With regard to special handling wastes, for the same year, about 3.6 thousand tons were generated. Moreover, 30 million tires and over 325 million L of used oil are generated per year, among other hazardous wastes, which could have been used as fuel.

There is an important need to find a viable option for the handling of waste materials and various technological options have been devised for treating them. Among these possibilities, we can mention biological, mechanical–biological, and thermal treatments.

Particularly, the waste thermal treatment can be defined as the conversion of solid wastes into gas, liquid, and solid products with the concurrent emission of energy in form of heat. Among these thermal processing technologies, we can mention:

- Incineration: It basically refers to the combustion of organic matter through the thermal oxidation process conducted in presence of oxygen within a temperature ranging from 800°C to 1800°C.
- Gasification: It is a thermal treatment conducted at high temperatures (around 2000°C) in which air (oxygen) is injected to produce mineralized waste and synthesis gas.
- Pyrolysis: It the thermal degradation of USW occurring at temperatures ranging from 400°C to 800°C in absence of oxygen in order to prevent gasification.
- Plasma: It is the process that permits the generation of energy and various exploitable products. However, we are still at the initial stage of its development because it is an expensive process. It is conducted at temperatures ranging from 3000°C to 20,000°C.

Currently, no sufficient practical information is available as regard to the operation of these technologies in Mexico in order to determine the technical feasibility of their instrumentation. However, it is known that, at experimental level, incineration, plasma, and plasma-arc assisted gasification have been applied, obtaining highly favorable results in the case of the last one, with weight reduction and volume reduction reaching up to 92% and 97%, respectively. Moreover, it is possible to generate and exploit the energy in the same process. However, the main drawback is that its operation is costly.

Simultaneously, another thermal method known as coprocessing has been developed. It is a treatment that uses hazardous USW requiring special handling (formulated or recovery fuel) as source of energy, replacing natural mineral resources, and fossil fuels such as coal, oil, and gas.

The difference between coprocessing and conventional incineration is that coprocessing takes advantage of the energy and the minerals present in wastes because of high temperatures reaching 2000°C; the permanence time of the gases at high temperatures is greater than 5 s, self-cleaning process of the gases because of an alkaline atmosphere, obtaining the integration of the ashes into the cement manufacturing process, while in the case of incineration, wastes generate ashes that must then be safely handled.

The Ley General Para la Prevención y Gestión Integral de los Residuos[12] [General Law for the Prevention and Integral Management of Wastes] (Official Gazette of the Federation, 2014), in Article 5, number four, states that *co-processing leads to an environmental safe integration of the waste generated by a known industry or source, through its use as alternative fuel for energy generation, which can be exploited in the production of goods and services.*

When the coprocessing activity is conducted under the criteria described in its definition "environmental safe integration," the expected result would be beneficial both for health and environment because wastes would be eliminated as a potential source of risk.

The abovementioned law indicates that *a distinction should be made among the wastes that, because of their characteristics, generation volumes, accumulation, environmental problems, economic and social impacts caused by their inadequate handling, could be subjected to co-processing. In turn, restrictions must be established as regard to the incineration or co-processing through combustion of wastes susceptible to be valorized through other processes, whenever available, environmentally efficient, and technologically and economically feasible. Only thus, actions susceptible to strengthen the valorization or treatment infrastructure of these wastes through other means will have to be promoted.*

5.2 CURRENT SITUATION OF THE USE OF COPROCESSING

As already mentioned, the inappropriate management of solid wastes in developing countries, together with the reduction of fossil fuel reserves make it necessary to develop new technical solutions and cooperation forms. This challenging development has led the Deutsche Gesellschaft für Technische Zusammenarbeit (GTZ), a company of international cooperation for the sustainable development of operations throughout the world, and Holcim, one of the world's leading cement and aggregate manufacturers, to exploit options for the use of wastes that otherwise would be useless, and sometimes troublesome, as valuable resource in an energy intensive industry.

Some deficiencies have been detected in waste handling, because:

- No all the developing countries have an integrated waste management strategy.
- Only a few developing countries have the adequate technical infrastructure for controlled and environmentally friendly waste disposal.
- Although in many cases there are laws regarding controlled waste management, said laws are frequently not adequately applied.
- Noncontrolled disposal is usually the most economical way and, frequently, the only form of waste disposal.
- The companies that generate industrial and commercial wastes tend not to be willing to pay much for adequate disposal.
- Legislators rarely give special attention to waste management and may be little informed with regard to the consequences on human health or the high cost related to the remediation of the damages caused by noncontrolled waste disposal.

In Figure 5.1, it can be seen that in 10 years (from 1995 to 2005) in the European Union, municipal solid wastes increased twice as much as population while the gross domestic product (GDP) increased a little more.

It is expected that municipal solid wastes volumes will increase twofold in the next 10 years despite the growing efforts made for recycling and reducing them (see Fig. 5.2). Although there are several alternative solutions for minimizing wastes, such as reduction, reuse, and recycling policies (the 3 Rs), currently over 80% of wastes are not used and they remain in dumping sites, landfills, or are illegally burned.

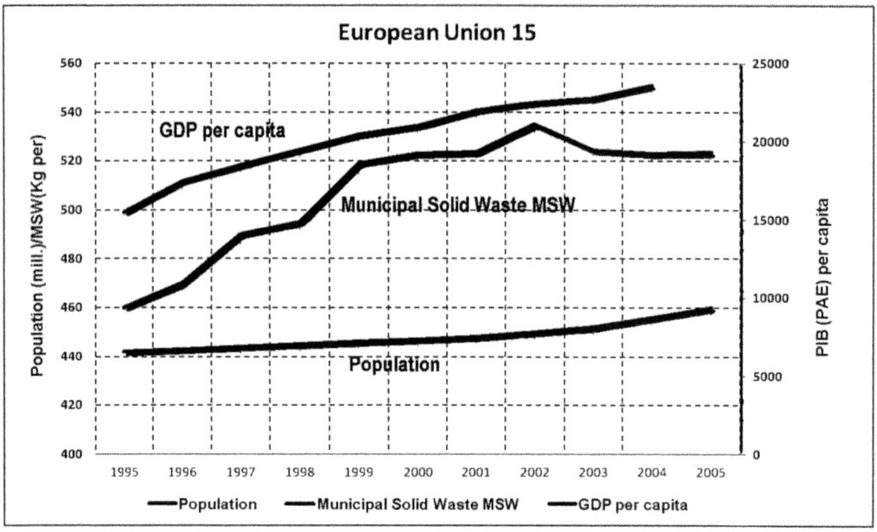

FIGURE 5.1 Correlation between population, and generation of municipal solid wastes in the European Union.

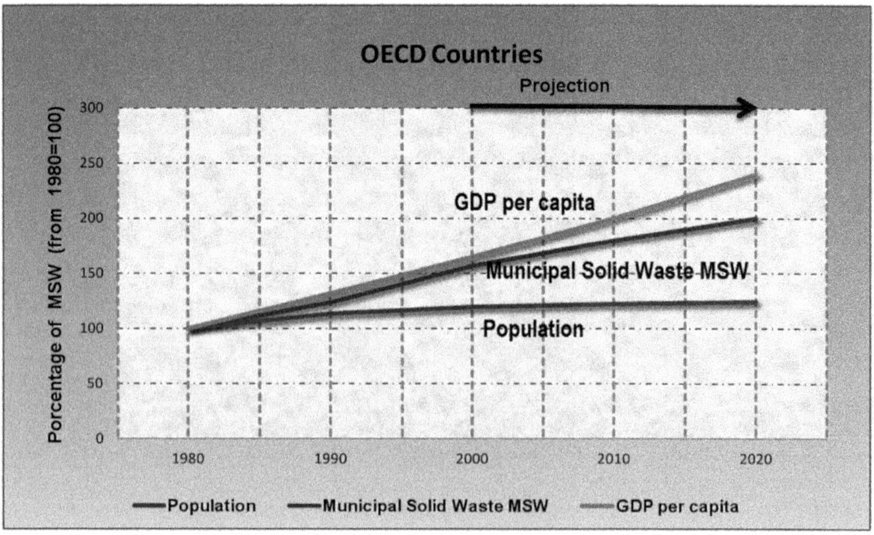

FIGURE 5.2 Correlation between GDP, population, and generation of municipal solid wastes in the OECD countries.

The members of the Organization for Economic Co-operation and Development (OCDE) are:

North America: Canada, United States, and Mexico.

Europe: Germany, Austria, Belgium, Denmark, Spain, Finland, France, Greece, Hungary, Ireland, Iceland, Italy, Luxemburg, Norway, The Netherlands, Poland, Portugal, United Kingdom, Czech Republic, Slovak Republic, Sweden, Switzerland, and Turkey.

Pacific: Australia, Japan, New Zealand, and Republic of Korea.

5.3 STUDY CASES IN DIFFERENT COUNTRIES

In 1999, Belgium faced an urgent public health problem that generated great interest: how to treat thousands of tons of potentially contaminated animal fat and meal. Federal authorities concluded that coprocessing in kilns of the cement industry was the best way to solve the crisis and the Belgium plants were summoned to collaborate. This process is supposed to be a safe and environmentally sound solution for the complete destruction of contaminants in the kiln as well as a reduction of emissions as a result of fuel substitution.[2]

A similar situation occurred in Italy in 2001. Italian cement plants, in collaboration with the authorities, were summoned to treat a large quantity of potentially contaminated animal meals.[2]

In 2007, the Spanish cement sector used 56.000 tons of waste tires as alternative fuel (AF), out of the 290.000 tons that were generated according to the estimates. On the other hand, in the same country, the cement sector took voluntarily part in the National Inventory of Dioxins and Furans conducted by the CIEMAT (belonging to the Secretariat of the Environment). It was concluded that the sector emission mean is 10 times smaller than the strictest legal limits. In the same way, no difference was evidenced between the emissions when conventional fossil fuels or waste coprocessing are used in partial fuel substitution.[2]

At Secil-Outão Intertox an itemized evaluation of the risks of coprocessing at Outão (Portugal) was conducted in order to determine the potential hazardous emissions that could represent a risk for human health and the ecosystem. It was calculated in the "worst possible scenario," simulating the accumulated effect of the least favorable situations in order to determine whether the emission levels could represent risk for human health and the environment.[2]

Currently, there is a greater awareness in many countries as regard to the use of coprocessing thermal technology, the legal frameworks have improved and finally the volume of coprocessed wastes in the cement plants has increased continuously. Four countries were selected for evaluating the functioning of said technology from 2004 to 2007, Figures 5.3 and 5.4 show that the Philippines, Morocco, and Chile increased the use of hazardous and nonhazardous wastes in their processes; Mexico remained at the forefront (see Figs. 5.3 and 5.4).

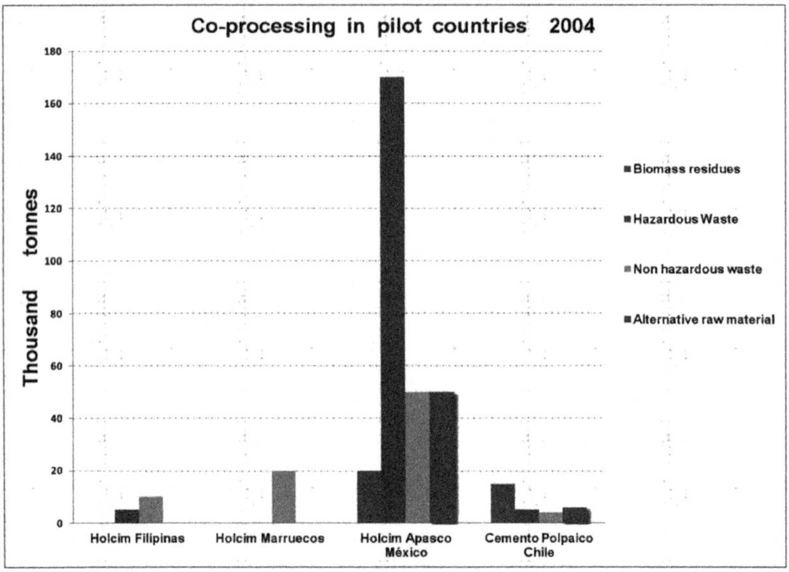

FIGURE 5.3 Pilot countries where coprocessing has been tested since 2004.

As can be seen in Figures 5.3 and 5.4, a good management of AF and alternative raw materials (AR/alternative raw materials, together, AFR), derives in: the use of AFR is advantageous for all the countries interested in integrating an alliance to avoid squandering materials that could be used in coprocessing. It does not increase the emissions of cement plants. Materials, products and emissions must be controlled at the source.

Figure 5.5 shows the progress of the use of the coprocessing technology in some continents from 2002 to 2007. It has to be highlighted that the European Community has been leading the trend. North America and Latin America have tried to follow this example while Asia and Africa, in 7 years, have not been able to adequately integrate wastes in coprocessing.

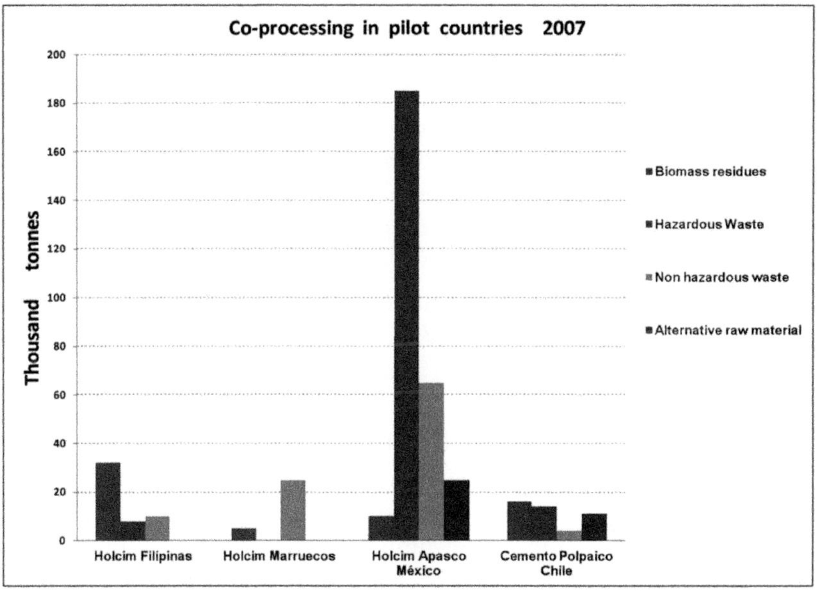

FIGURE 5.4 Pilot countries where coprocessing has been tested in 2007.

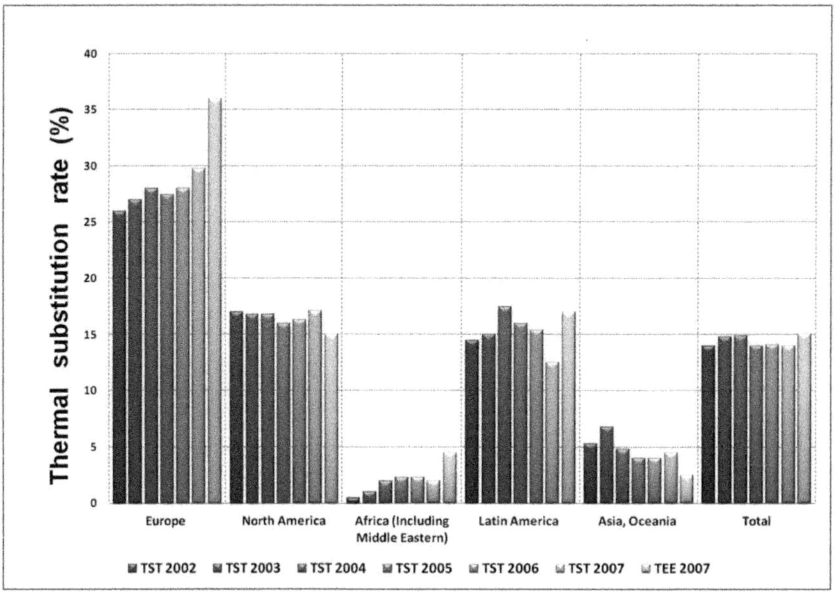

FIGURE 5.5 Use of alternative fuels and alternative raw materials in coprocessing.

The results showed that worldwide, 10 million tons of wastes and residues are coprocessed every year. Countries such as Japan, Thailand, Bangladesh, Egypt, Philippines, Indonesia, the United States of America, Canada, and members of the European Union have used coprocessing as an environmentally sustainable and economically viable alternative in their production processes. The cement industry in Germany, Belgium, Austria, and Switzerland currently substitutes over 40% of its fuel needs by wastes especially conditioned for said purpose, reaching over 80% in Holland. In Latin America, in countries such as Argentina, Brazil, Chile, Costa Rica, Colombia, Guatemala, Venezuela, Puerto Rico, Panama, Nicaragua, Mexico, and the Dominican Republic, among others, substitution ranges from 7% to 18%.

According to an estimate of European Cement Research Academy, the World Business Council Sustainable Development and the International Energy Agency, in 2030, in *developed countries* AF will represent between 40% and 60% of their consumption while in developing countries it will represent between 25% and 35% in 2050 (see Fig. 5.6).

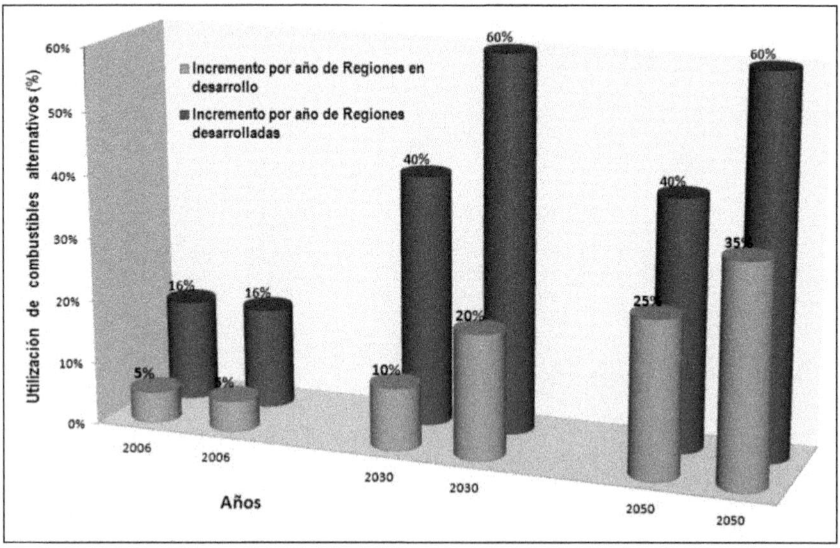

FIGURE 5.6 Estimated use of alternative fuels.

In *Mexico*, out of 34 cement plants, 32 use waste coprocessing as AF or raw materials. Figure 5.7 shows a list of the companies established in

Mexico and the type of waste they incorporate. In most of the cases, a substitution capacity by AF of 30% by weight is reported. However, said percentage may vary depending on the plant.

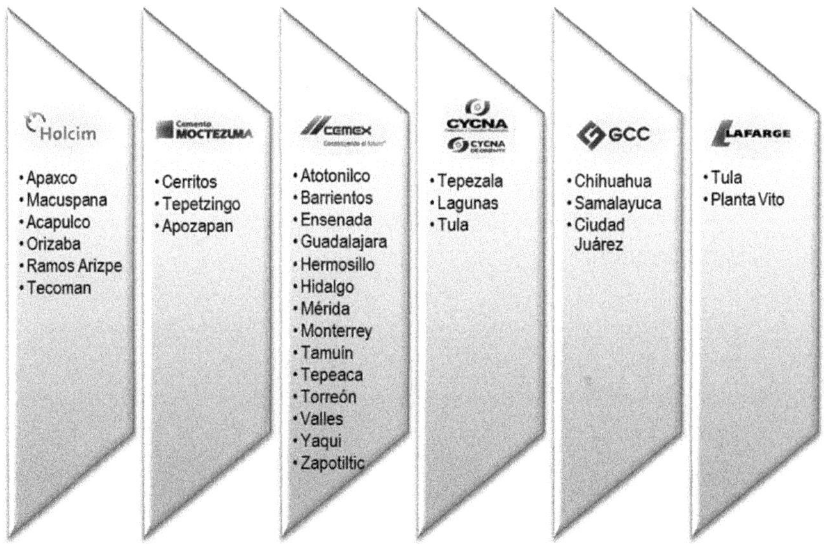

FIGURE 5.7 Coprocessing companies in Mexico.

CEMEX coprocesses over 90% of the wastes generated at its cement plants, through environmental care operations. In 2011, 632 tons of these by-products were coprocessed.[3]

Three thousand six hundred tons of waste paper and cardboard, treatment plant sludge, and airport residues are available. This figure is an underestimate since it considers only Mexico City Airport wastes.

In total 68% of hotel generated wastes can be used as AF, while in Mexico City, 91% of tires are not used.[10]

Petroleum wastes can also be used as coprocessing AF or raw materials. Since most of them contain oil, they have a large calorific power. This is also the case of solids wastes contaminated with hydrocarbons, oily sludge, and used lubricant oils. However, they have not been used at 100% yet.

This is mainly due to the following causes:

- Limited infrastructure for the proper handling of the wastes.

- Noncontrolled disposal which is the cheapest and most irresponsible way of getting rid of wastes.
- A large number of companies do not declare the real generation of their hazardous wastes.
- An important part of wastes is not disposed off according to the most sustainable environmental alternatives because wastes are still sent to confining places or practices are used that do not recycle, reuse, or recover materials and energy. As already mentioned, as regard to the sustainable management of wastes, there is an international strategy that indicates several alternatives to ensure that wastes are disposed off in the safest environmental manner depending on their characteristics.

However, in Mexico, the pyramid leading to the sustainable development is inverted compared to the European Community, North America, and Latin America (see Fig. 5.8).

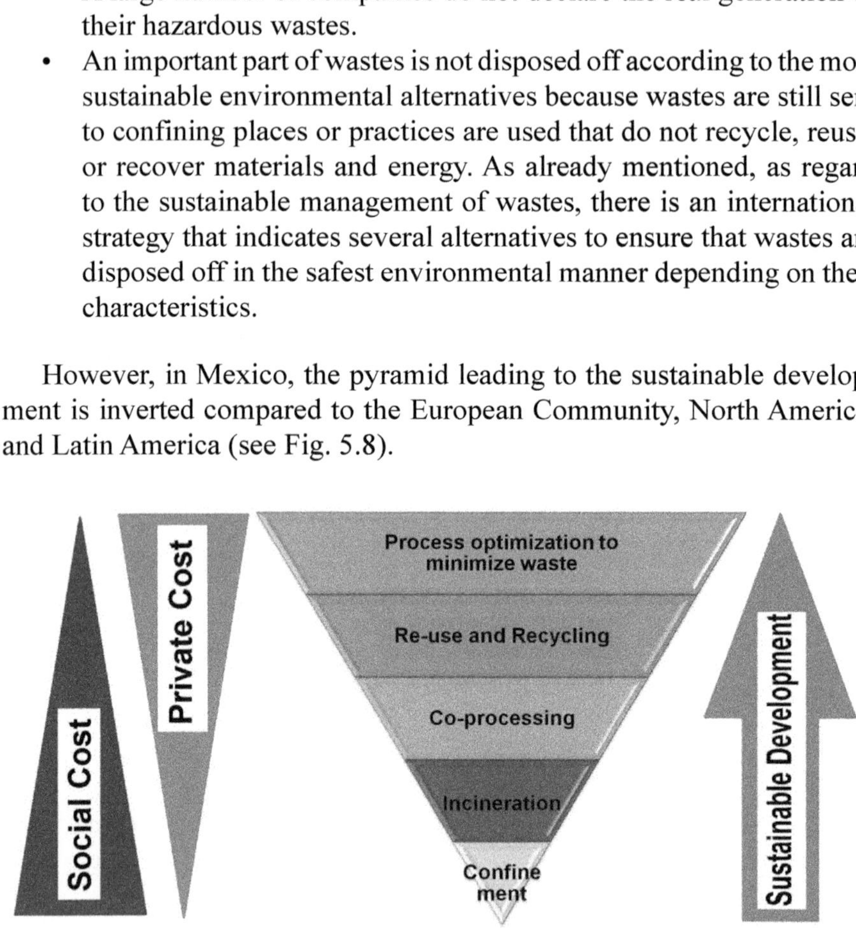

FIGURE 5.8 Pyramid in Mexico leading to sustainable development.

5.4 MAIN CHARACTERISTICS OF WASTE USE PROCESS

Although the practice varies from one plant to the other, the cement manufacturing may consume significant quantities of wastes as fuel and nonfuel raw materials. This consumption reflects the process characteristics in the clinker kilns, ensuring the complete rupture of the raw materials in the oxides comprising it and the recombination of said oxides in clinker minerals. The main characteristics of the process for the use of wastes and hazardous wastes to feed the kiln, through appropriate feeding points, can be summarized as follows:[7]

- Maximum temperatures of about 2000°C (main burner, flame temperature) in rotary kilns;
- Gas retention times of around 8 s at temperatures higher than 1200°C in rotary kilns;
- Material temperature of about 1450°C in the sinterization zone of the rotary kilns;
- Oxidizing gas atmosphere in rotary kilns;
- Gas retention time in secondary burners of about 2 s at temperatures higher than 850°C; in the precalciner, the retention times are conveniently longer and the temperatures higher;
- The solid waste temperature is 850°C in the secondary burner and in the calciner;
- Uniform combustion conditions for load fluctuations due to the high temperatures during sufficiently long retention times;
- Destruction of contaminant organisms because of the high temperatures during sufficiently long retention times;
- Absorption of gaseous components such as HF, HCl, and SO_2 in alkaline reagents;
- High retention capacity for heavy metal bound to particles;
- Short retention times of exhaust gases in the temperature range of the formation of PCDD/PCDF;
- Concurrent material recycling and energy recovery through the complete use of fuel ashes as clinker components;
- Lack of generation of specific wastes of the product because of the complete use of the material in the clinker matrix (although some cement plants discard kiln dust or deviated dust);
- Chemical–mineralogical incorporation of nonvolatile heavy metals to the clinker matrix.

In summary, a solid waste coprocessor works as follows: (1) before being coprocessed, the preselected wastes pass through an analysis phase in a laboratory; an analysis is performed in order to determine humidity, calorific power, chlorine, sulfur, heavy metals, etc; (2) the selected wastes pass then to the pretreatment phase wherein they are triturated to reduce their size; a triturator is used to obtain a size of about 5 × 5 cm and they are fed to a size selector; (3) the wastes are then taken to the kiln on the conveyor belt and finally the feeding of the materials to the cement kiln is controlled through the plant operation center (see Fig. 5.9).

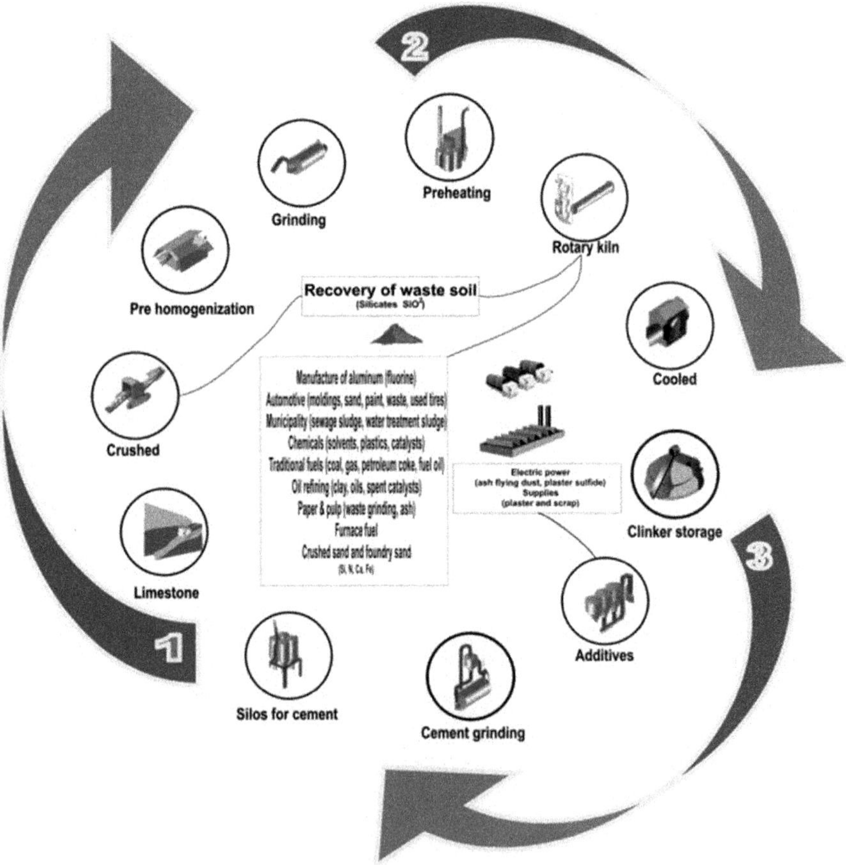

FIGURE 5.9 Flow diagram of the functioning of a coprocessor in a cement plant.

5.5 MATERIALS USED IN COPROCESSING

There is a huge variety of wastes and materials; thus, four main types of wastes can be classified as shown in Table 5.1.

TABLE 5.1 Coprocessable Wastes.

Wastes type	Uses
Liquid wastes	Used oils, solvents, polluted water, water base paints, solvent base, antifreeze, etc.
Solid wastes	Impregnated with hydrocarbon, solids coming from various industrial processes, plastics, greases, air, and gasoline filters, packaging materials, expired or out of specification materials and drugs, etc.
Sludges	Soil impacted with hydrocarbons, sludges from water treatment plants; oil coming from tank cleaning; phosphates coming from various manufacturing processes, resins, paint residues, etc.
Other wastes	Used tires (whole or triturated), materials rich in silicon, iron, calcium and/or aluminum, used catalyzers, wood meal, etc.

5.5.1 HAZARDOUS WASTES APPROPRIATE FOR COPROCESSING IN CEMENT KILNS

There is a wide range of hazardous wastes that are appropriate for coprocessing, but because of the fact that the emissions of cement kilns are specific for each installation, there is no uniform response with regard to the type of wastes that can be used in a given plant. Wastes selection depends on several factors such as the nature of wastes, their hazard characteristics, the available waste management operations, kiln functioning, raw material and fuel composition, waste feeding points, exhaust gases cleaning process, resulting clinker quality, general environmental impact, probability of formation and release of persisting organic pollutants, concrete waste management considerations, fulfillment of the standard, and acceptance by the government and the population.[7,8,15,19]

The following variables are some of the parameters that have to be taken into account for waste selection purposes:

i) Alkaline (sodium, potassium, etc.), sulfur, and chloride content: An excessive feeding of these compounds may cause accumulation and blocking in the kiln. When they cannot be captured in cement clinker or kiln dust, it may be necessary to install derivation to eliminate the compounds in excess from the kiln preheater and precalcinator systems.
ii) The high alkaline content may also limit the recycling of kiln dust in the kiln itself;

iii) Calorific power: It is a basic parameter of energy supplied to the process;
iv) Water content: The total humidity contents may affect productivity, efficiency, and increase energy consumption. Waste water content must be taken into account together with fuel and conventional raw materials water content;
v) Ash content: The ash content affects the chemical composition of the cement and makes it possibly necessary to adjust the raw material mixture composition;
vi) Exhaust gases flow speed and wastes feeding speed: A sufficiently long residence time is necessary to allow the destruction of organic compounds and prevent the production of an incomplete combustion because of waste overload;
vii) Functioning stability (e.g., duration and frequency of disconnections due to CO excess) and state (liquid and solid), preparation (crushed and milled), and residue homogeneity;
viii) Emissions: Organic compounds content—Organic constituents are associated with CO_2 emissions and can generate the emission of CO as well as other products of incomplete combustion, if wastes are introduced in inappropriate zones or in unstable functioning conditions.

Chloride content: Chlorides can combine with alkalis to form fine particles that are difficult to control. In some cases, chlorides have been combined with the ammonia present in the limestone used as raw material. This causes the release of highly visible flows of fine particles having a high ammonium chloride content.

Metal content: The nonvolatile behavior of most of the heavy metals makes it possible that most of them pass directly through the kiln system and are incorporated to the clinker. The volatile metals introduced are recycled partially internally through evaporation and condensation till a balance is reached; the rest is emitted in the exhaust gases. Thallium, mercury, and their compounds are highly volatiles; cadmium, lead, selenium, and their compounds are volatile to a lesser extent. It must be taken into account that dust control devices only capture the fraction of heavy metals and their compounds that is bound to the particles. Wood treated with preservatives containing copper, chromium, and arsenic also requires special considerations with regard to the efficiency of the exhaust gases cleaning system. Mercury is a highly volatile metal that, depending on the temperature of the exhaust gases, is found either bound to particles or as vapor in air contamination control equipments.[7]

5.5.2 COPROCESSING PRINCIPLES IN CEMENT MANUFACTURING

In order to avoid situations in which poor planning may cause an increase of contaminant emissions or the impossibility to agree on the priority of adopting preferable environmental management practices, the companies GTZ GmbH and Holcim Group Support Ltd. developed a group of general principles.[8] These principles (see Table 5.2) offer a comprehensive and concise summary of the basic considerations for planners and persons interested in coprocessing projects.

TABLE 5.2 General Principles of the Coprocessing of Wastes and Hazardous Wastes in Cement Kilns.

Principle	Description
Management hierarchy has to be respected.	Wastes must be coprocessed in cement kilns when there are no other ecological and economic methods to recover a larger quantity of solids.
	Coprocessing must be considered as an integral part of waste management.
	Coprocessing must adhere to the Basel and Stockholm Conventions as well as to other relevant international environmental agreements.
Additional emissions and negative impact on human health must be avoided.	The negative effects of environmental contamination on human health must be avoided or kept to a minimum.
	The atmospheric emissions from waste coprocessing in cement kilns cannot be statistically higher than the ones not coming from waste coprocessing.
Cement quality must not be modified.	The product (clinker, cement, and concrete) must not be used as deposit for heavy metals.
	The product must not have a negative impact on the environment (e.g., the one that can be determined through leachate assays).
	The quality of the product must allow recovery at the end of its useful life.

TABLE 5.2 *(Continued)*

Principle	Description
The companies performing coprocessing must be qualified.	Ensure the fulfillment of all the laws and standards.
	Have good fulfillment records as regard to environment and safety.
	Have dedicated in situ personnel, processes, and systems for the protection of environment, health and safety.
	Be able to control the contributions to the production process.
	Maintain good relation with the public and the other parties involved in the local, national, and international schemes of waste management.
Coprocessing application must take national circumstances into account.	The requirements and specific needs of each country must be reflected in the standards and procedures.
	Its application must allow the development of the necessary capacities and the establishment of institutional agreements.
	The introduction of coprocessing must be consistent with other change processes in the waste management infrastructure of a country.

i) In systems equipped with an appropriate alkali derivation system, exhaust gases can be released through an independent exhaust chimney or through the main kiln chimney. According to the Environmental Protection Agency (EPA) of the United States (1998), the same hazardous air contaminants are found in the main chimneys and in the chimneys of alkali derivation. Upon installing an alkali derivation system, it is necessary to supply also the derivation gas chimney with an appropriate system, similar to the one mandatory for the main chimney, for controlling the gases emitted to the atmosphere[15];

ii) A high sulfur content of the raw materials, fuel, and wastes may cause the release of sulfur dioxide, SO_2;

iii) Clinker, cement, and final product quality;

iv) Large quantities of phosphate may prolong setting time;

v) Large quantities of fluor will affect setting time and hardness development;

vi) Large quantities of chlorine, sulfur, and alkalis affect the global quality of the product;

vii) Thallium and chromium content may negatively affect cement quality and cause allergic reactions in sensitive users. Chromium leachate from concrete scraps may be more frequent than leachate from other metals.[21] Limestone, sand, and clay contain chromium, and thus chromium content in cement will be unavoidable and highly variable. The Norwegian National Institute of Labor Health[11] reviewed several studies as regard to allergy to chromates, especially the ones related to building industry workers, and found that the main sources of chromium in cement were the raw materials, the bricks of the kiln, and the chromated steel mills. The relative contribution of these factors may vary depending on the chromium content of the raw materials and the manufacturing conditions. Other less important sources include both conventional and AF.[7] Eczema caused by cement may be provoked by the exposition to wet cement having a high pH, which induces the appearance of an irritant contact dermatitis, and through an immunological reaction to chromium causing an allergic contact dermatitis.[11] If there is a skin contact possibility, cement containing, when hydrated, a percentage of soluble chromium(VI) greater than 0.0002% of the total cement dry weight, or preparations containing said cement, may not be used or sold in the European Union. Since the main source of chromates is the raw material used, in order to reduce the chromium(VI) quantity in cement, it is necessary to add a reducing agent to the finished product. The main reducing agents used in Europe are ferrous sulfate and tin sulfate.[7]

viii) Oligoelements leachate: Heavy metals are present in all raw materials, both conventional and nonconventional. Despite this fact, in some analysis conditions, the concentrations of other metals, besides chromium, in the concrete leachate may be around the levels found in drinking water.[8]

In principle, the following wastes should not be processed together in cement kilns:

a) Radioactive or nuclear wastes;
b) Electrical and electronic wastes (e-waste);
c) Whole batteries;
d) Corrosive wastes, including mineral acids;

e) Explosives;
f) Cyanide containing wastes;
g) Asbestos containing wastes;
h) Medical infectious wastes;
i) Chemical or biological weapons contemplated for destruction;
j) Wastes containing mercury or contaminated by it;
k) Unknown or unpredictable composition wastes, including nonclassified municipal wastes.

Each installation can also exclude other wastes depending on the local circumstances.

In general terms, the use of these wastes is not recommended because of health and safety concerns, the potentially negative impact on the functioning of the kilns, clinker quality, and atmospheric emissions, and when a preferable alternative waste management option is available.

Feeding the kiln with wastes containing mercury or contaminated by it should be avoided and kept to the lowest possible levels. In all the cases, a limit value should be established for mercury emissions since limiting the quantity of mercury in the wastes does not ensure a low emission of this metal from the kiln.

CEMEX is the first Mexican company to have developed a strategy for coprocessing AF denominated fraction of inorganic solid urban waste (FIRSU), at the Atotonilco, Ensenada, Guadalajara, Huichapan, Mérida, Monterrey, Tepeaca, and Zapotiltic plants, as part of an environmental sustainability program and innovation projects.[3]

In 2011, CEMEX, substituted 17% of the conventional fuels through the use of various AF, including FIRSU.[3] The combustion of USW permits an 85–90% volume reduction and USW ash is used as raw material for the production of cement clinker.[4]

FIRSU is obtained from the selection of trash containing nonfermentable matter colloquially called inorganic matter such as paper, plastics, and textiles that cannot be recycled because of their condition. These waste materials are compacted and ground, and then used as AF in cement kilns. Since the ash obtained from FIRSU calcination is compatible with the raw material, it is incorporated to the product elaboration process, closing the coprocessing cycle.[3]

As already mentioned, some special management wastes can also be used in coprocessing; for example, in 2011, 3.6 thousand tons of paper, cardboard, treatment plant sludge, and airports wastes were generated.[10]

Besides, 68% of hotel-generated wastes can be used as AF, whereas 91% of the tires in Mexico City are not used.[10]

Oil wastes can also be used as fuel or alternative material in coprocessing, due to the fact that a large part of them contains oil contributing a high calorific power, such as hydrocarbon contaminated solid wastes, oily sludges, and used lubs. However, they have not being used at 100% for this purpose.[10]

Wastes that are not appropriate for coprocessing in cement plants are hospital and sanitary residues, batteries, and untreated municipal wastes.

5.6 CALORIFIC CAPACITY OF SOME WASTE MATERIALS

According to the characteristics of the production process, the cement industry can coprocess: (1) AF having a large calorific power, such as used oils (Table 5.3); (2) AFR containing mineral components that are appropriate for clinker or cement production, such as contaminated soils; (3) materials supplying calorific power and concurrently bringing mineral components, such as paper industry sludge and used tires (Table 5.4).

Most USW can be used in kilns as AF sources because their minimum calorific value is 10.7 MJ/kg—which is lower than the calorific value of commonly used fuels (Queiroz et al., 2012).

Table 5.3 shows the calorific capacity of commonly used fuels and Table 5.4 shows the calorific capacity of some waste materials.

TABLE 5.3 Calorific Capacity of Commonly Used Fuels.

Fuel	Calorific capacity (MJ/kg)
Coke	33.488
Natural gas	48
Fuel oil	44
Coal	29
Lignite	20
Other materials	
Polyethylene	46
Heavy fuel oil	42
Tar	40
Animal fat	38

TABLE 5.3 *(Continued)*

Fuel	Calorific capacity (MJ/kg)
Rubber	36
Residual oils, refinery wastes	30–40
Petroleum coke	33
Used tires	28–32
Tar coal (low ash content)	27
Substitution liquid fuel	20–30
Fluff	18–22
Filling gas	16–20 (in Nm^3)
Animal meal	18
Dry wood, rice husk (10% H_2O)	16
Impregnated wood meal (25% H_2O, CSS, Seneffe)	10–13
Wastewater dry sludges (10% H_2O)	10

The calorific content may vary depending on the waste used. However, it is not only the calorific power that matters but also the mineral contents of the residue (ash) that can be used as alternative raw material in clinker production (clinker is the kiln product that is ground to manufacture cement).

TABLE 5.4 Calorific Capacity of Urban Solid Wastes (USW).

Waste type	Calorific capacity (MJ/kg)		
	As collected, lower calorific capacity	Dry, normal, without water	Dry, ash-free, higher calorific capacity
Mixed food	4.2	13.9	16.7
Fats	37.4	38.2	39.1
Fruit	4	18.6	19.2
Meat	17.6	28.9	30.4
Mixed paper	15.7	17.6	18.7
Newspapers	18.5	19.7	20
Cardboard	26.2	27.1	27.4
Mixed plastics	32.7	33.4	37.1
Polyethylene	43.4	43.4	43.9
Polystyrene	38.1	38.1	38.1

TABLE 5.4 *(Continued)*

Waste type	Calorific capacity (MJ/kg)		
	As collected, lower calorific capacity	**Dry, normal, without water**	**Dry, ash-free, higher calorific capacity**
Polyurethane	26	26	27.1
PVC	22.5	22.5	22.7
Textiles	18.3	20.4	22.7
Garden wastes	6	15.1	15.1
Mixed wood	15.4	19.3	19.3
Glass	0.2	0.2	0.15
Metals	0.7	0.7	0.7
Household urban solid Wastes	11.6	14.5	19.3
Commercial urban solid Wastes	12.8	15	
Urban solid wastes	10.7	13.4	

5.7 COPROCESSING BENEFITS

The most direct benefit is the energy contained in the AF that is used in the cement plants and substitute fossil fuel.[13] Thus, the dependency on fossil fuel is reduced and savings are generated through resource conservation. The demand for fossil fuels eliminated in this form depends, among on the factors, on the calorific power and water content of the AF.

Moreover, substitution fuels can have a lower carbon content (in mass) than the fossil fuels, and the AFR, that do not need a significantly greater quantity of heat (and fuel), can bring part of the CaO necessary for the production of clinker from a source different from $CaCO_3$.[20] Another direct benefit of waste coprocessing in cement manufacturing is thus the potential reduction of CO_2 emissions. The integration of coprocessing in cement kilns is a global waste management strategy that offers a potential reduction of the global net emissions of CO_2 compared to a scenario in which wastes are burned in an incinerator without energy recovery.[2,5]

The use of alternative materials to substitute traditional raw materials reduces the exploitation of natural resources and the environmental footprint of these activities.[2,22]

The savings in terms of investment costs derived from the use of a preexisting kiln infrastructure to carry out the coprocessing of wastes that cannot be reduced or recycled in any other way, derive from the absence of need to invest in specialized incinerators or dumps.[8,13] Contrary to what occurs in specialized waste incinerators, the ashes of the hazardous wastes that are processed jointly in cement kilns are incorporated to the clinker in such a way that final products requiring further handling are not generated.

5.7.1 COPROCESSING AS AN ECONOMICALLY VIABLE ALTERNATIVE

A direct calculation of the production values related to the Balsa Nova plant, Paraná, gives an approximate idea of the use of coprocessing at a cement plant. Table 5.5 shows an example of the savings generated by the replacement of 15% of traditional fuel (petroleum coke) by fuel obtained from waste (Queiroz et al., 2012).

TABLE 5.5 Evaluation of the Savings Generated by the Replacement of 15% of Traditional Fuel by Fuel from Waste.

Description	Unit	Value
Annual cement production	Tons	1,500,000
Total number of kiln operating days	days/year	330
Daily clinker production	Tons	4545
Energy consumption/kg of clinker	Kcal (MJ)	800 (3.4)
Requested daily energy	10^9 kcal/day (10^6 MJ/day)	3.6 (15.1)
Minimum calorific value of petroleum coke	kcal/kg (MJ)	8200 (34.3)
Daily fuel consumption	tons/day	443,459
Daily fuel saving through the replacement of coke by waste (15%)	tons/day	70,296,652
Cost of petroleum coke	US$/tons	61
Daily fuel savings (coke)	US$/day	4287.69
Annual fuel savings	US$/year	1,414,938

On the other hand, following the same reasoning, a 15% substitution of 3,630,000,000 kcal (15,184 MJ)/day is equivalent to 545,450,000 kcal (2282 MJ)/day. Using a calorific power source offering less than 3,000 kcal (0.01 MJ)/kg would require 1818 tons of wastes per day. If the plant charges US$ 100.00/ton of incinerated wastes, the daily income would be US$ 18,180.00, and the annual income would reach US$ 5,999,400.00 (Queiroz et al., 2012).

Thus, adding together the reduced use of fossil fuels and the burning of wastes, with the above hypothetical values, kiln coprocessing savings would amount to US$ 7,414,337.70 per year. Moreover, 59,994 tons of waste would have been burned during 1 year (Queiroz et al. 2012).

In the Table 5.6, other benefits from coprocessing are mentioned.

TABLE 5.6 Summary of Coprocessing Related Benefits.

Economic benefits	Environmental benefits	Social benefits
Generation of direct and indirect jobs.	Reduction of environmental pollution through lower greenhouse effect gases emissions.	Supply of a complementary tool to society for waste handling purposes.
Reduction of the necessary investments because existing installations are used, such as cement plants.	Technical and environmental safe removal of industrial wastes.	Contribution to public health because disease propagation is avoided.
Alternative energy source.	Absence of generation of waste at the end of the coprocessing process and no wastes deposited in dumps.	
Saving of material that were previously obtained from nature such as fossil fuel and mineral resources.	No wastes deposited in dumps.	

5.8 CONCLUSIONS

The use of waste as AF source reduces the energy dependence on traditional fuels. The use of AFR offers various benefits; among them are the reduced need to tap natural resources and an improved environmental footprint.

Cement plants recycle several types of wastes generated both by themselves as well as by other productive sectors. In this way, the need for new raw materials and the generation of wastes can be avoided.

Solid alliances between public and private sectors are key to reaching the maximum benefits from wastes coprocessing in cement kilns. There is a clear distribution of tasks and responsibilities. Innovative and specialized techniques are available that will be further developed by the private sector, while both the private sector and the public sector should ensure that environmental standards are preserved and health and safety regulations are enforced. In this context, the private sector is already a step ahead and there is a strong need to continue strengthening the institutional capacity and the human resources in the public sector in order to ensure that it is qualified to fulfill its task. The cooperation between GTZ and Holcim can be considered successful, despite the fact that both companies have totally different activities.

ACKNOWLEDGMENT

The coauthors are grateful to CEMEX for the interest shown with regard to the conduction of this research.

KEYWORDS

- **coprocessing**
- **reuse**
- **recycling**
- **solid wastes**
- **cement**

REFERENCES

1. Supino, S.; Ornella, M.; Mario, T.; Daniela, S. Sustainability in the EU Cement Industry: The Italian and German Experiences. *J. Clean. Prod.* **2016,** *112*, 430–442.
2. CEMBUREAU; Oficemen; CEMA Fundación and SUSTAINABLE ENERGY EUROPE. Producción sostenible de cemento, la recuperación de residuos como combustibles y materias primas alternativas en la industria cementera (Sustainable Cement Production, Waste Recovery as Alternative Fuels and Raw Materials in

the Cement Industry). 2009. http://www.fundacioncema.org/Uploads/docs/ProduccionSostenibleCemento.pdf (accessed Feb 15, 2014).

3. CEMEX. Inclusion, Sustainable Development Report, 2012, 1–116. http://www.cemexmexico.com/DesarrolloSustentables/files/CemexMexicoInformeDesarrollo-Sustentable2012.pdf (accessed March 22, 2014).
4. Choy, H.; Ko, K.; Cheung, H.; Fung C.; Hui, W.; Porter F. y Mckay, G. Municipal Solid Waste Utilization for Integrated Cement Processing with Waste Minimization, A Pilot Scale Proposal. *Process Safety Environ. Protec.* **2004,** *82*(B3), 200–207.
5. EA (Environment Agency of England and Wales). Substitute Liquid Fuels (SLF) Used in Cement Kilns–Life Cycle Analysis. Research and Development Technical Report P274. Bristol: Environment Agency, 1999b, p 109.
6. EIPPCB (European Integrated Pollution Prevention and Control Bureau). Integrated Pollution Prevention and Control, Reference Document on the General Principles of Monitoring. European Commission, Joint Research Centre, Institute for Prospective Technological Studies. Sevilla. 2013. p 56. ftp://ftp.jrc.es/pub/eippcb/doc/mon_bref_0703.pdf (accessed June 15, 2014).
7. EIPPCB (European Integrated Pollution Prevention and Control Bureau). Reference Document on Best Available Techniques in the Cement, Lime and Magnesium Oxide Manufacturing Industries. European Commission, Joint Research Centre, Institute for Prospective Technological Studies. Sevilla. 2010, p 9. http://www.ebrd.com/downloads/policies/environmental/construction/cement-production.pdf (accessed June 11, 2014).
8. GTZ-Holcim. Guidelines on Co-processing Waste Materials in Cement Production. The GTZ–Holcim Public Private Partnership. 2006, p 135. http://www.coprocem.com/trainingkit/documents/diverse/guideline_coprocem_v06-06.pdf (accessed June 9, 2014).
9. GTZ-Holcim. La alianza estratégica GTZ–Holcim para el co-procesamiento de residuos en la producción de cemento (The Strategic Alliance GTZ–Holcim for Waste Co-processing in Cement Production). D&PG—Diseño y Producciones Gráficas Santiago de Chile. 2009, p 12.
10. INECC-SEMARNAT, Diagnóstico básico para la gestión integral de los residuos 2012, Versión extensa (Basic Diagnosis for Integral Waste Management 2012) Full version, pages 201, http://www.inecc.gob.mx/descargas/publicaciones/495.pdf. (accessed Feb 20, 2014).
11. Kjuus, H.; Lenvik, K.; Kjærheim, K.; Austad, J. Epidemiological Assessment of the Occurrence of Allergic Dermatitis in Workers in the Construction Industry Related to the Content of Cr (VI) in Cement. National Norwegian Institute of Work Hygiene. 2003, p 47. http://www.wbcsd.org/web/projects/cement/tf3/nioh-study_chromium_allergic_dermatitis.pdf (accessed June 21, 2014).
12. Ley General Para la Prevención y Gestión Integral de los Residuos (DOF, 2014) [General Law for the Prevention and Integral Management of Wastes (Federal Gazette of the Federation, 2014)], p 1–48.
13. Murray, A.; Price, L. Use of Alternative Fuels in Cement Manufacture: Analysis of Fuel Characteristics and Feasibility for Use in the Chinese Cement Sector. China Energy Group, Ernest Orlando Lawrence Berkeley National Laboratory, U. S. Department of Energy, 2008, p 62.

14. Queiroz Lamas; Fortes Palau y Rubens de Camargo. Waste Materials co-processing in Cement Industry: Ecological Efficiency of Waste Reuse. *Renew. Sustain. Energy Rev*. 2004, *19*, 200–207.
15. PNUMA, (United Nations Environmental Program). Guidelines on best available Techniques and Provisional Guidance on Best Environmental Practices Relevant to Article 5 and Annex C of the Stockholm Convention on Persistent Organic Pollutants: Cement Kilns Incinerating Hazardous Wastes. Experts Group on Best Available Techniques and Best Environmental Practices. Geneva: PNUMA, 2007; p 109.
16. PNUMA, (United Nations Environmental Program). Technical Guidelines on the Environmentally sound co-processing of hazardous Wastes in Cement Kilns. Conferences of the Parties to the Basel Convention on the Control of Cross-border Movements of Hazardous Wastes and Their Elimination. Cartagena: UNEP, 2012; p 62.
17. Rojas-Valencia M. N. *Construyendo ciudades sustentables: experiencias de Pekín y la Ciudad de México. Capítulo: Tecnologías para atender la situación de los residuos sólidos urbanos y agua en la Ciudad de México. (Building Sustainable Cities: Experiences in Beijing and Mexico City. Chapter: Technologies for Handling Urban Solid Wastes and Water in Mexico City;* 1st ed. ISBN: 978-607-02-2951-0. 2012; pp 197–211.
18. SEMARNAT, Rubro 3. Co-procesamiento de residuos sólidos industriales, 2014 (Section 3. Industrial Solid Wastes Co-processing, 2014).
19. Van Oss, H. G.; Padovani, A. C. Cement Manufacture and the Environment. Part II: Environmental Challenges and Opportunities. *J. Ind. Ecol*. **2003,** *7*(1), 93–126.
20. Van Oss, H. G. Background Facts and Issues Concerning Cement and Cement Data. Open-File Report 2005-1152. U. S. Department of the Interior, U. S. Geological Survey. 2005. p 80. http://pubs.usgs.gov/of/2005/1152/2005-1152.pdf (accessed June 20, 2014).
21. Van der Sloot, H. A.; Van Zomeren, A.; Stenger, R.; Schneider, M.; Spanka, G.; Stoltenberg-Hansson, E.; Dath, P. Environmental Criteria for Cement Based Products, ECRICEM. Executive Summary. Energy Research Centre of The Netherlands (ECN). ECN Report N° ECN-E--08-011. 2008, p 28. http://www.ecn.nl/docs/library/report/2008/e08011.pdf (accessed June 24, 2014).
22. WBCSD (World Business Council for Sustainable Development). Guidelines for the Selection and Use of Fuels and Raw Materials in the Cement Manufacturing Process. Cement Sustainability Initiative (CSI). Geneva: WBCSD. 2005, p 38. http://www.wbcsdcement.org/pdf/tf2_guidelines.pdf (accessed June 13, 2014).

CHAPTER 6

FEEDSTOCK RECYCLING OF AUTOMOBILE SHREDDER RESIDUE

JUMA HAYDARY* and DALIBOR SUSA

Institute of Chemical and Environmental Engineering, Faculty of Chemical and Food Technology, Slovak University of Technology, Radlinského 9, 81237 Bratislava, Slovakia

Corresponding author. E-mail: juma.haydary@stuba.sk

CONTENTS

ABSTRACT

In this work, the potential of automobile shredder residue (ASR) for feedstock recycling was studied. Composition of ASR was determined by the separation of a 10 kg representative sample into its basic components. Thermogravimetric (TG) and differential scanning calorimetric (DSC) analyses of individual components as well as of the representative sample of ASR were provided by a simultaneous TG/DSC analyzer. Elemental analysis of mixed ASR and individual components of ASR was provided by an elemental analyzer. The bomb calorimetric method was used to determine the higher heating value (HHV) of all individual components and also of mixed ASR. A model for the determination of the following parameters of mixed ASR was developed: kinetics of thermal decomposition, heat of thermal decomposition, proximate and elemental analyses, and HHV. The model enables the determination of the abovementioned parameters by entering the content of individual types of waste present in ASR. In a second series of experiments, the ASR samples were pyrolyzed in a laboratory screw-type reactor at reactor temperature ranging from 550°C to 800°C. Individual product yields as well as composition of pyrolysis gases were estimated.

6.1 INTRODUCTION

Recycling of automobile shredder residue (ASR, auto fluff) by pyrolysis or gasification can be an economically and environmentally effective method for its disposal. The product of a typical EU automobile shredder contains around 70 wt. % of metals separated by magnetic separators and other 5–6 wt. % represent nonferrous metals. After the separation of metals, the remaining part is a complex mixture of materials including plastics, foam, textiles, rubber, glass, and others. In the European Union, about 2–2.5 million tons of this waste is produced every year.[1,2]

At the present time, ASR is usually landfilled but the European draft Directive 2000/53/EC[3] forces the development of alternative solutions requiring 95% of end of live vehicles to be reused/recovered and 85% to be reused/recycled by January 1, 2015. Since auto fluff is a complex mixture and usually it is contaminated with rust, dirt, and a variety of fluids, its recyclability poses a challenge to shredding operators. Consequently, the need to explore new and innovative ways of ASR recycling, or of recovering valuable resource materials from this waste, is an urgent environmental and economic issue.

Feedstock recycling methods such as pyrolysis and gasification represent an economically and ecologically acceptable solution for the disposal of this waste. For this reason, significant attention has been dedicated to thermal processing of ASR in the last decade.

Lin et al.[4] presented an analysis of catalytic gasification of ASR for the generation of high-purity hydrogen in a lab-scale fixed-bed downdraft gasifier using 15 wt. % NiO/Al_2O_3 catalyst at 760–900 K. Based on laboratory data and according to the calculations, approximately 220 kg/h of ASR is gasified to generate 100 kW of electric power. Horri and Lida[5] presented a process referred to as gasification and dry distillation of ASR. At certain conditions, the gas heating value of about 15.5 MJ/kg was achieved.

Pyrolysis of ASR was studied using different technologies at different scales. Day et al.[6] applied a commercial screw kiln unit. ASR was pyrolyzed into 26% of gas, 21% of oil, 10% of water, 11% of magnetic solids, 25% of fines, and 7% of other solids. The main gaseous components were CH_4, H_2, CO, CO_2, C_2H_2, C_2H_4, and other hydrocarbons. Quality and composition of the gas, liquid, and solid products is described in their work. A laboratory scale reactor using 10–15 g samples was employed by Santini et al.[7] Using laboratory tests, they compared the conversion of raw ASR with that of ASR treated by flotation. Conversion of raw ASR was 20% and 60% for treated samples. Different aspects of ASR pyrolysis were studied also in.[8–11]

Results achieved by different authors show significant differences. Vermeulen et al.[12] collected and compared the results of a large number of works in a review on thermal processing of ASR. Also Harder and Fordon[13] studied a large number of works devoted to the pyrolysis of ASR in their critical review. The results presented in these works show significant differences in the amount and quality of the obtained products. The reason is that ASR is a heterogeneous mixture of all materials found in cars and the composition of ASR varies depending on the producer, type of shredded vehicles, used shredding technology, fractionation of the shredded cars, and many other factors. However, design of a thermal process (incineration, gasification, or pyrolysis) requires the knowledge of the characteristics of the raw materials and allows the variability of parameters only within a defined range.

The energy content of ASR, proximate and elemental composition, and kinetics of thermal decomposition are crucial input data for the design of a thermal process. However, all these parameters depend on the composition of ASR. The aim of this work was to develop an experimentally based model for the calculation of these parameters and a material balance of the ASR pyrolysis process.

6.2 MATERIALS AND METHODS

The material used in this work was a fraction of ASR generally separated from inorganic material and dust also called light fraction. The used ASR contained mainly plastics, rubber, textile, foam, and other organics. Composition of the ASR was estimated by the separation of a 10 kg (20–50 mm) representative sample into its basic components by hand separation. Figure 6.1 shows an illustration of the ASR used for further experiments. The mass fraction of individual fluff components is shown in Table 6.1.

FIGURE 6.1 Shredded ASR sample.

Samples of each category were homogenized by grounding of around 200 g of individual waste materials to particles of sizes lower than 1 mm. Each category of waste was studied separately. In addition, a mixed sample of ASR with the composition shown in Table 6.1 was prepared. The following experimental methods and procedures were applied:

- Simultaneous thermogravimetric (TG)/differential scanning calorimetric (DSC) measurements using a simultaneous thermal analyzer (Netzsch STA 409 PC Luxx, selb, Germany). Experimental conditions were met at the linear heating rates of 10°C/min in the nitrogen flow of 60 mL/h. The samples were heated from 20°C to 800°C. At this temperature, they were maintained for around 30 min and consequently were combusted by entering the oxygen to the system. Samples of individual categories of ASR and also the sample of the mixed ASR with the mass of around 20 mg were used in the TG/DSC measurements. Data obtained from TG experiments were used for the determination of kinetic parameters of thermal decomposition and also for the estimation of proximate composition (moisture, volatiles, fixed carbon, and ash content) of individual samples.

TABLE 6.1 Composition of Light Fraction of ASR After Separation of Inorganic Material and Wires Coated with PVC.

Category	w_i [kg/kg]
Rubber	0.46
Foam	0.02
Plastics	0.47
Textile	0.02
Other OM	0.03

- Measurement of higher heating value (HHV) using a FTT isoperibolic calorimetric bomb (Fire Testing Technology Limited). Combustion of the sample took place in a calorimetric bomb under oxygen atmosphere at 30 bars. The sample mass was around 1 g. Benzoic acid was used as a standard material. Each measurement was repeated at least three times, the permissible variation between the measurements was 0.2 MJ/kg.
- Elemental analysis using an elemental analyzer Vario Macro Cube ELEMENTAR. A CHNS (carbon, hydrogen, nitrogen, sulfur) module with the combustion tube temperature of 1150°C and the reduction tube temperature of 850°C was used. The mass of the sample was about 1 g.
- Pyrolysis experiments were carried out in a laboratory pyrolysis reaction unit in nitrogen atmosphere. A description of the apparatus can be seen in Figure 6.2. The equipment consisted of a screw-type reactor electrically heated by a tube furnace; ASR sample was fed

into the reactor and moved down the reactor by means of a rotating screw powered by an electric motor. Residence time of the particles in the reactor can be modified by varying the screw rotation speed using a computer. After the reaction solid char was deposited at the end of this reactor and volatile product gases passed upwards a series of coolers where liquid was collected. The remaining noncondensable gases flew to a set of ice baths before entering gas chromatography apparatus. The entire system was blanketed under nitrogen. The concurrent flow of inert gas was regulated by a flow meter. The pyrolysis experiments were performed in the temperature range of 550–800°C.

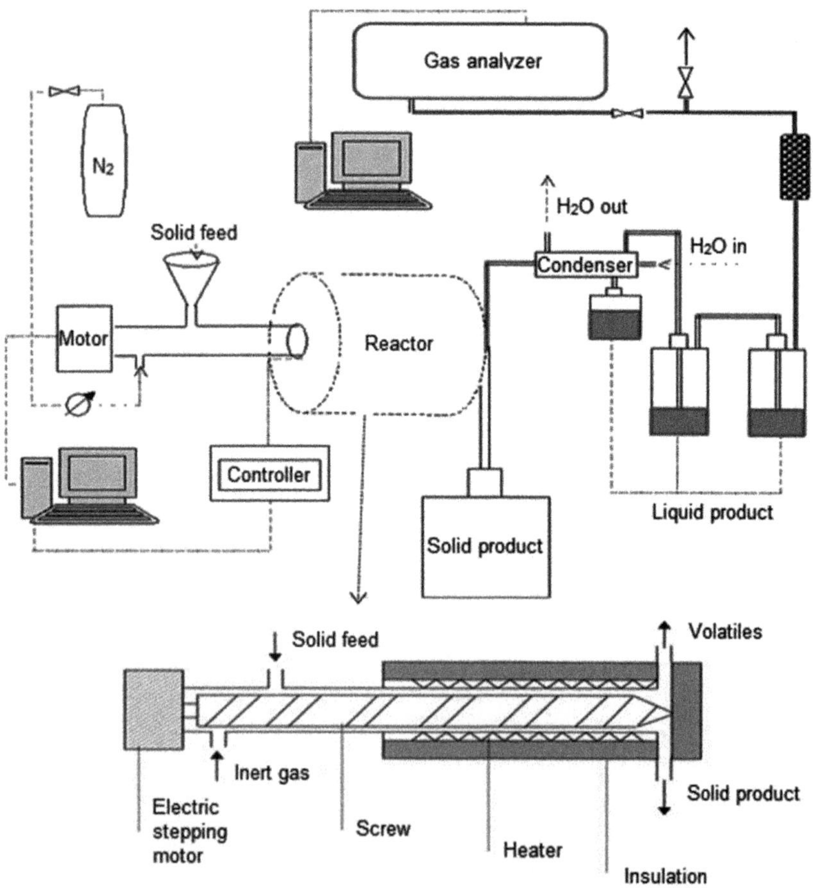

FIGURE 6.2 Laboratory pyrolysis unit.

6.3 RESULTS AND DISCUSSIONS

Thermal characteristics of an ASR mixture were estimated and compared using two methods:

- Direct measurement of the parameter using a homogenized mixed sample with composition presented in Table 6.1.
- Calculation of the parameter using parameters of individual components and an additive formula (eq 6.1)

$$P_{ASR} = \sum_{i=1}^{k} w_i P_i \qquad (6.1)$$

where P represents a general indication of a parameter and w the mass fraction of component i *(i=rubber, foam, plastics, textile, other OM)*.

Table 6.2 shows the proximate and elemental composition of a mixed ASR calculated based on the content of the individual category and obtained directly. The data calculated based on ASR components are close to those estimated directly.

TABLE 6.2 Proximate and Elemental Composition of ASR in wt. %.

Sample	Moisture	Volatile matter	Fixed carbon	Ash	N	C	H	S	O[a]
ASR (based on the components)	0.55	70.62	9.90	18.93	1.41	58.03	6.84	0.89	14.02
ASR (direct measurement)	0.50	71.11	10.00	18.00	1.42	58.88	6.75	0.90	13.10

[a]Calculated from the difference to 100%.

HHV of ASR was estimated using two different approaches: based on HHV of individual components (eq 6.1) and direct measurement of a mixed sample. The value of HHV based on the individual component was 28.4 MJ/kg. By direct measurement of the mixed sample, a value of 28.12 MJ/kg was measured. The values of HHV for ASR in literature vary from 5 to 30 MJ/kg depending on the composition of used ASR.

Thermal decomposition is an endothermic process; in some cases, also an exothermic oxidation process takes place because of the presence of oxygen in the raw material. However, overall heat of the reaction is usually endothermic. The Netzsch STA 409 PC Luxx analyzer enables simultaneous recording of TG and DSC curves. DSC data were used for the determination of the overall heat of thermal decomposition of individual samples. The values of the heat of thermal decomposition vary between 130 and 650 J/g for individual material categories of ASR. For the mixed ASR sample, the value of 304.9 J/g was calculated by eq 6.1 and the value of 301.5 J/g was estimated directly from the DSC curve of the mixed sample.

TABLE 6.3 Higher Heating Value and Endothermic Heat of Thermal Decomposition of ASR.

Sample	HHV [MJ/kg]	Heat of thermal decomposition (endothermic) [J/g]
ASR (based on the components)	28.40	307.9
ASR (direct measurement)	28.12	301.5

6.3.1 KINETIC MEASUREMENTS

TG measurements of individual material categories of ASR were used for the determination of kinetic parameters of thermal decomposition. Figure 6.3 shows an example of the measured TG/DSC curves.

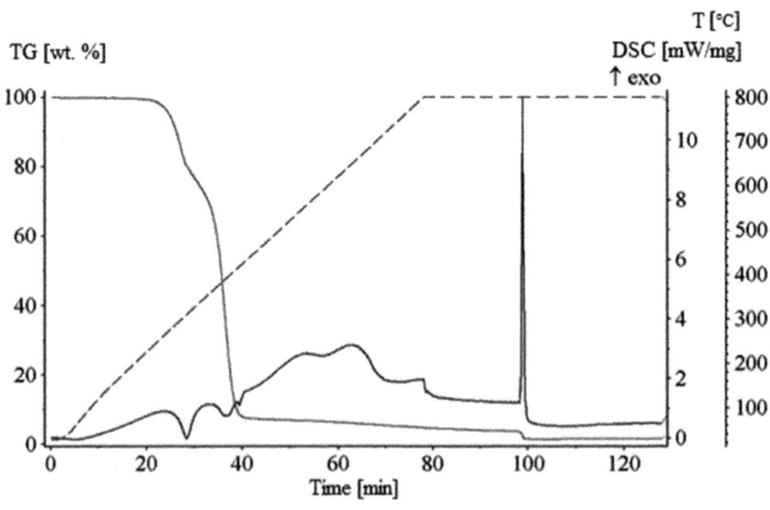

FIGURE 6.3 Example of TG/DSC curves.

For a one-step thermal decomposition, the rate equation using the Arrhenius law is:

$$\frac{d\alpha}{dt} = A\exp\left(-\frac{E}{RT}\right)(1-\alpha)^n \quad (6.2)$$

where A is the apparent preexponential factor, E is the apparent activation energy, T is the temperature, n is the reaction order, t is the time, and R is the gas constant. Conversion α was calculated according to eq 6.3:

$$\alpha = \frac{m_0 - m}{m_0 - m_{final}} \quad (6.3)$$

where m_0, m, and m_{final} correspond to the initial, actual, and final sample mass, respectively.

Two clearly separated peaks on the DTG curves predetermine that a two-step decomposition model will provide more accurate description of the process. For a multistep thermal decomposition, the total decomposition rate can be calculated as a sum of individual decomposition reactions:

$$\frac{d\alpha}{dt} = \sum_{i=1}^{n} A_i \exp\left(-\frac{E_i}{RT}\right)\left(1-\alpha_i\right)^{n_i} \quad (6.4)$$

Parameters A, E, and n for each decomposition step were obtained from a set of kinetic experiments on the dependence of the reaction rate versus temperature by fitting the calculated and experimental data (Fig. 6.4). The used objective function was:

$$f = \sum_{i=1}^{k}\left(\frac{d\alpha_{\exp}}{dt} - \frac{d\alpha_{cal}}{dt}\right)^2 \quad (6.5)$$

The estimated kinetic parameters for all material categories are presented in Table 6.4.

The total decomposition rate of mixed ASR can be calculated additively using the rate of thermal decomposition of individual material categories as follows:

$$\frac{d\alpha_{ASR}}{dt} = \sum_j w_j \frac{d\alpha_j}{dt} = \sum_j w_j \left(\sum_{i=1}^{n} A_i \exp\left(-\frac{E_i}{RT}\right)\left(1-\alpha_i\right)^{n_i}\right)_j \quad (6.6)$$

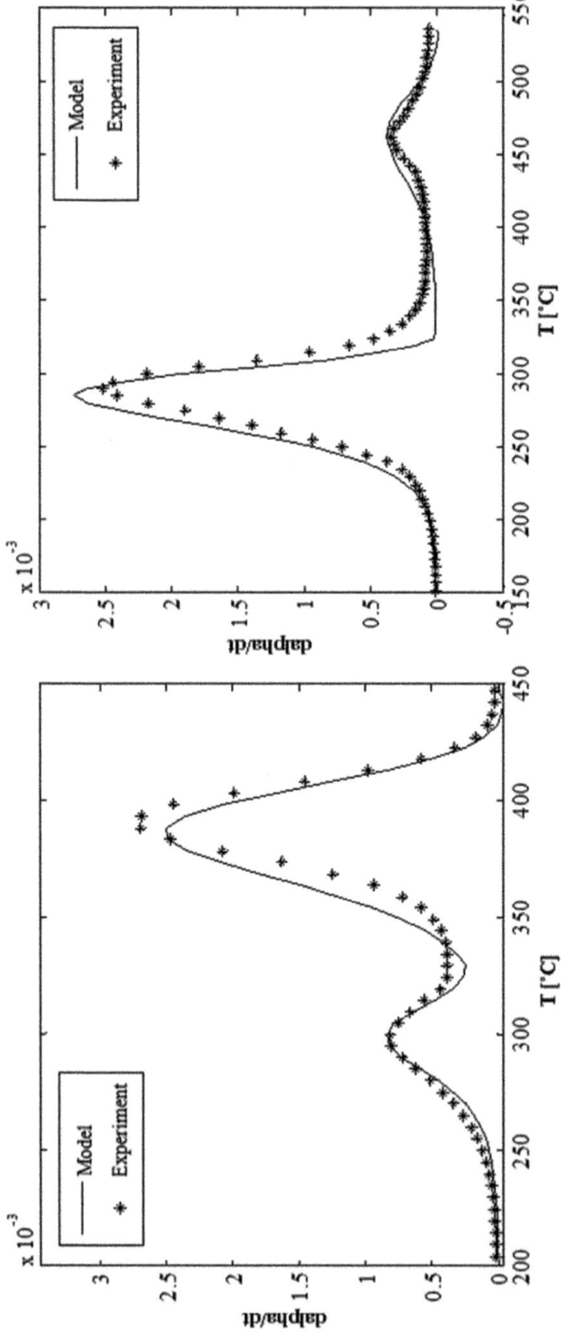

FIGURE 6.4 Fitting of DTG curves of foam (left) and plastic (right) samples for the determination of kinetic parameters.

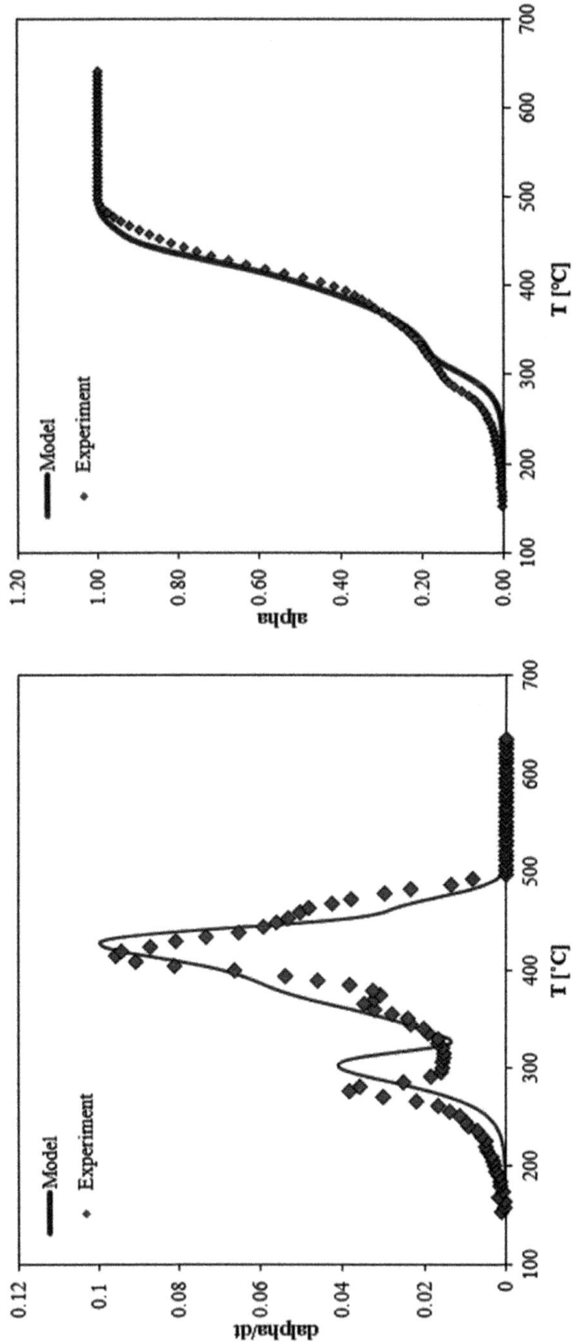

FIGURE 6.5 Model and experimental TG and DTG curves of ASR mixed sample.

TABLE 6.4 Kinetic Parameters of Thermal Decomposition of ASR Components and Mixed ASR.

Sample	A_1 [s^{-1}]	E_1 [J/mol]	n_1 [–]	A_2 [s^{-1}]	E_2 [J/mol]	n_2 [–]
Rubber	2.50×10^{11}	1.70×10^5	1.70	2.88×10^{11}	1.83×10^5	1.01
Plastics	3.96×10^7	1.38×10^5	1.07	4.20×10^{11}	1.95×10^5	1.08
Foam	1.00×10^{12}	1.63×10^5	1.10	3.48×10^{14}	2.10×10^5	1.00
Textile	3.61×10^{14}	2.22×10^5	1.00	–	–	–
Other OM	1.86×10^{10}	1.33×10^5	1.05	1.04×10^{11}	1.86×10^5	1.06
Mixed ASR	9.20×10^{10}	1.47×10^5	2.02	8.61×10^{10}	1.79×10^5	1.00

6.3.2 LABORATORY SCALE PYROLYSIS OF ASR

Samples of mixed ASR milled to particles with size around 4 mm were pyrolyzed in a laboratory pyrolysis unit shown in Figure 6.2 at temperatures ranging from 550°C to 800°C. The average residence time of particles in the reactor was 100 s. Solid and liquid pyrolysis yields (Fig. 6.6) were determined in each experiment by weighing the obtained amount and calculating the corresponding percentage. The gas yields were determined by difference and, therefore, include all the experimental errors and inaccuracies. Based on TG measurements of all material categories, the temperature of 550°C is sufficient to achieve total ASR conversion. Product yields of solid, gaseous, and liquid fractions at various temperatures and residence times from our experimental runs are shown in Figure 6.7. An increase in the amount of gases and the consequent decrease in the amount of liquids can be seen. This decline in the liquid fraction and increase in the gaseous yields were observed also by other authors and it can be explained by the stronger thermal cracking processes taking place at higher temperatures[14,15] as well as by the occurrence of secondary reactions that are involved both in liquid and gas fractions.

ASR pyrolysis gases are mainly composed of light hydrocarbons, hydrogen, and methane together with some CO, CO_2, and H_2S. The content of the main components of gas at different temperatures shows (Fig. 6.8) that the higher the pyrolysis temperature is, the more hydrogen and CO is present in the gas. Maximum amount of methane was observed at temperatures of around 750°C, when its degradation starts. The content of remaining hydrocarbons, marked in this publication as CH_x decreases in the whole temperature range. This drop in the light hydrocarbon and the increase in the methane and hydrogen production with temperature are due to the mechanism of the pyrolysis process itself. As the temperature increases, heavier hydrocarbons are cracked to components with lower molecular weight.

FIGURE 6.6 Liquid and solid pyrolysis products.

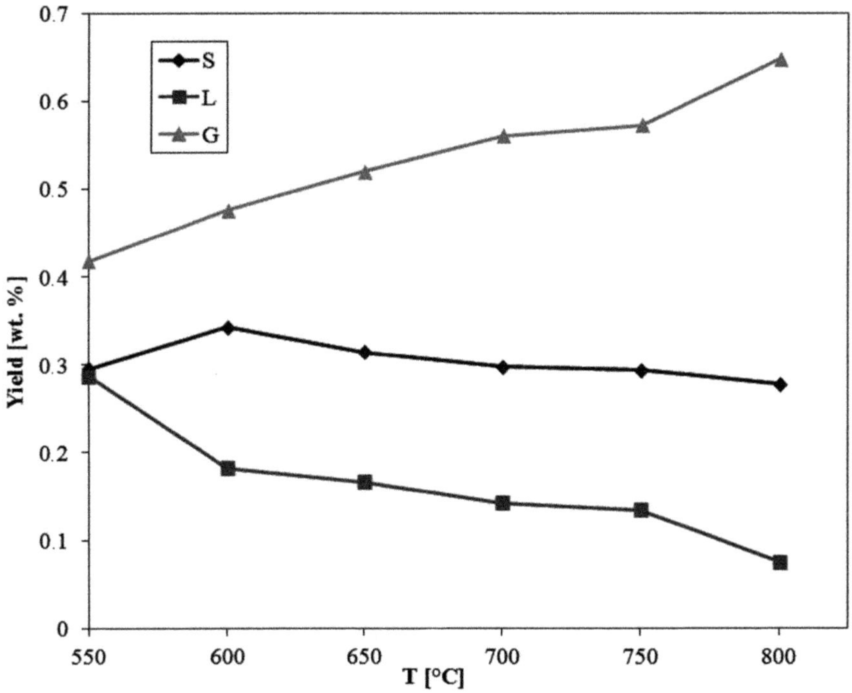

FIGURE 6.7 Product yields of ASR pyrolysis in a screw-type reactor.

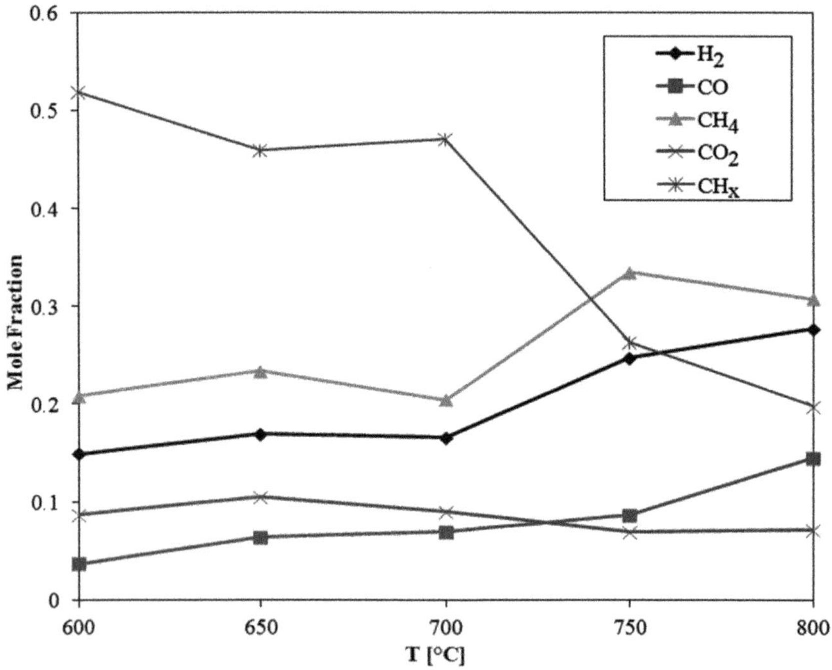

FIGURE 6.8 Composition of the gas product of ASR pyrolysis.

6.4 CONCLUSIONS

The light fraction of ASR, predominantly consisting of rubber, plastics, textile, and polyurethane foam, has a big potential to be recycled by feedstock recycling methods such as pyrolysis and gasification. Proximate and elemental composition as well as the heat content of ASR designates it as a suitable material for thermal processing to recover valuable materials and energy. Basic parameters needed for the design of a thermal process can be estimated using the parameters of individual components of ASR. All, proximate and elemental analyses, HHV, heat of thermal cracking, and the rate of the reaction calculated based on the individual components, are in a very good agreement with the data estimated for a mixed ASR sample. In a laboratory scale pyrolysis unit with a screw-type reactor, the yield of solid product was 30% at 550°C and it increased to 34% at 600°C; at higher temperatures (800°C), it decreased to 28%. The yield of liquid product decreased from 28% at 550°C to only 7% at 800°C. The gas yield increased from 42% at 550°C to 65% at 800°C. By increasing the temperature, the content of hydrogen and CO in the gas fraction increased, whereas the content of light hydrocarbons, except for methane, decreased. The content of methane showed a maximum level at 750°C.

ACKNOWLEDGMENT

This work was supported by the Grant VEGA No. 1/0757/13 from the Slovak Scientific Grant Agency and the OP Research and Development of the project National Centrum of Research and Application of Renewable Sources of Energy, ITMS 26240120016, cofinanced by the Fund of European Regional Development.

KEYWORDS

- **recycling**
- **automobile shredder residue**
- **bomb calorimetric method**
- **reuse**

REFERENCES

1. Zorpas, A. A.; Inglezakis, V. J. Automotive Industry Challenges in Meeting EU 2015 Environmental Standard. *Technol. Soc.* **2012,** *34*, 55–83.
2. Fiore, S.; Ruffino, B.; Zanetti, M. C. Automobile Shredder Residues in Italy: Characterization and Valorization Opportunities. *Waste Manage.* **2012,** *32*, 1548–1559.
3. Kanari, Ndue; Pineau, J-L; Seit, Shallari. End-of-Life Vehicle Recycling in the European Union. *Jom.* **2003,** *55*, 15–19.
4. Lin, K. S.; Chowdhury, S.; Wang, Z. P. Catalytic Gasification of Automotive Shredder Residues with Hydrogen Generation. *J. Power Sour.* **2010,** *195*, 6016–6023.
5. Horii, M.; Iida, S. Gasification and Dry Distillation of Automobile Shredder Residue (ASR). *JSAE Rev.* **2001,** *22*, 63–68.
6. Day, M.; Cooney, J. D.; Shen, Z. Pyrolysis of Automobile Shredder Residue: An Analysis of the Products of a Commercial Screw Kiln Process. *J. Anal. Appl. Pyrol.* **1996,** *37*, 49–67.

7. Santini, A.; Passarini, F.; Vassura, I.; Serrano, D.; Morselli, L. Auto Shredder Residue Recycling: Mechanical Separation and Pyrolysis. *Waste Manage.* **2012,** *32*, 852–858.
8. Zolezzi, M.; Nicolella, C.; Ferrara, S.; Iacobucci, C.; Rovatti, M. Conventional and Fast Pyrolysis of Automobile Shredder Residues (ASR). *Waste Manage.* **2004,** *24*, 691–699.
9. Joung, H. T.; Seo, Y. C.; Kim, K. H. Distribution of Dioxins, Furans, and Dioxin-like PCBs in Solid Products Generated by Pyrolysis and Melting of Automobile Shredder Residues. *Chemosphere* **2007,** *68*, 1636–1641.
10. Donaj, P.; Yang, W.; Błasiak, W.; Forsgren, C. Recycling of Automobile Shredder Residue with a Microwave Pyrolysis Combined with High Temperature Steam Gasification. *J. Hazard. Mater.* **2010,** *182*, 80–89.
11. Patierno, O.; Cipriani, P.; Pochettj, F.; Giona, M. Pyrolysis of Automotive Shredder Residues: A Lumped Kinetic Characterization. *Chem. Eng. J.* **1998,** *70*, 157–163.
12. Vermeulen, I.; Caneghem, J. Van; Block, C.; Baeyens, J.; Vandecasteele, C. Automotive Shredder Residue (ASR): Reviewing Its Production from End-of-life Vehicles (ELVs) and Its Recycling, Energy or Chemicals' Valorization. *J. Hazard. Mater.* **2011,** *190*, 8–27.
13. Harder, M. K.; Forton, O. T. A Critical Review of Developments in the Pyrolysis of Automotive Shredder Residue. *J. Anal. Appl. Pyrol.* **2007,** *79*, 387–394.
14. Leung, D. Y. C.; Wang, C. L. Kinetic Study of Scrap Tyre Pyrolysis and Combustion. *J. Anal. Appl. Pyrol.* **1998,** *45*, 153–169.
15. Li, S. Q.; Yao, Q.; Chi, Y.; Yan, J. H.; Cen, K. F. Pilot-scale Pyrolysis of Scrap tires in a Continuous Rotary Kiln Reactor. *Ind. Eng. Chem. Res.* **2004,** *43*, 5133–5145.

CHAPTER 7

MODIFICATION AND REUSE OF POLYVINYL CHLORIDE USING POLYANILINE AND CONDUCTING FILLERS

RENJANADEVI B.[1*] and K. E. GEORGE[2]

[1]*Department of Chemical Engineering, Government Engineering College, Thrissur, India*

[2]*Albertian Institute of Science and Technology Engineering College, Kalamassery, Kochi, Kerala, India*

[*]*Corresponding author. E-mail: renjanab@gmail.com*

CONTENTS

ABSTRACT

Plastics, especially polyvinyl chloride (PVC), are used in enormous volumes and a possible way to reduce its waste is to recycle and reuse by value addition. In this study, waste PVC from blood bag processing industry is collected and it is then transformed into value added plastics by two different technologies. In one method, PVC is made more conducting by melt blending it with conducting polymers such as polyaniline (PAni) and conducting fillers such as acetylene black. Second method involves the production of nanocomposites of PVC with nanofillers such as nanosilica. The present work involves melt blending and compression molding of the samples and the determination of the mechanical and electrical properties of PVC. The electrical properties were determined in the DC field using cavity perturbation technique. Acetylene black is found to improve the electrical properties of PVC in the DC field. By melt blending with PVC, doped PAni sample showed very good electrical properties. The improvement in mechanical strength indicates that the addition of these conducting materials and nanofillers does not adversely affect the mechanical properties of the material.

7.1 INTRODUCTION

Polyvinyl chloride (PVC) is a considered to be the most versatile material. Applications of PVC are substantial which ranges from building applications to toys. PVC pipes and conduits account for 30% of the PVC consumption. It is also used in medical products, tubings, and blood bags.[1,2] PVC is an amorphous polymer. Fillers are incorporated into PVC to modify its physical properties. Examples of fillers include calcium carbonate, clay talc, silica, and glass fibers. There is a wide scope for recycling of plastics in developing countries due to several factors such as lower labor costs, fewer laws to control the standards of recycled materials, lower cost of raw materials, etc.

PVC is a polar polymer and it is a good base for adding fillers such as acetylene black for improving the electrical conductivity.[3] Since PVC is an amorphous polymer, it can also accommodate a fairly good amount of filler compared to other commodity plastics. Hence nanofillers and conventional fillers may be added to enhance the mechanical strength of PVC.[4] The tensile strength of PVC/nano-$CaCO_3$ nanocomposites at various weight ratios was studied.[5] At a weight ratio of 95/5, the nanocomposite exhibits a slightly higher tensile strength than neat PVC, indicating that the nanoparticles have enhanced the strength of the PVC matrix. The strength of particulate composites is determined not only by particle size and particle/matrix interfacial adhesion but also by particle loading. Various trends of the effect of particle loading on composite strength have been observed.

For many years, finely divided carbon black has been a valuable addition to electrically conductive materials and the resulting composite materials exhibit a wide spectrum of conductivity depending on the loading of carbon black. A study was conducted by Krishnan Rajeshwar et al., regarding the applicability of composites of carbon black as an electronically conductive polymer, polypyrrole, in environmental pollution abatement.[6] Honey John et al. prepared PAni–PVC composites and the dielectric properties of the composites were studied using a HP8510 vector network analyzer.[7] The microwave absorption of the composites were reported at different frequency bands, that is, S, C, and X bands (2–12 GHz). Conducting polymer composites with some suitable composition of one or more insulating materials lead to desirable properties. These materials are especially important owing to their bridging role between the world of conducting polymers and that of nanoparticles.

A new conducting graft copolymer PVC-g–PAni, that is processable and having desirable mechanical properties, was prepared by electrochemical method using a precursor polymer, poly (vinyl 1,4-phenylenediamine)(PVPD). [8] The electrochemical properties of graft copolymer were compared with those of PVC/PAni composites. PVC was used for the backbone polymer as it can provide good mechanical properties and has reactive chlorine atoms for graft sites even though their reactivity is low.

Many attempts have been made to prepare polymer nanocomposites that generate a class of materials with novel properties. The applications of such materials include automotive, biomedical, engineering, and aerospace applications. In a study conducted by Laska et al., PVC blends were made by mechanical mixing of PVC with PAni which exhibited very good mechanical properties.[9]

7.2 MATERIALS

Waste PVC obtained from blood bag manufacturing unit was used for the study. Acetylene Black, having a size of 41 nm was supplied by

Travancore Electrochemicals. Acetylene black has very good conductivity, and it is highly structured and has high surface area. A wide spectrum of conductivity can be obtained depending upon the loading of carbon black. Aniline, ammonium peroxodisulfate, and hydrochloric acid (HCL) used for synthesis of PAni was supplied by E. Merck (P) Ltd., Mumbai.

Precipitated silica was supplied by Degussa, Germany with surface area of 234 m^2/g and particle size of 20 nm. Nanosilica used was Ultracil VN-3 supplied by United Silica Industrial Ltd. Taiwan with particle size of 19 nm. Silane coupling agents: triethoxysilyl propyl tetra sulfide (TESPT) was obtained from Degussa Corporation, Germany.

7.3 EXPERIMENTAL

7.3.1 PREPARATION OF PAni

PAni was prepared by using Chemical oxidative polymerization[10] with ammonium per sulfate initiator in the presence of 1 M HCI at room temperature for 4 h. The PAni formed was dried in a hot air oven at 50–60°C for 6 h and then it was ground into a fine powder.

7.3.2 PREPARATION OF PAni–PVC NANOCOMPOSITE

Polyaniline–PVC nanocomposite used in this study was prepared by melt blending method. The materials were dried in an oven at 50°C for 1 h before melt blending. Brabender Plasticorder was used in this study for mixing polymer melt with polyacrylonitrile (PAN) and acetylene black. Mixing of samples of different compositions was carried out at a temperature of 135°C for about 5 min. The composites were prepared by compression molding.

7.3.3 DOPING

The samples containing PAni were cut into thin strips and dipped in 1 M HCl solution for 24 h. These samples were washed with water, dried, and used for studying the electrical properties. The samples for electrical properties were prepared by using varying compositions (2–40 parts) of PAni and acetylene black (from 2 to 20 parts) in 30 g of PVC. The samples for measuring mechanical properties were prepared by using varying compositions (2–40 parts) of PAni and acetylene black (from 2 to 20 parts) in 30 g of PVC.

7.3.4 MEASUREMENT OF MICROWAVE PROPERTIES

The DC and AC conductivity of PAN/PVC and PVC–acetylene black composites and PAN–zinc sulfide composites were studied in the microwave field[11,12] using cavity perturbation techniques[13] at S-band frequency (2 GHz).

The experimental set up consists of an HP 8510 vector network analyzer, sweep oscillator, S-parameter test set, and the rectangular cavity resonator. The measurements were done at 25°C in a band at 2.97 GHz.

7.3.5 THEORY

The cavity resonates at different frequencies depending on the dimension of the cavity.

The basic principle involved in this technique is that the introduction of the dielectric sample through the nonradiating slot perturbs the field within the cavity resonator. The field perturbation is given by Kupfer et al.[14]

$\sigma = 2\pi f_s \varepsilon_o \varepsilon_r$ Here f_s is the resonant frequency of the loaded cavity. ε_o is the permittivity of vacuum.

The real and imaginary parts of the complex permittivity known as dielectric constant and dielectric loss of the material were given by the following equations.

$$(\varepsilon_{r'}) = 1 + \frac{(f_o - f_s)}{2f_s}\left(\frac{V_c}{V_s}\right) \tag{7.1}$$

$$\varepsilon_{r''} = \left(\frac{V_c}{4Vs}\right)\frac{(Q_t - Q_s)}{Q_t Q_s} \tag{7.2}$$

The loss tangent is given b, $\tan\delta = \dfrac{\sigma + \omega\varepsilon_r''}{\omega\varepsilon'}$ (7.3)

Here, $\sigma + \omega\varepsilon_{r''}$ is the effective conductivity of the medium. When the conductivity due to free charge is negligibly small (good dielectric), the effective conductivity is due to electric polarization and is reduced to $\sigma = 2\pi f_s \varepsilon_0 \varepsilon_{r''}$.

The efficiency of heating is usually compared by means of a comparison coefficient, which is defined as $J = \frac{1}{\varepsilon_r' \tan\delta}$. The absorption coefficient is the measure of absorption of electromagnetic waves when it passes through the medium is given by

$$\alpha_f = \frac{\varepsilon_r' f}{nc} \tag{7.4}$$

and defined as the effective distance of penetration of an electromagnetic wave into the material which is found out using the equation

$$\delta_f = \frac{1}{\alpha_f} \tag{7.5}$$

where δ_f is the skin depth.

The electrical conductivity, σ, in the DC field measured by Keithly 236 source measurement unit can be found out by using the equation

$$\sigma_{dc} = \frac{t}{AR} \tag{7.6}$$

where 't' is the thickness of sample, A is the area of cross section of the sample through which current flows.

$$R = V/I \tag{7.7}$$

where 'I' is the applied current and V is the voltage drop.

7.4 RESULTS AND DISCUSSION

7.4.1 VARIATION OF DIELECTRIC PROPERTIES

The variation of dielectric properties of the PVC/PAni and PVC mixed with other conductive fillers are depicted in the following figures.

The microwave field, dielectric loss of the PVC samples filled with acetylene black, PAni, and acetylene black/PAni mixture increases with their composition in PVC (Fig. 7.1). When two matrices of different dielectric constants are mixed, interfacial polarization happens and it will increase dielectric loss. The interfacial polarization also depends on the geometrical size and shape. Doping provides additional H^+ and Cl^- ions. This causes higher dielectric loss. Doped samples also show higher relaxation phenomenon. Undoped PAni is nonpolar, and in nonpolar materials, the orientation polarization is absent and the polarizability arises from two effects: electronic and atomic polarization. This polarization leads to dielectric loss and conductivity. Figure 7.2 indicates the DC conductivity of samples containing PAni and other conductive fillers. It is clear from Figure 7.2 that the samples containing acetylene black, that is, PVC/acetylene black and PVC/PAni +acetylene black are more conductive than PVC/PAni samples. The DC conduction in PAni is due to polaron, bipolaron, and solitons. But in acetylene black conduction occurs due to tunneling. The charge carriers are believed to be transported between the phases by tunneling mechanism. Hence, the conductivity increases with the composition of conducting materials in PVC.

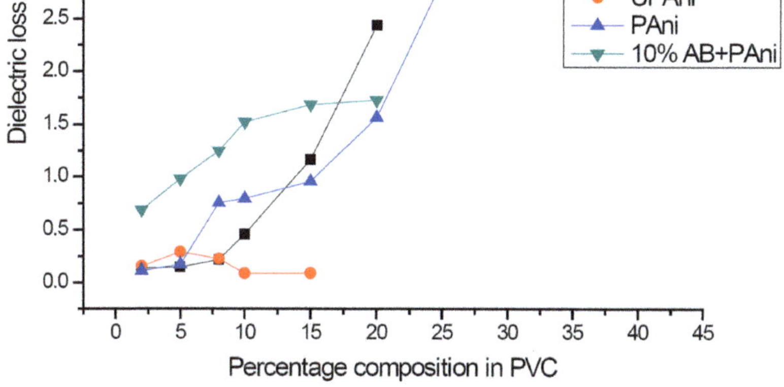

FIGURE 7.1 Dielectric loss of PVC with conducting fillers.

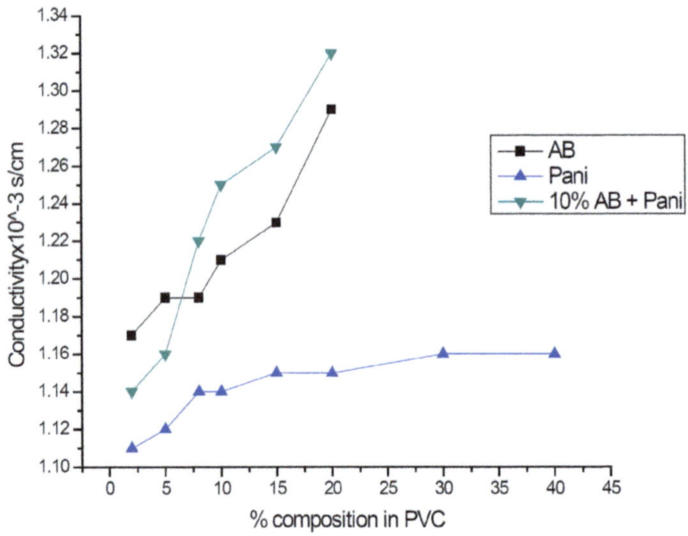

FIGURE 7.2 DC conductivity of PVC with conducting fillers.

From Figure 7.3, it is evident that the conductivity of PVC/acetylene black, PVC/PAni, and PVC/PAni + acetylene black samples increases with the composition of conducting materials in PVC. In the composition range of 5–30 wt%, the conductivities of all samples are high enough

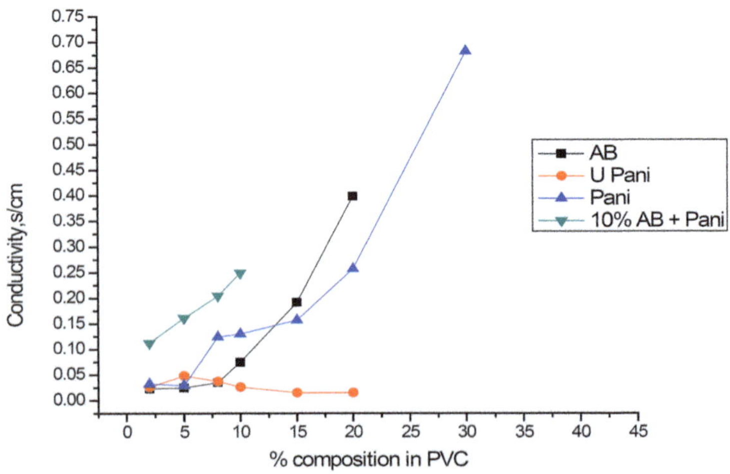

FIGURE 7.3 Variation of conductivity of PVC with conducting fillers.

for important applications. But for PVC-undoped PAni samples, conductivity decreases after showing a marginal increase. Conductivity is directly related to dielectric loss. Hence, they show the same behavior as that of dielectric loss.

7.4.2 VARIATION OF MECHANICAL PROPERTIES OF PVC COMPOSITES

The mechanical properties of PVC/PAni and PVC/acetylene black, and PVC/silica were strongly influenced by the composition. A lower composition, which is below 15 wt%, was found to produce a significant change in the tensile properties of the composite formed.

From Figure 7.4, it is clear that the tensile strength of PVC is increased by the addition of conducting materials up to a medium composition. This is due to the dilution effect. At higher composition of fillers, filler–filler interaction is more than the filler–polymer matrix interaction and hence the mechanical strength decreases. The PVC–PAni + acetylene black samples have comparatively lower strength.

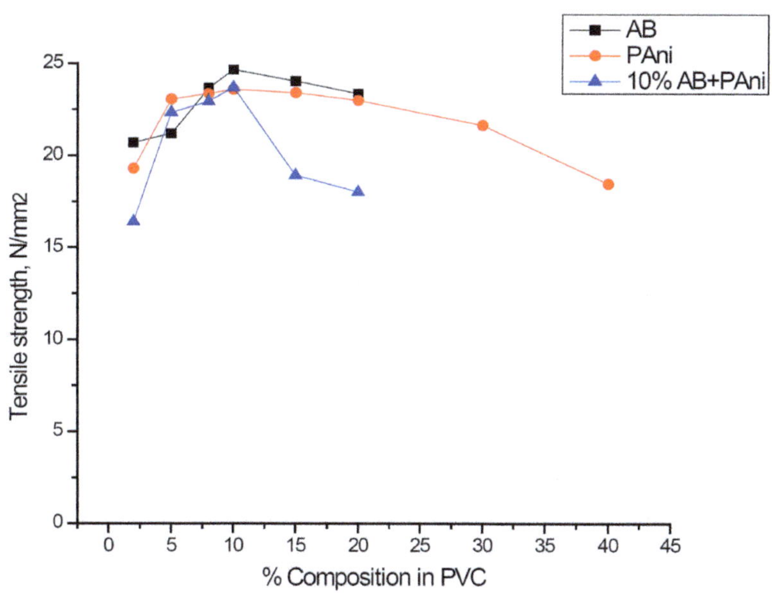

FIGURE 7.4 Tensile strength of PVC with conducting fillers.

Figure 7.5 shows the variation of tensile strength of PVC/silica nanocomposites. When compared to conductive fillers, nanosilica provides good mechanical strength at lower compositions. Maximum value of tensile strength is recorded for nanosilica mixed with silane coupling agent. The coupling agent improves the adhesion between the filler and the PVC matrix by forming covalent bonds. At higher compositions, coupling agent is not enough. Therefore, the value decreases after the initial increase at higher compositions.

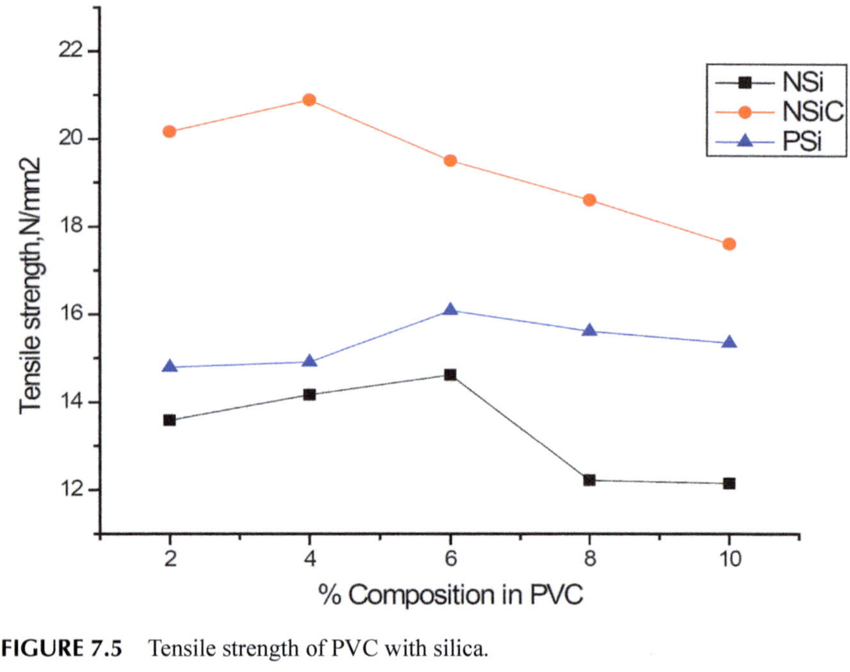

FIGURE 7.5 Tensile strength of PVC with silica.

7.5 CONCLUSION

The microwave and DC conductivity of PVC/acetylene black, PVC/PAni, and PVC/PAni + acetylene black samples increases with the composition of conducting materials in PVC. The doped samples of PAni (30%) show very good electrical properties. The improvement in tensile strength implies that the mechanical properties of PVC are also enhanced by the incorporation of fillers. In the case of PVC filled with silica, use of coupling agents provides better adhesion between the matrix and fillers and this causes improved performance of the nanocomposites.

KEYWORDS

- **PVC**
- **PAni**
- **acetylene black**
- **nanosilica**
- **nanocomposite**
- **electrical properties**
- **mechanical properties**

REFERENCES

1. Crawford, R. J. *Plastics Engineering,* 3rd Ed.; Elsevier: US, 1998; 1–7.
2. Jain and Jain. *Engineering Chemistry*, 9th Ed.; Dhanpat Rai Publishing Company: India, 1998, 234–238.
3. Pant, H. C.; Patra, M. K.; Vashistha, P.; Manzoor, K.; Vardera, S. R.; Kumar, N. In *Study of Dielectric Behaviour in Conducting Polymer–Semiconductor Composite.* Annual General Meeting and Theme Symposium on Novel Polymeric Materials, Mumbai, India (accessed Feb 11–13, 2003).
4. Shao-Yun, F; Xi-Qiao, F.; Bernd, L.; Yiu-Wing, M. Effects of Particle Size, Particle/Matrix Interface Adhesion and Particle Loading on Mechanical Properties of Particulate–Polymer Composites. *Composites Part B* **2008,** *39*, 933–961.
5. Wang, M.; Berry, C; Braden, M.; Bonfield, W. Young's and Shear Moduli of Ceramic Particle Filled Polyethylene. *J. Mater. Sci. Mater. Med.* **1998,** *9*, 621–624.
6. Krishnan, R.; Wesley, A. W.; Goeringer, S.; Gerspacher, M. Carbon Black and Carbon Black-conducting Polymer Composites for Environmental Applications. The University of Texas at Arlington. Texas 76019-0065.
7. John, H.; Thomas, R. M.; Jacob, J.; Mathew, K. T.; Joseph, R. Conducting Polyaniline Composites as Microwave Absorbers. *Polym. Compos.* **2007**, 25(5), 588–592.
8. Laska, J.; Zak, K.; Pron, A. Conducting Blends of Polyaniline with Conventional Polymers. *Synth. Metals* **1997,** *84*, 117–118.
9. MacDiarmid, A. G.; Chiang, J. C.; Richter, A. F.; Somasiri, N. L. D.; Epestein, A. J. Polyanilines: Synthesis and Characterization of the Emeraldine Oxidation State by Elemental Analysis. *Conduct. Polym.*, Springer: Dordrecht**, 1987,** 105–120. (Alacacer, L. Ed. Dordrecht).

10. Banerjee, P.; Mandal, B. M. Conducting Polyaniline Nanoparticle Blends with Extremely Low Percolation Thresholds. *Macromolecules* **1995,** *28*, 3940 (1995).
11. Subramaniam, C. K.; Kaiser, A. B.; Gilberd, P. W.; Wessling, B. Electronic Transport Properties of Polyaniline/PVC Blends, *J. Polym. Sci. Part B Polym. Phys.* **1993,** *31*, 1425.
12. Mathew, K. T.; Raveendranath, U. *Sensors Update*; Baltes, H.; Gopel, W.; Hesse, J. Eds.; Wiley-VCH: Weinheim Pordrecht, 1998; p 105.
13. Kupfer, K.; Kraszewski, A.; Knochel, R. *Sensors Update: Microwave Sensing of Moist Materials, Food and Other Dielectrics;* Baltes, H.; Gopel, W.; Hesse, J., Eds.; Wiley-VCH: Germany, 2000; Vol. 7, p 186.

CHAPTER 8

RECYCLABILITY OF EPDM RUBBER BY CHEMICAL MODIFICATION

L. M. POLGAR[1,2] and F. PICCHIONI[1,2*]

[1]*Department of Chemical Engineering, University of Groningen, Nijenborgh 4, 9747 AG Groningen, The Netherlands*

[2]*Dutch Polymer Institute (DPI), Eindhoven, The Netherlands*

[*]*Corresponding author. E-mail: f.picchioni@rug.nl*

CONTENTS

ABSTRACT

The current climate within the chemical industry is characterized by the development of new or improved polymeric products, displaying new distinctive properties such as (thermally) reversible cross-linking. For polymeric materials, this is often achieved by chemical modification of existing (commercial) polymers. In this short review, we present and briefly discuss the newest development on the chemical modification of ethylene propylene diene random copolymers by focusing on the most promising routes form an industrial perspective.

8.1 INTRODUCTION

Ethylene propylene diene (EPDM) rubbers have many fascinating properties, such as outstanding ozone-, weather-, and low temperature resistance. Extending the technological applications of existing polymers by their chemical modification represents a convenient route to yield improved and cheaper materials as compared to developing completely new ones.[1–6] Functionalization reactions can improve mechanical, chemical, and environmental properties by introducing controlled amounts of functional groups onto the polymer backbone.[7]

Current trends towards sustainability and *cradle-to-cradle* products render the recyclability of cross-linked elastomers, an interesting target property. Especially for rubbers, recycling is currently of crucial importance, even if factually unfeasible according to a *cradle-to-cradle* strategy. The reason is that the currently used cross-linking methods, which are usually irreversible chemical reactions such as sulfur vulcanization and peroxide curing, prohibit reusal as raw material.[8,9]

During the last decades, a lot of effort has therefore been put in the devulcanization of cross-linked rubbers.[10–12] Although reclaiming sulfur-cured natural rubber is now a common technology, less successful attempts have been reported for EPDM rubbers, which can be found in all nontire automotive parts, roofing, and window profiles among others.[13,14] The lack of chemical functionalities in EPDM is one of the reasons behind its wide range of applications. Unfortunately, this also results in lower compatibility with polar polymers as well as with fillers. Introducing adequate functional groups might help overcoming this problem.[15–17] In this context, chemical modification reactions represent an absolute necessity and have been widely studied in the last three decades. This work aims at providing an overview of the most commonly employed modification routes for EPDM, with primary focus on modification strategies applicable at industrial level.

8.2 COMMONLY APPLIED RUBBER MODIFICATIONS

The carbon–carbon double bonds that are present in the rubber (originating from the diene monomeric units) are often, although not always, employed as reactive moieties towards other functional groups.[18] Some modification methods do not require these unsaturations and would just as well work with ethylene–propylene rubber (EPM—copolymers of ethylene and propylene) rubbers (Fig. 8.1). The disadvantage of such methods can be the occurrence of polymer degradation due to chain cleavage.

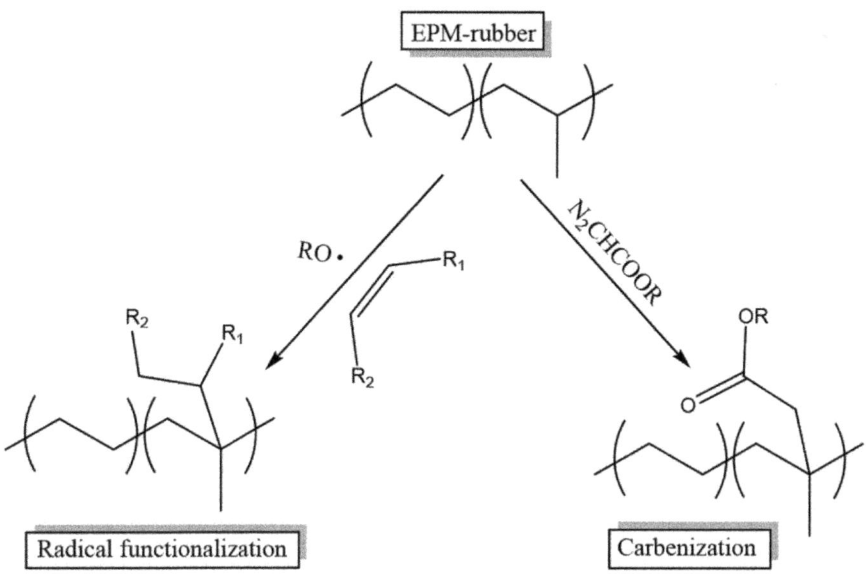

FIGURE 8.1 Commonly employed modification reactions on saturated (EPM) rubbers.

Most other modification methods require the presence a certain amount of carbon–carbon double bonds (Fig. 8.2). Besides, the methods that are feasible on saturated systems prefer to react via different pathways if double bonds are present. These unsaturations can be introduced into the rubbers by copolymerization with dienes. Due to polymer branching, however, the amount of diene that can be incorporated is limited to roughly 1–5 wt%. For EPDM rubbers, the most widely used dienetermonomer is ethylidene norbornene (ENB). ENB groups have an anomalously high reactivity due to bond angle distortion and the associated loss of ring strain upon addition. Other commercially used dienes are dicyclopentadiene (DCPD) and vinyl-norbornene (VNB). EPDM rubbers maintain their excellent stability compared to more unsaturated rubbers because the double bonds that remain after polymerization are not part of the rubber backbone.

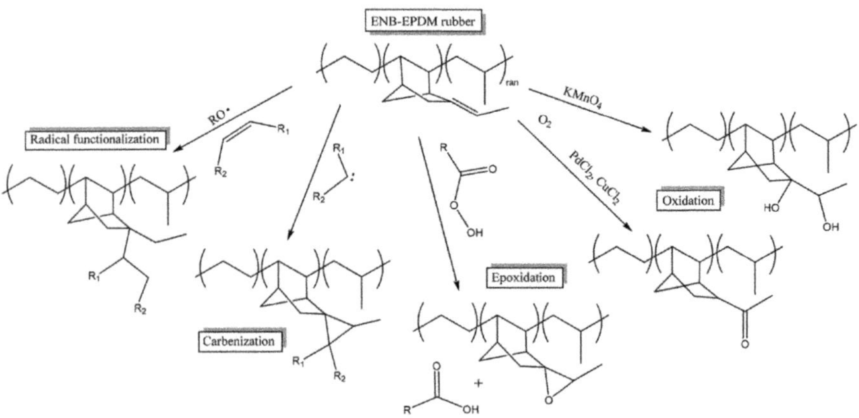

FIGURE 8.2 Commonly employed modification reactions on unsaturated (EDPM) rubbers.

8.2.1 *FREE RADICAL PROCESSING*

Radical processing is the most commonly applied modification techniques for rubbers. It is generally assisted by coagents and is initiated in the presence of peroxides that easily decompose into free radicals. By subtracting hydrogens from the polymer backbone, they create radical moieties on the polymer, resulting in a reactive sites that are susceptible to other reactions, for example, addition to double bonds.[19] Free radical reactions offer a lot of possibilities due to the large number of functional substrates which can be used in combination with radical initiators.[7] Numerous peroxides and hydroperoxides, commonly employed as radical initiators, are also available. They can be selected for individual processes based on their reactivity, selectivity, processing temperature, and solubility in the molten polymer; as such modification reactions generally take place via reactive extrusion.[20,21]

The benefit of using peroxides is that the reactions proceed without the presence of carbon–carbon double bonds and therefore, enables modification reactions on otherwise unreactive saturated polymers such as EPM rubber. The simplicity and cost-effectiveness make functionalization by melt grafting an attractive method to prepare reactive polymers.[22] Organic peroxides are therefore frequently used in the modification of polymer melts to obtain polymers which have physical and mechanical properties different from the original materials.[23,24]

Free radical reactions suffer several disadvantages, such as low specificity, degradation, and undesired cross-linking of the macromolecules.[7] Problems encountered with peroxides include reduced efficiencies and unwanted side reactions resulting in surface degradation and volatiles.[25,26] The addition of a comonomer can increase the extent of functionalization by promoting chain extension and preventing β-scission. Although styrene and methylstyrene are often used for this purpose, furan or thiophene derivatives have also been investigated as coagents.[27,28]

Maleinization is the most common radical polymer modification technique applied to EPDM rubbers.[29–37] Maleinized rubbers exhibit hydrophilicity, polarity, and improved adhesion and binding capabilities, making them suitable for printing and coating applications, compatibilizing agents, and impact modifiers in polymer blends.[2,38–42] On the other hand, rubber products grafted with maleic anhydride (MA) are quite sensitive to moisture and fast hydrolysis may occur at atmospheric conditions.

The functionalization reaction starts with hydrogen abstraction from the polymer backbone and formation of a macroradical, which may subsequently react with a MA, resulting in the desired grafting of up to 3 wt%.[35–37,43–45] This functionalization can be realized by chemical treatment, photoirradiation or high-energy irradiation.[46,47] On an industrial scale, however, maleinization is initiated with peroxides.[37,38] The reaction is usually carried out via continuous reactive extrusion rather than in solution. Grafting takes mostly place on the tertiary carbon atoms, which yield the more stable radicals upon hydrogen abstraction by the primary radicals (*vide supra*). Suppressants are required as the random covalent grafting by a radical mechanism that is accompanied by degradation and leads to undesirable rheological changes in the product due to unwanted side reactions such as chain scission, cross-linking, disproportionation, and coupling.[42,48] For EPDM, mostly single anhydride ring grafting takes place on both tertiary carbons and terminal unsaturations formed by β-scission. However, unsaturated, oligomeric, and polymeric grafts can also be formed.[38]

The isostructural analogues of MA, such as maleimides, maleates, anhydrydes, succinic acid, esters, and imides can be grafted in a similar fashion.[46] Other unsaturated compounds, such as diethyl maleate,[49,50] diethyl fumarate,[49,50] several amino-, hydroxyl-, and glycidyl (meth)acrylates,[22,51–57] benzotriazoles,[58] acrylic acid,[59] methyl esters of itaconic acid,[60] oxazoline,[22,61,62] furan an thiophene derivatives,[26,28] Polar ester groups,[7] p-phenylen-bis-maleamic acid,[63] amino groups,[64] styrenes,[65] and methyl esters of unsaturated fatty acids were also applied as functionalizing agents for radical processing in several studies.

8.2.2 CARBENIZATION

Carbenes are reactive towards both alkenes and alkanes and, therefore, of particular interest for the modification of polymers with a low degree of unsaturations such as EPDM rubbers.[66] Carbenes are extremely reactive as their lifetimes are generally shorter than 1 s.[7,67] Carbene insertion to C–H bonds allows for functionalization with well-defined groups while avoiding undesired side reactions.[7] Carbenization reactions can be used to introduce groups with a diverse chemical functionality by using diazocarbonyl compounds such as polar ester groups into the polymer backbone.[7,68,69]

Carbenization reactions can be performed without solvent, in the molten state, and proceed via a one-step mechanism in which the carbine species react in a singlet state after the emission of nitrogen gas. These compounds can also be used to induce a certain amount of photo cross-linking.[66] For EPDM rubbers, complete decomposition of the additives and a controlled degree of functionalization between 5 and 10 wt% can be achieved.[7]

8.2.3 EPOXIDATION

Rubber modification by epoxidation of carbon–carbon double bonds in the polymer is a promising method for improving properties such as its miscibility with polar polymers for composites and its adhesion to other materials, expanding their applications to the field of polymer alloys.[70,71] Furthermore, epoxidation of natural rubber is known to improve solvent and wear resistance, and its gas barrier property compared to the virgin natural rubber. The introduction of epoxy groups also increases the miscibility with epoxy resin. Epoxidized rubbers can therefore be used as toughening agents for epoxy resins, which are brittle at room temperature.[72]

Conventional epoxidation reactions, which utilize peroxide agents, usually result in a significant amount of side reactions. Current epoxidation techniques, on the other hand, require relatively easy solventless reactive processing conditions, preferably involving a catalyst.[73] The reactivity of EPDM rubbers is higher than that of other unsaturated materials such as styrene–butadiene rubber.[6] The degree of substitution that can be obtained is relatively high (1–5 wt%, depending on the diene content).[74] In addition to adhering to polar materials such as brass and nylon, epoxydized unsaturated rubbers offer a potential reactivity for further reactive blending.[6]

8.2.4 OXIDATION

The radical oxidation of unsaturated polymers proceeds as autoxidation under the influence of light and air and will eventually result in decomposition.[75] A more controlled radical oxidation can be achieved with the use of catalysts. Generally, these are cobalt based, although many alternatives have been also reported.[76,77] Such reactions result in the conversion of carbon–carbon double bonds into alcohols. Similarly, the Wacker oxidation specifically yields ketone groups.[75] Another way to achieve carbonyl groups is by bond cleavage methods, such as ozonolysis. Oxidative cleavage, however, results in shortening of the polymers and can only yield end-group functionalities.[78] A more desirable method is oxosynthesis, which denotes the reaction of olefins with carbon monoxide and hydrogen mixtures to yield alcohol or ketone functionalized products.[79] These methods could very well be used to convert pending carbon–carbon double bonds into oxygenized forms.

8.3 CONCLUSIONS

Functionalization or modification of rubbers can improve mechanical, chemical, and environmental properties, or it can introduce new properties in a material such as the recyclability of cross-linked elastomers. The lack of chemical functionalities in EPDM rubbers limits the possibilities to modify these materials accordingly. Introducing adequate functional groups by chemical modification might help overcoming this problem. Several modification routes are commonly applied to EPDM in industry. The most well-known are radical processes that are normally used for the peroxide cross-linking of EPDM rubbers. Radical reactions can be used to graft a large array of functional groups onto the rubber backbone, even without available carbon–carbon double bonds. The largest disadvantage of radical reactions is the lack of control which may cause numerous side reactions, usually leading to partial degradation and undesired cross-linking. Carbenization and epoxidation on the other hand are more controlled, but do require the presence of carbon–carbon double bonds. The advantage is that the formed carbene and epoxy functionalities are extremely reactive towards a large number of chemical reactions and can therefore very well be used to (reversibly) cross-link the functionalized elastomers in a second reaction step. Lastly, the oxidation reaction is very versatile and can be chosen to selectively introduce alcohol, ketone, aldehyde, or even acid functionalities, which may again be used in any desired second-degree chemical reaction.

KEYWORDS

- **ethylene propylene diene**
- **EPDM rubber**
- **recycle**
- **reuse**
- **carbenes**

REFERENCES

1. Brosse, J. C.; Campistron, I.; Derouet, D.; El Hamdaoui, A.; Houdayer, S.; Reyx, D.; Ritoit-Gillier, S. Chemical Modifications of Polydiene Elastomers: A Survey and Some Recent Results. *J. Appl. Polym. Sci.* **2000,** *78*(8), 1461–1477.

2. Thames, S. F.; Gupta, S. Synthesis and Characterization of Maleinized-silylated Low Molecular Weight Guayule Rubber (MASiGR). *J. Appl. Polym. Sci.* **2001,** *81*(3)754–766.
3. Lei, Y. *et al.* Thiol-containing Ionic Liquid for the Modification of Styrene–Butadiene Rubber/Silica Composites. *J. Appl. Polym. Sci.* **2012,** *123*(2), 1252–1260..
4. Dick, J. S. *How to Improve Rubber Compounds: 1500 Experimental Ideas for Problem Solving*; Carl Hanser Verlag GmbH & Co. KG, 2004;1–420.
5. Zhang, Y.; Chen, X. Z.; Zhang, Y.; Zhang, X. Z. Preparation of Epoxidized Rubber using a Reactive Processing Technique. I. Synthesis and Characterization of Epoxidized Polybutadiene Rubber. *J. Appl. Polym. Sci.* **2001,** *81*(12), 2987–2992.
6. Zhang, Y.; Chen, X. Z.; Zhang, Y.; Zhang, Y. X. Epoxidation of Unsaturated Rubbers with Alkyl Hydroperoxide Using Molybdenum Compounds as Catalyst in the Reactive Processing Equipment. *Macromolecul. Mater. Eng.* **2001,** *286*(8), 443–448.
7. Aglietto, M.; Alterio, R.; Bertani, R.; Galleschi, F.; Ruggeri, G. Polyolefin Functionalization by Carbene Insertion for Polymer Blends. *Polymer* **1989,** *30*, 1133–1136.
8. Chino, K.; Ashiura, M. Themoreversible Cross-linking Rubber Using Supramolecular Hydrogen-bonding Networks. *Macromolecules* **2001,** *34*, 9201–9204.
9. van der Mee, M. A. J. Thermoreversible Cross-linking of Elastomers. A Comparative Study Between Ionic Interactions, Hydrogen Bonding and Covalent Cross-links; Technische Universiteit Eindhoven: Eindhoven, **2007**,1–177.
10. Dijkhuis, K. A. J. Recycling of Vulcanized EPDM-rubber Mechanistic Studies into the Development of a Continuous Process Using Amines as Devulcanization Aids. *Enschede, The Netherlands,* **2008,** 169.
11. Verbruggen, M. Devulcanization of EPDM Rubber. Doctoral Dissertation, Ph.D. Thesis, University of Twente, Enschede, The Netherlands, September 14, 2007.
12. Sutanto, P. Development of a Continuous Process for EPDM Devulcanization in an Extruder. University Library Groningen, Netherlands, **2006**.
13. Sutanto, P.; Picchioni, E.; Janssen, L. P. B. M.; Dijkhuis, K. A. J.; Dierkes, E. K.; Noordermeer, J. W. M. State of the Art: Recycling of EPDM Rubber Vulcanizates. *Int. Polym. Proc.* **2006,** *21*(2): 211–217.
14. Dijkhuis, K. A. J.; Babu I.; Lopulissa, J. S.; Noordermeer, J. W. M.; Dierkes, W. K. A Mechanistic Approach to EPDM Devulcanization. *Rubber Chem. Technol.* **2008,** *81*.
15. Kolarik, J.; Agrawal, G. L.; Krulis, Z.; Kovar, J. Dynamic Mechanical-behavior of Binary Blends Polyethylene EPDM Rubber and Polypropylene EPDM Rubber. *Polym. Compos.* **1986,** *7*(2) 190–208.
16. Ruksakulpiwat, Y.; Sridee, J.; Suppakarn, N.; Sutapun, W. Improvement of Impact Property of Natural Fiber-polypropylene Composite by Using Natural Rubber and EPDM Rubber. *Compos. B-Eng.* **2009,** *40*(7) 619–622.
17. Tall, S.; Karlsson, S.; Albertsson, A. C. EPDM Elastomers as Impact Modifiers for Contaminated, Recycled HDPE. *Polym. Polym. Compos.* **1997,** *5*, 417–422.
18. Thomsen, A. D.; Malmstrom, E.; Hvilsted, S. Novel Polymers with a High Carboxylic Acid Loading. *J. Polym. Sci. A-Polym. Chem.* **2006,** *44*(21) 6360–6377.
19. Li, X.; Tabil, L. G. Panigrahi, S. Chemical Treatments of Natural Fiber for use in Natural Fiber-reinforced Composites. *J. Polym. Environ.* 2007, 15(1) 25–33.

20. Xie, H. Q.; Seay, M.; Oliphant, K.; Baker, W. E. Search for Nonoxidative, Hydrogen-abstracting Initiators Useful for Melt Grafting Processes. *J. Appl. Polym. Sci.* **1993,** *48*, 1199–1208.
21. Hettema, R.; J. Van Tol, J.; Janssen, L. P. B. M. In-situ Reactive Blending of Polyethylene and Polypropylene in Co-rotating and Counter-rotating Extruders. *Polym Eng. Sci.* **1999,** *39*(9), 1628–1641.
22. Liu, N. C.; Xie, H. Q.; Baker, W. E. Comparison of the Effectiveness of Different Basic Functional-group for the Reactive Compatibilization of Polymer Blends. *Polymer* **1993,** *34*(22) 4680–4687.
23. Lambla, M. Reactive Extrusion—A New Tool for the Diversification of Polymeric Materials. *Macromol. Symp.* **1994,** *83*, 37–58.
24. Yu, D. W.; Xanthos, M.; Gogos, C. G. Structure Properties Comparison of Peroxide Modified and Irradiated LDPE/PP Blends. *Abstr. Pap. Am. Chem. Soc.* **1992,** *204*.
25. Zielinska, A. J. Cross-linking and Modification of Saturated Elastomers Using Functionalized Azides. 2011.
26. Augier, S.; Coiai, S.; Gragnoli, T.; Passaglia, E.; Pradel, J.; Flat, J. Coagent Assisted Polypropylene Radical Functionalization: Monomer Grafting Modulation and Molecular Weight Conservation. *Polymer* **2006,** *47*.
27. Coiai, S.; Augier, S.; Pinzino, C.; Passaglia, E. Control of Degradation of Polypropylene During Its Radical Functionalisation with Furan and Thiophene Derivatives. *Polym. Degrad. Stab.* **2010,** *95*, 298–305.
28. Augier, S.; Coiai, S.; Passaglia, E.; Ciardelli, F.; Zulli, F.; Andreozzi, L.; Giordano, M. Structure and Rheology of Polypropylene with Various Architectures Prepared by Coagent-assisted Radical Processing. *Polym. Int.* **2010,** *59*.
29. Guldogan, Y.; Egri, S.; Rzaev, Z. M. O.; Piskin, E. Comparison of Maleic Anhydride Grafting onto Powder and Granular Polypropylene in the Melt by Reactive Extrusion. *J. Appl. Polym. Sci.* **2004,** *92*.
30. Chang, D.; White, J. L. Experimental Study of Maleation of Polypropylene in Various Twin-screw Extruder Systems. *J. Appl. Polym. Sci.* **2003,** *90*, 1755–1764.
31. Li, Y.; Xie, X. M.; Guo, B. H. Study on Styrene-assisted Melt Free-radical Grafting of Maleic Anhydride onto Polypropylene. *Polymer* **2001,** *42*, 3419–3425.
32. Samay, G.; Nagy, T.; White, J. L. Grafting Maleic-Anhydride and Comonomers onto Polyethylene. *J. Appl. Polym. Sci.* **1995,** *56*, 1423–1433.
33. Lu, B.; Chung, T. C. Synthesis of Maleic Anhydride Grafted Polyethylene and Polypropylene, with Controlled Molecular Structures. *J. Polym. Sci. A-Polym. Chem.* **2000,** *38*, 1337–1343.
34. Machado, A. V.; Covas, J. A.; van Duin, M. Effect of Polyolefin Structure on Maleic Anhydride Grafting. *Polymer* **2001,** *42*(8), 3649–3655.
35. Mehrabzadeh, M.; Kasaei, S.; Khosravi, M. Modification of Fast-cure Ethylene-propylene Diene Terpolymer Rubber by Maleic Anhydride and Effect of Electron Donor. *J. Appl. Polym. Sci.* **1998,** *70*(1), 1–5.
36. van Duin, M. Grafting of Polyolefins with Maleic Anhydride: Alchemy or Technology? *Macromol. Symp.* **2003,** *202*(1) 1–10 (Wiley-VCH Verlag).
37. Burlett, D. J.; Lindt, J. T. Reactive Processing of Rubbers. *Rubber Chem. Technol.* **1993,** *66*(3), 411–434.

38. Heinen, W.; Rosenmoller, C. H.; Wenzel, C. B.; deGroot, H. J. M.; Lugtenburg, J.; vanDuin, M. C-13 NMR Study of the Grafting of Maleic Anhydride onto Polyethene, Polypropene, and Ethene–Propene Copolymers. *Macromolecules* **1996,** *29*, 1151–1157.
39. Greco, R.; Malinconico, M.; Martuscelli, E.; Ragosta, G.; Scarinzi, G. Role of Degree of Grafting of Functionalized Ethylene Propylene Rubber on the Properties of Rubber-Modified Polyamide-6. *Polymer* **1987,** *28*(7), 1185–1189.
40. Cimmino, S.; Coppola, F.; Dorazio, L.; Greco, R.; Maglio, G.; Malinconico, M.; Mancarella, C.; Martuscelli, E.; Ragosta, G. Ternary Nylon-6 Rubber Modified Rubber Blends: Effect of the Mixing Procedure on Morphology, Mechanical and Impact Properties. *Polymer* **1986**, *27*(12), 1874–1884.
41. Kayano, Y.; Keskkula, H.; Paul, D. R. Fracture Behaviour of some Rubber-toughened Nylon 6 Blends. *Polymer* **1998,** *39*(13), 2835–2845.
42. Vicente, A. I.; Campos, J.; Bordado, J. M.; Ribeiro, M. R. Maleic Anhydride Modified Ethylene-diene Copolymers: Synthesis and Properties. *React. Funct. Polym.* **2008,** *68*(2), 519–526.
43. van Duin, M.; Dikland, H. A Chemical Modification Approach for Improving the Oil Resistance of Ethylene–Propylene Copolymers. *Polym. Degrad. Stab.* **2007,** *92*(12), 2287–2293.
44. Oostenbrink, A. J.; Gaymans, R. J. Maleic-anhydride Grafting on EPDM Rubber in the Melt. *Polymer* **1992,** *33*(14), 3086–3088.
45. Barra, G. M. O.; Crespo, J. S.; Bertolino, J., R.; Soldi, V.; Pires, A. T. N. Maleic Anhydride Grafting on EPDM: Qualitative and Quantitative Determination. *J. Braz. Chem. Soc.* **1999,** *10*, 31–34.
46. Rzayev, Z. M. O. Graft Copolymers of Maleic Anhydride and Its Isostructural Analogues: High Performance Engineering Materials. *Int. Rev. Chem. Eng.* **2011,** 3. 1105.1260.
47. Bhattacharya, A.; Misra, B. N. Grafting: A Versatile Means to Modify Polymers: Techniques, Factors and Applications. *Prog. Polym. Sci.* **2004,** *29*(8), 767–814.
48. Wu, C. H.; Su, A. C. Suppression of Side Reactions During Melt Functionalization of Ethylene Propylene Rubber. *Polymer* **1992,** *33*, 1987–1992.
49. Ruggeri, G.; Aglietto, M.; Petragnani, A.; Ciardelli, F. Some Aspects of Polypropylene Functionalization by Free-radical Reactions. *Eur. Polym. J.* **1983,** *19*(10–11), 863–866.
50. Ciardelli, F.; Aglietto, M.; Passaglia, E.; Ruggeri, G. Molecular and Mechanistic Aspects of the Functionalization of Polyolefins with Ester Groups. *Macromol. Symp.* **1998,** *129*, 79–88.
51. Huang, H.; Liu, N. C. Nondegradative Melt Functionalization of Polypropylene with Glycidyl Methacrylate. *J. Appl. Polym. Sci.* **1998,** *67*, 1957–1963.
52. Chen, L. F.; Wong, B.; Baker, W. E. Melt Grafting of Glycidyl Methacrylate onto Polypropylene and Reactive Compatibilization of Rubber Toughened Polypropylene. *Polym. Eng. Sci.* **1996,** *36*, 1594–1607.
53. Citovicky, P.; Chrastova, V.; Mejzlik, J.; Majer, J.; Benc, G. Modification of Polypropylene with 2,3[epoxypropyl] Methacrylate. *Collect. Czech. Chem. Commun.* **1980,** *45*(8), 2319–2328.
54. Song, Z.; Baker, W. E. Melt Grafting of Tert-butylaminoethyl Methacrylate Onto Polyethylene. *Polymer* **1992,** *33*(15), 3266–3273.

55. Song, Z. Q.; Baker, W. E. In-situ Compatibilization of Polystyrene Polyethylene Blends Using Amino-methacrylate-grafted Polyethylene. *J. Appl. Polym. Sci.* **1992,** *44*, 2167–2177.
56. Song, Z. Q.; Baker, W. E. Grafting of 2-(Dimethylamino)Ethyl Methacrylate on Linear Low-density Polyethylene in the Melt. *Angew. Makromol. Chem.* **1990,** *181*, 1–22.
57. Mousavi-Saghandikolaei, S. A.; Frounchi, M.; Dadbin, S.; Augier, S.; Passaglia, E.; Ciardelli, F. Modification of Isotactic Polypropylene by the Free-radical Grafting of 1,1,1-trimethylolpropane Trimethacrylate. *J. Appl. Polym. Sci.* **2007,** *104*, 950–958.
58. Pradellok, W.; Vogl, O.; Gupta, A. Functional Polymers.14. Grafting of 2(2-Hydroxy-5-Vinylphenyl)2H-Benzotriazole onto Polymers with Aliphatic Groups. *J. Polym. Sci. A-Polym. Chem.* **1981,** *19*, 3307–3314.
59. Oromehie, A. R.; Hashemi, S. A.; Meldrum, I. G.; Waters, D. N. Functionalisation of Polypropylene with Maleic Anhydride and Acrylic Acid for Compatibilising Blends of Polypropylene with Poly(Ethylene Terephthalate). *Polym. Int.* **1997,** *42*, 117–120.
60. Yazdani-Pedram, M.; Vega, H.; Quijada, R. Melt Functionalization of Polypropylene with Methyl Esters of Itaconic Acid. *Polymer* **2001,** *42*, 4751–4758.
61. Vainio, T.; Hu, G. H.; Lambla, M.; Seppala, J. Functionalization of Polypropylene with Oxazoline and Reactive Blending of PP with PBT in a Corotating Twin-screw Extruder. *J. Appl. Polym. Sci.* **1997,** *63*, 883–894.
62. Vainio, T.; Hu, G. H.; Lambla, M.; Seppala, J. V. Functionalized Polypropylene Prepared by Melt Free Radical Grafting of Low Volatile Oxazoline and Its Potential in Compatibilization of PP/PBT Blends. *J. Appl. Polym. Sci.* **1996,** *61*, 843–852.
63. Garcia-Martinez, J. M.; Cofrades, A. G.; Areso, S.; Laguna, O.; Collar, E. P. Grafting of P-phenylen-bis-maleamic Acid into Polypropylene in Melt. *J. Appl. Polym. Sci.* **1998,** *69*, 931–939.
64. Immirzi, B.; Lanzetta, N.; Laurienzo, P.; Maglio, G.; Malinconico, M.; Martuscelli, E.; Palumbo, R. Acid and Base Functionalization of Ethylene-propylene Rubbers.1. Grafting of Tertiary Amino-groups and Interpolymer Network Formation. *Makromol. Chem.-Macromol. Chem. Phys.***1987,** *188*, 951–960.
65. Picchioni, F.; Goossens, J. G. P.; van Duin, M.; Magusin, P. Solid-state Modification of Isotactic Polypropylene (iPP) via Grafting of Styrene. I. Polymerization Experiments. *J. Appl. Polym. Sci.* **2003,** *89*, 3279–3291.
66. Blencowe, A.; Blencowe, C.; Cosstick, K.; Hayes, W. A Carbene Insertion Approach to Functionalised Poly(ethylene oxide)-based Gels. *React. Funct. Polym.* **2008,** *68*(4), 868–875.
67. Aglietto, M.; Bertani, R.; Ruggeri, G.; Fiordiponti, P.; Segre, A. L. Functionalization of Polyolefins: Structure of Functional-groups in Polyethylene Reacted with Ethyl Diazoacetate. *Macromolecules* **1989,** *22*(3), 1492–1493.
68. Jellema, E.; Jongerius, A. L.; Reek, J. N. H.; de Bruin, B. C1 Polymerisation and Related C–C bond Forming "Carbene Insertion" Reactions. *Chem. Soc. Rev.* **2010,** *39*(5), 1706–1723.
69. Choong, C.; Foord, J. S.; Griffiths, J.; Parker, E. M.; Baiwen, L.; Bora, M.; Moloney, M. G. Post-polymerisation Modification of Surface Chemical Functionality and Its Effect on Protein Binding. *New J. Chem.***2010,** *36*(5), 1187–1200
70. Baker, C. S. L.; Gelling, I. R.; Newell, R. Epoxidized Natural-rubber. *Rubber Chem. Technol.* **1985,** *58*(1), 67–85.

71. Gallucci, R. R.; Going, R. C. Preparation and Reactions of Epoxy-modified Polyethylene. *J. Appl. Polym. Sci.* **1982,** *27*(2), 425–437.
72. Krishnan, P. S. G.; Ayyaswamy, K.; Nayak, S. K. Hydroxy Terminated Polybutadiene: Chemical Modifications and Applications. *J. Macromol. Sci. A-Pure Appl. Chem.* **2013,** *50*, 128–138.
73. Palomeque, J.; Lopez, J.; Figueras, F. Epoxydation of Activated Olefins by Solid Bases. *J. Catal.* **2002,** *211*(1), 150–156.
74. Hoyle, C. E.; Lee, T. Y.; Roper, T. Thiolenes: Chemistry of the Past with Promise for the Future. *J. Polym. Sci. A-Polym. Chem.* **2004,** *42*(21), 5301–5338.
75. Koeckritz, A.; Martin, A. Oxidation of Unsaturated Fatty Acid Derivatives and Vegetable Oils. *Eur. J. Lipid Sci. Technol.***2008,** *110*(9), 812–824.
76. Oyman, Z. O.; Ming, W.; van der Linde, R. Catalytic Activity of a Dinuclear Manganese Complex (MnMeTACN) on the Oxidation of Ethyl Linoleate. *Appl. Cat. A-Gen.* **2007,** *316*(2), 191–196.
77. Micciche, F.; van Haveren, J.; Oostveen, E.; Laven, J.; Ming, W. H.; Oyman, Z. O.; van der Linde, R. Oxidation of Methyl Linoleate in Micellar Solutions Induced by the Combination of Iron(II)/Ascorbic Acid and Iron(II)/H2O2. *Arch. Biochem. Biophys.* **2005,** *443*(1–2), 45–52.
78. Poyatos, M.; Mata, J. A.; Falomir, E.; Crabtree, R. H.; Peris, E. New Ruthenium(II) CNC-Pincer Bis(Carbene) Complexes: Synthesis and Catalytic Activity. *Organometallics* **2003,** *22*, 1110–1114.
79. Drent, E.; Mul, W. P.; Budzelaar, P. H. M. Teaching a Palladium Polymerization Catalyst to Mono-oxygenate Olefins. *Comments Inorg. Chem.* **2002,** *23*(2), 127–147.

CHAPTER 9

WETTING STUDIES AS ENABLERS FOR RECYCLING AND REUSE OF MICROFLUIDIC DEVICES

V. MADHURIMA*

Department of Physics, Central University of Tamil Nadu, Neelakudi Campus, Thiruvarur 610101, Tamil Nadu, India

**E-mail: madhurima@cutn.ac.in*

CONTENTS

ABSTRACT

Good health care to the growing world population is a prime concern for all. Miniaturization of analytical instruments and drug delivery systems has gained importance in order to enable medical resources to reach the poorest of people and the remotest of places. Miniaturization of devices helps in reducing both the volume of analytes used per person and the quantity of materials used. However, disposal of these small-scaled devices is not easy and reuse is also very difficult, unlike those used with bulk analytes. The possibility of recycling and reusing microfluidic device is explored here, by understanding the wetting properties of materials associated with these devices. Two methods will be discussed in detail with respect to (1) changing the properties of substrates and (2) altering the surface tension of the concerned liquids.

9.1 INTRODUCTION

Integration of sensors and actuators to CMOS technology has been responsible for the "more than Moore" state of electronics [1] where in the size of the devices has been greatly reduced and hence more number of components per unit device areas has been possible. This integration has been brought about by both monolithic integration and hybrid integration through the machining of Si and other materials to form sensors and actuators. Such devices go by the broad name of MEMS—micro electro mechanical systems. The need for machining CMOS compatible materials arose from the fact that actuators draw the maximum power in most circuits. Integrating sensors and actuators on the same platform as the electronic circuit leads to a dramatic reduction in power consumption,[2] leading to a large-scale increase in the number of electronic devices used.[3] Size reduction of electronic devices combined with their lowered power consumption has meant a lowering of the footprint of most devices ranging from mobile phones to printers all the way to analytical chemical instruments, which in turn has lead to a rapid proliferation of such devices.

This rapid proliferation comes with its associated problem of disposal of this large number of devices. All the waste from electrical and electronic appliances and devices that are at the end of use are termed e-waste.[4] The usual procedure of e-waste disposal involves the collection of the devices, sorting and separation followed by recycling those materials which can be recycled and disposing those which cannot be. E-wastes have, on the one hand, toxic materials and on the other, valuable materials. The high toxicity of e-wastes when they are disposed in an uncontrolled fashion has lead them to be classified as hazardous according to the Basel convention.[5]

In the Indian context, porous borders, non-availability of stringent regulations and the participation of private sector in e-waste management had seen a proliferation of e-waste imports into the country, often illegally.[4] Of the many methods suggested to curb this menace, is to reduce the wastage at source. Recycling e-waste is known to be a labor-intensive task,[6] with the returns in terms of reusables being much less compared to the effort needed to retrieve them. This calls for new strategies to handle e-waste recycling and management.

9.1.1 INTRODUCTIONTOMICROFLUIDICS—DIFFERENCEIN DROPLET-AND RHEOLOGY-BASED DEVICES

Here we take the specific case of MEMS devices that have a fluidic component. Such devices are called microfluidics and they have a ubiquitous existence as seen in inkjet printers,[7] insulin pumps,[8] lab-on-chips[9] and other such devices.[10] While it is a fact that the larger components of e-wastes come from mobile phones and personal computers, the contribution from devices such as lab-on-chips is increasing due to the pressing need for point-of-care biochemical analysis for larger population. The proliferation of these small-scale devices cannot be curbed and hence strategies for their reuse become imperative.

Typically in microfluidic devices, liquid(s) are converted into droplets, moved, mixed, reacted, or detected. The handling of liquids can occur in micron/sub-micron sized chambers or channels. The advantage of microfluidics comes from their high sensitivity, high speed of operation, high throughput, and low cost of analysis.[11]

The strategy being suggested for reuse is to be able to empty, rinse, clean and refill these devices so that the wastage is minimized. In order to be able to achieve this, the nature of drop-formation, drop detachment and fluid flow in them needs to be understood carefully. This, in turn, needs an understanding of the solids being used to make the devices, the liquids flowing through them and the actual interaction between such solids and liquids.

Microfluidics relies largely on micromachining technologies, which are highly material specific. The earliest devices were silicon-based followed by glass-based devices. While these materials are good for electrophoresis, they are expensive for micro-fabrication. Elastomers and plastics—both of which are relatively inexpensive and allow rapid prototyping of the structures were the next set of materials to be explored for microfluidics.[11] Current microfluidic technologies rely on polymeric materials, especially polydimethylsiloxane (PDMS) and cellulose.[12] Lab-on-chip is now being replaced by lab-on-paper, especially for point of care diagnosis.[13]

9.2 MICROFLUIDIC DEVICES

Microfluidic devices are based either on ejection of droplets from reservoir chambers, flow of liquids through channels or a combination of both. There are a large number of reviews of drop formation and transport in microfluidic channels.[14–17] We shall consider here, briefly, some salient points of the scaling rules and liquid-surface interaction.

In general, the formation of droplets over a surface in bulk is usually a three phase equilibrium condition with the liquid droplet being in thermodynamical equilibrium with the gas (air) on one hand and the solid surface on the other. The equilibrium condition is given by Young's equation:

$$\gamma_{SG} = \gamma_{SG} + \gamma_{SG} \cos \theta$$

where Υ_{ij} is the surface energy between i and j. S, L, and G stand for the solid, liquid and gas phases and θ is the contact angle.[18] The important point to note here is that as the size of the drop decreases, the effects due to gravity also decrease and the surface effects are dominant. One important scaling phenomenon is that of the surface area to volume ratio. As the dimensions of the device decrease, the surface area increases with respect to the volume.

For a generic length scale L, body forces such as gravitational forces scale as L^3 while surface forces scale as L^2 and surface tension as L (since surface tension is a function of the line of contact between the drop and the surface). Consider the action of an external for on a system that is a combination of body force, surface force and a line force, such as the case of a bulk water droplet pinching off and falling due to gravitational effects. The force balance can be given as:

$$m\frac{dv}{dt} = \rho\left(L^3\right) + \tau\left(L^2\right) + \xi\left(L\right)$$

where ρ, τ, and ξ are the volume, surface and line forces respectively. If the length scale of the device (L) reduces by a factor of 10, the bulk effects, such as due to gravity, will reduce by an order of 1000 and that of surface by 100 and line forces by 10. Hence, while drop detachment for larger volumes is assisted by gravity and that of small volumes is a challenge since all the energy to overcome the strong molecular interactions has to be supplied externally.

Drop manipulation at micron/sub-micron scales thus involves two phases—solid and liquid, and their interactions, as understood by Young's equation. Fluid flow at these scales is still governed by classical hydrodynamical equations such as Navier–Stokes, but with appropriate scaling rules.[19] Drop/fluid manipulation at these scales can be achieved by many methods[20] such as (1) application of Constant/Switching Electric Fields—this is a process called Electrowetting wherein the wetting of a (generally) hydrophobic surface is altered by the application of either a constant (DC) or switching (AC) electric field between the liquid and the substrate.[21,22] (2) Magnetic effects—Magnetic nanoparticles are added to the fluid which is then subject to an external magnetic field so as to enable the nano fluid to move.[23] (3) Change in surface tension[24]—The (Gibbs-) Marangoni effect is the flow of a fluid due to surface tension gradients since the tangential stress across an interface is proportional to the gradient of surface tension. This in turn implies that any system with a surface tension gradient cannot be under static equilibrium. (4) Use of surfactants—binary liquids, etc. The Marangoni effect can be induced though methods such as addition of surfactants and addition of a different liquid.[25] (5) Temperature gradient—Fluid flows can be induced by a temperature gradient. Such localized heating can be caused either by ohmic heaters[26] or through shining Laser light.[27] (6) Mechanical methods—Pneumatic pumps using micro mechanical valves and gauges patterned out of the substrate material.[28] (7) Geometry mediated—One method is to modify the surface roughness,[29] that is, have a smooth/rough transition, so as to induce flow and the other is to pattern the channels such that one channel feeds the other channel and by careful control of junctions, collapse/mixing of liquids is possible.

Of the very many available techniques to control drop formation and flow at micro/sub-micron scales, two methods of taken up here for further elucidation are (1) changing the properties of substrates and (2) altering the surface tension of the concerned liquids.

9.3 ALTERING SURFACE PROPERTIES OF SOLIDS

Altering surface properties such as surface energy and roughness in order to control their wetting properties have been used in designing materials such as self-cleaning surfaces.[30] This is an idea borrowed from nature. Many natural surfaces, such as the lotus leaf, are known to be water repellant. A drop of water on such a surface rapidly assumes a spherical shape and will tend to roll off. This is known as the lotus leaf effect[31] and its mimicry on artificial surfaces relies on the production of a hierarchical structure of roughness. On the other hand, natural surfaces such as the petal of a rose exhibit spherical water drops although they do not roll off. Both these effects are known to be caused by hierarchical roughness of the surface.[32] The rose petal effect is said to be in a Wenzel state wherein the capillary forces of the liquid trapped between the pillars holds the drop without rolling (Fig. 9.1a). The lotus leaf effect is explained as being in the Cassie–Baxter State (Fig. 9.1b) where the drops sit atop the pillared structure.[33] Here, r refers to the roughness factor and f to the fraction of the solid surface that is wetted by the liquid:

$$\cos\theta_W = r_W \left\{ \frac{Y_{SV} - Y_{SL}}{Y_{LV}} \right\} = r_W \cos\theta_Y \text{ Wenzel State equation}$$

$$\cos\theta_{CB} = r_f f \cos\theta_Y + f - 1 \text{ Cassie–Baxter state}$$

Controlling the roughness of surfaces can be achieved in a multitude of ways such as (1) plasma treatment,[34] (2) chemical treatment,[35] (3) directional growth of substrate materials[36] and (4) mechanical treatment of surfaces,[37] etc. By altering the surface roughness liquids can be made to be more sticky non-sticky over any surface.

This control of surface properties is crucial for micron/sub-micron sized devices since the intermolecular interaction strength far exceeds that due to gravitational forces. Any control of fluid motion at these scales depends entirely on the control of the intermolecular forces between the liquid and solid. It is evident from the previous discussion that the wetting of a solid surface by a liquid depends on (1) the cohesive forces in solid and in the liquid, (2) the adhesive forces between the molecules of the solid and that of the liquid, and (3) the roughness of the solid surface.

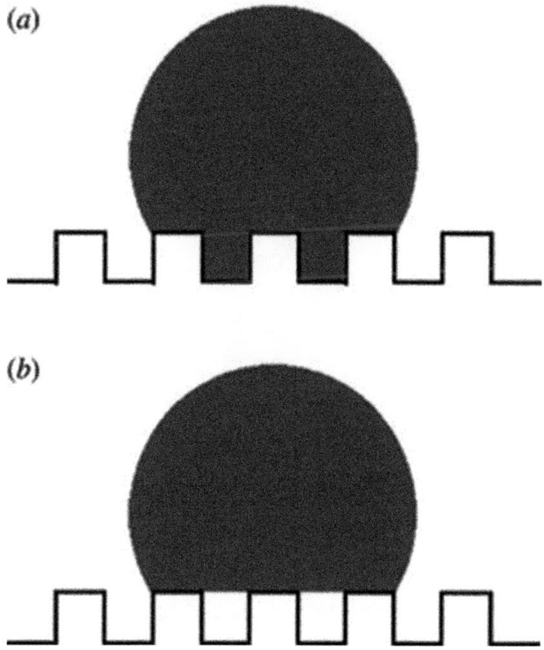

FIGURE 9.1 (a) Wenzel state and (b) Cassie–Baxter state.

The intermolecular forces within a solid are relatively more difficult to manipulate. The cohesive forces in a liquid can be easily altered by the addition of a second component of liquid, or a surfactant or any such additive. The net wetting of a solid by a liquid can be altered by also controlling the surface roughness of the solid substrate material.

In the process of device production, materials such as oxides are often coated over hard substrates such as glass. The roughness of the underlying material often propagates to the material layer. This, in turn, alters the wetting of the top layer by liquids. An example of this process can be seen in the wetting of Zn on borosilicate glass and quartz by a water drop.[38] The contact angle of water over Zn varies from ~110 to ~150 depending on the roughness of the surface. This is further affected by the exposure to ultraviolet radiation, which can be used as the next level for wetting management, as seen from Figure 9.2.

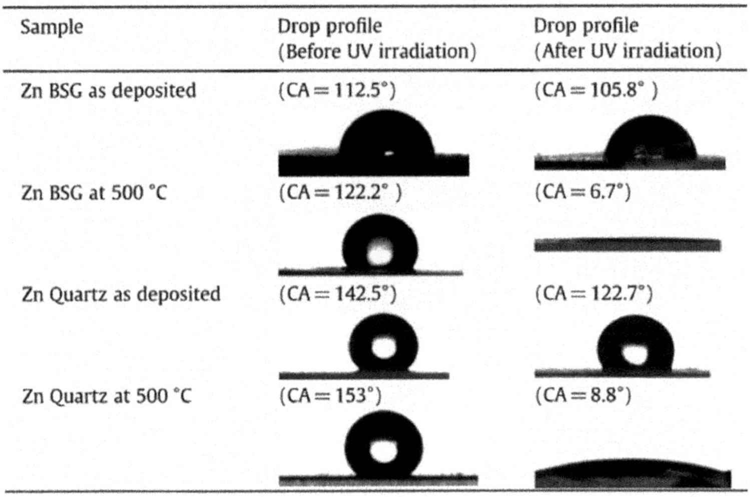

FIGURE 9.2 Contact angles (CA) of as deposited (As-dep) Zn films and films annealed to 500°C on BSG and quartz substrates.[38] (Reprinted with permission from Shaik, U. P.; Kshirsagar, S.; Krishna, M. G.; Tewari, S. P.; Purkayastha, D. D.; Madhurima, V. Growth of Superhydrophobic Zinc Oxide Nanowire Thin Films. *Mater. Lett.* **2012,** *75*, © 2012, Elsevier.)

9.4 CONTROLLING SURFACE TENSION OF LIQUIDS

The relatively easier control over the formation of drops and the flow of liquids can be achieved through the alteration of the surface tension of a liquid flowing through the microchannels. This can be done by the addition of another component of liquid, or surfactants. The former is easily achievable in microfluidics[39] given that the fabrication of channels wherein liquids can intermix and react has been firmly established. Specific channels can be opened and closed by means of piezoelectric valves to let in a second (or third) component of liquid, thus altering its surface tension. Given that it is possible to mix liquids, it is imperative to understand how the surface tension varies as a function of concentration.

Consider here the case of mixing of water with 1,4 dioxane.[40] This binary system is of interest since it exhibits hydrogen bonds, which are the typical bonds seen in most liquids used in microfluidics and are also the kind of bonds that come into effect at micron/sub-micron scales. Strong intermolecular bonds such as covalent and ionic bonds are responsible for the mechanical strength and stability of molecules while weak molecular bonds such as the hydrogen bond and van der Waals bond are responsible for their dynamics.

In the case of 1,4 dioxane with water, it is seen that the adhesion energy of the system over various surfaces varies as a function of the concentration of the components (Fig. 9.3). This goes to show that by injecting a second liquid into a fluid stream, it is possible to alter the "stickiness" of liquids to surfaces and hence aid the cleaning of the surfaces. This concept is of particular use for biological liquids, wherein due to the presence of an abundance of hydrogen bonds, the "stickiness," or the adhesiveness, is larger than non-hydrogen bonded liquids. It is interesting to note that the adhesive energy changes in a non-linear manner with concentration. A small addition of a second component of liquid is sufficient to alter the adhesive energy in a drastic manner. This non-linearity is due to steric and conformal effects.[40]

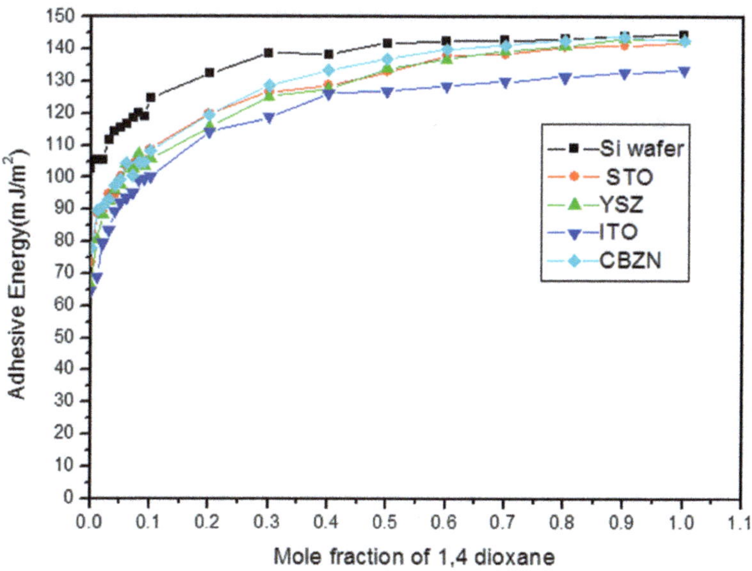

FIGURE 9.3 Adhesive energy of the binary system of 1,4 dioxane + water for various concentrations, over five substrates—Silicon wafer (Si), Strontium Titanate (STO), Yttrium Stabilized Zirconia (YSZ), Indium Tin Oxide (ITO), and Calcium Bismuth Zinc Niobate (CBZN)[40] (Reprinted with permission from Madhurima, V.; Purkayastha, D. D.; Rao, N. V. S. Wettability, FT-IR and Dielectric Studies of 1,4 Dioxane and Water System. *J. Colloid Interface Sci.* **2011,** *357*, © 2011, Elsevier.)

In order to further this idea, it can be observed from Figure 9.4, that the contact angle, and hence the adhesive property of a binary system of aniline and alcohol also shows remarkable concentration dependence. This system is an important in the study of bioMEMS since it is a prototype for

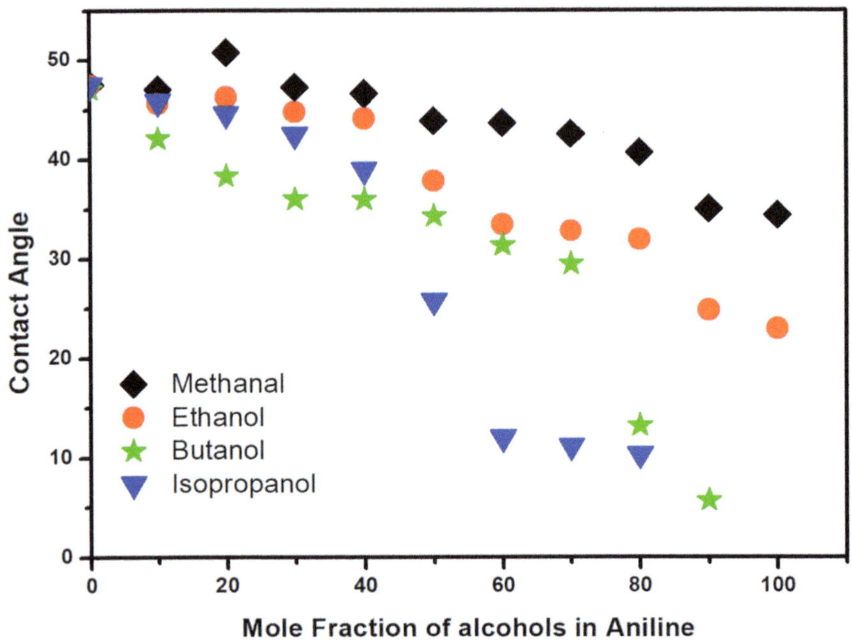

FIGURE 9.4 The variation of contact angle and surface tension with mole fraction of alcohols in aniline over indium tin oxide surfaces.

the N–H–O hydrogen bonds that are seen in biological systems. A similar concentration dependent change is observed. The substrates shown in Figures 9.3 and 9.4 are typical of those used in microelectronic devices. These go to prove that by the addition of very small volumes of a second, appropriate liquid, sufficient control can be achieved over the molecular interactions in a liquid.

9.5 SUMMARY AND CONCLUSION

There is an urgent need to explore novel methods for dealing with micron/sub-micron sized electronic devices. A large class of such devices involves microfluidic components.[42] Since bulk effects such as gravity have little or no role to play at these scales, it is important to understand the molecular interactions that occur, especially those between solid and liquid surfaces, to be able to control the formation of drops and the flow of liquids in micron/sub-micron sized fluidic devices. While a large volume of literature in this field has focused on altering surface properties, it is suggested here that the control of surface tension of the liquids can be a potential tool for the reuse of microfluidic devices through their rinsing, cleaning and refilling.

KEYWORDS

- **wetting studies**
- **microfluidic devices**
- **recycling**
- **reuse**
- **e-waste**
- **microfluidic device**

REFERENCES

1. Lantz, M.; Hagleitner, C.; Despont, M.; Vettiger, P.; Cortese, M.; Vigna, B. Sensors and Actuators on CMOS Platforms, Chapter 5. In *More than Moore: Creating High Value Micro/Nanoelectronics Systems*; Zhang, G. Q., Roosmalen, A. V., Eds.; Springer: US, 2010; p 109.
2. Varadan, V. K.; Vino, K. J.; Jose, K. A. In *RF MEMS and Their Applications*; John Wiley and Sons: Hoboken, 2003; p 125.
3. http://www.itnewsonline.com/news/CE-and-Healthcare-Industries-Helping-the-MEMS-Market-to-grow-at-a-19-Pct-Annual-Rate/35132/8/1 (accessed Oct 22, 2014).
4. Ram K. and Sampa S., E-Waste in India. Research Unit (LARRDIS). Rajya Sabha Secreteriat, June, 2011.
5. Widmera, R.; Oswald-Krapf, H.; Sinha-Khetriwal, D.; Schnellmannc, M.; Boni, H. Global Perspectives on E-Waste. *Env. Impact. Ass. Rev.* **2005,** *25*, 436–458.
6. Joseph, K. In *Electronic Waste Management in India—Issues and Strategies*, Proceedings of Sardinia 2007, Eleventh International Waste Management and Landfill Symposium, S. Margherita di Pula, Cagliari, Italy; Oct 1–5, 2007.
7. Maluf, N.; Williams, K. In *Introduction to Microelectromechanical Systems Engineering*; Artech House: Norwood, 2004.
8. Schetky, L. McD.; Jardine, P.; Moussy, F. In *A Closed Loop Implantable Artificial Pancreas Using Thin Film Nitinol MEMS Pumps*, Proceedings of SMST-2003, Pacific Grove, California, May 5–8, 2003, p 555.
9. Papautsky, I. Hot Embossing for Lab on Chip Applications. In *Bio-MEMS: Technologies and Applications*; CRC Press: Boka Raton, 2006; p 117.

10. Zimmerman, W. B. J. In *Microfluidics: History, Theory and Applications*, Springer Science and Business: Berlin, Germany, 2006.
11. Ren, K.; Zhou, J.; Wu, H. Materials for Microfluidic Chip Fabrication. *Acc. Chem. Res.* **2013,** *46*, 2396–2406.
12. Herold, K. E.; Rasooly, A. In *Lab-on-Chip Technology: Fabrication and Microfluidics*; Horizon Scientific Press: Poole, 2009.
13. Zhao, W.; van den Berg, A. Lab on Paper. *Lab Chip.* **2008,** *8*, 1988–1991.
14. Baroud, C. N.; Gallaire, F.; Dangla, R. Dynamics of Microfluidic Droplets. *Lab Chip.* **2010,** *10*, 2032–2045.
15. Garstecki, P.; Ganan-Calvo, A. M.; Whitesides, G. M. Formation of Bubbles and Droplets in Microfluidic Systems. *Bull. Pol. Acad. Sci.* **2005,** *53*, 361–372.
16. Rosenfeld, L.; Lin, T.; Derda, R.; Tang, S. K. Y. Review and Analysis of Performance Metrics of Droplet Microfluidics Systems. *Microfluid. Nanofluidics.* **2014,** *16*, 921–939.
17. Cristini, V.; Tan, Y.-C. Theory and Numerical Simulation of Droplet Dynamics in Complex Flows—A Review. *Lab Chip.* **2004,** *4*, 257–264.
18. Neumann, A. W.; David, R.; Zuo, Y. In *Applied Surface Thermodynamics;* 2nd Ed.; CRC Press: Boka Raton, 2010.
19. Hadjiconstantinou, N. G. The Limits of Navier-Stokes Theory and Kinetic Extensions for Describing Small-scale Gaseous Hydrodynamics. *Phys. Fluid.* **2006,** *18*, 111301.
20. Lee, C.-Y.; Chang, C.-L.; Wang, Y.-N.; Fu, L.-M. Microfluidic Mixing: A Review. *Int. J. Mol. Sci.* **2011,** *12*, 3263–3287.
21. Pollack, M. G.; Fair, R. B.; Shenderov, A. D. Electrowetting-based Actuation of Liquid Droplets for Microfluidic Application*Appl. Phys. Lett.* **2000,** *77*, 1725.
22. Paik, P.; Pamula, V. K.; Pollack, M. G.; Fair, R. B. Electrowetting-based Droplet Mixers for Microfluidic Systems. *Lab Chip.* **2003,** *3*, 28–33.
23. Xia, N.; Hunt, T. P.; Mayers, B. T. Combined Microfluidic—Micromagnetic Separation of Living Cells in Continuous Flow. *Biomed Microdevices* **2006,** *8*. 299–308.
24. Darhuber, A. A.; Troian, S. M. Principles of Microfluidic actuation by Modulation of Surface Stresses. *Ann. Rev. Fluid. Mech.* **2005,** *37*, 425–455.
25. Baret, J.-C. Surfactants in Droplet-based Microfluidics. *Lab Chip* **2012,** *12*, 422–433.
26. Darhuber, A. A.; Valentino, J. P.; Davis, J. M.; Troian, S. M.; Wagner, S. Microfluidic Actuation by Modulation of Surface Stresses. *Appl. Phys. Lett.* **2003,** *82*, 657–659.
27. Kotz, K. T.; Noble, K. A.; Faris, G. W. Optical Microfluidics. *Appl. Phys. Lett.* **2004,** *85*, 2658–2660.
28. Zhang, C.; Xing, D.; Li, Y. Micropumps, Microvalves, and Micromixers within PCR Microfluidic Chips: Advances and Trends. *Biotechnol. Adv.* **2007,** *25*, 483–514.
29. Sun, C.; Zhao, X.-W.; Han, Y.-H.; Gu, Z.-Z. Control of Water Droplet Motion by Alteration of Roughness Gradient on Silicon Wafer by Laser Surface Treatment. *Thin Solid Films* **2008,** *516*, 4059–4063.
30. Krishna, M. G.; Madhurima, V.; Purkayastha, D. D. Metal Oxide Thin Films and Nanostructures for Self-cleaning Applications: Current Status and Future Prospects. *Eur. Phys. J. Appl. Phys.* **2013,** *62*, 30001.
31. Marmur, A. The Lotus Effect: Superhydrophobicity and Metastability. *Langmuir* **2004,** *20*, 3517–3519.

32. Bhushan, B.; Jung, Y. C.; Koch, K. Micro-, Nano- and Hierarchial Structures for Superhydrophobicity, Self-cleaning and Low adhesion. *Phil. Trans. R. Soc. A.* **2009,** *367*, 1631–1672.
33. Whyman, G.; Bormashenko, E.; Stein, T. The Rigorous Derivation of Young, Cassie—Baxter and Wenzel Equations and the Analysis of the Contact Angle Hysteresis Phenomenon. *Chem. Phys. Lett.* **2008,** *450*, 355–359.
34. Morra, M.; Occhiello, E.; Garbassi, F. Contact Angle Hysteresis in Oxygen Plasma Treated Poly (tetrafluoroethylene). *Langmuir* **1989,** *5*, 872–876.
35. Cazabat, A. M.; Stuart, M. A. C. Dynamics of Wetting: Effects of Surface Roughness. *J. Phys. Chem.* **1986,** *90*, 5845–5849.
36. Purkayastha, D. D.; Pandeeswari, R.; Madhurima, V.; Krishna, M G. Metal Buffer Layer Mediated Wettability of Nanostructured TiO_2 Films. *Mat. Lett.* **2013,** *92*, 151–153.
37. Arifvianto, B.; Mahardika, S. M.; Dewo, P.; Iswanto, P. T.; Salim, U. A. Effect of Surface Mechanical Attrition Treatment (SMAT) on Microhardness, Surface Roughness and Wettability of AISI 316L. *Mat. Chem. Phys.* **2011,** *125*, 418–426.
38. Shaik, U. P.; Kshirsagar, S.; Krishna, M. G.; Tewari, S. P.; Purkayastha, D. D.; Madhurima, V. Growth of Superhydrophobic Zinc Oxide Nanowire Thin Films. *Mater. Lett.* **2012,** *75*, 51–53.
39. Baret, J.-C. Surfactants in Droplet-based Microfluidics. *Lab Chip.* **2012,** *12*, 422–433.
40. Madhurima, V.; Purkayastha, D. D.; Rao, N. V. S. Wettability, FT-IR and Dielectric Studies of 1,4 Dioxane and Water System. *J. Colloid Interface Sci.* **2011,** *357,* 229–233.
41. Lalnunsiama, J.; Madhurima, V. Wettability Studies of ITO Substrates Using Binary Mixture of Aniline and Alcohols. *Environ. Eng. Res.* **2012,** *17*, S49–S52.
42. Watanabe, M. Refreshable Microfluidic Channels Constructed using an Inkjet Printers. *Sensors Actuators B* **2007,** *122*, 141–147.

CHAPTER 10

PROPOSAL FOR RECYCLING ORGANIC AND INORGANIC MATERIALS IN AN ECOLOGICAL PARK

M. NEFTALÍ ROJAS-VALENCIA* and
ALFREDO GALICIA MARTÍNEZ

Coordination of Environmental Engineering, Institute of Engineering, National Autonomous University of Mexico, Post Box 70-472, Coyoacán 04510, D. F. Mexico, Mexico

**Corresponding author. E-mail: nrov@iingen.unam.mx*

CONTENTS

ABSTRACT

The objective of this work was to design, within the installations of an ecological park, the building of a reception and storage center for the storage of economically valuable solid wastes, and the setting up of a composting plant for the treatment of the organic fraction. From a generation study conducted in an ecological park in operation, the volumetric weights of the generated wastes were obtained, the most abundant ones being polyethylene terephthalate (better known as PET), 38.75 kg/m^3; food wastes, 110.5 kg/m^3; and gardening wastes, 2443 kg/m^3. These data were used as basis to determine the size of a reception and storage center and composting plant for the ecological park planned to be built in Tuzandepetl, Veracruz. PET wastes and plastics were added and a volume of 2.4 m^3 in 3 months was obtained. This volume was used to design the four specific areas into which the reception and storage center was divided, each area being 2-m long, 1.5-m wide and 2-m high, plus 1-m for ventilation purposes. A movable partition was proposed in order to accommodate the eventual need of a larger space for a given type of wastes. As regard to the composting plant, the method applied in the Food and Agriculture Organization of the United Nations (FAO) workshop-composting techniques was used, in which compost density is established at 250 kg/m^3 and the daily volume of organic wastes (2.25 m^3) was obtained to design a weekly mound being 1.5-m long, 0.8-m wide, and 1.2-m high, forming a 16-m-long pile in 3 months. For both plants, a 10-cm-thick concrete slab was proposed as well as an area for waste reception. The building materials will be obtained from the recycling of tree felling wastes, excavation, and constructions carried out in the park. These areas will help promote awareness among the visitors.

10.1 INTRODUCTION

The parks are public green areas. These areas generally have abundance of trees and plants, grass and various facilities (such as benches, playgrounds, fountains, and other equipment) that allow enjoying the leisure and rest. Ecological, is an adjective that refers to what is linked to the ecology. The term ecology, in its broadest sense, mentions the interactions that maintain living beings to the environment.[1–5]

By definition, an ecological park is a place characterized by the special care received by the species living in it. The purpose of an ecological park is to protect the ecosystem in which it develops. These regions also serve as recreation and allow to the public about the nature of a particular place. Sustainable parks can help promote the environmental culture of reutilization and recycling, leaving an impact on visitors and workers who, in turn, may transmit an environmental awareness onto their children, the environmental managers of the future.[6–9] The importance and success of the above does not depend only on waste awareness courses but also on practical and concrete implementations, and thus, parks integrating a reception and storage center and a composting plant in a global solid waste management program are essential.[10–13]

Depending on its formal characteristics, an ecological park can be variously classified and receives different denominations. Within the group of protected zones, it is possible to talk about national park, sustainable park, or ecological reserve. The specificity of each name depends on the regulations of the country in question.[14,15]

In Mexico, few parks can be considered sustainable or ecological. Among them, the Xochimilco Park,[16] Luis Donaldo Colosio Park, Xochitla Park, Huayamilpas Park, and "Sendero del Abuelo" park can be mentioned, as well as the Jaguaroundi Park which is one of the most recent.

The purpose of this study was to design a specific reception and storage center for the Tuzandepetl ecological park. The reception and storage center is an appropriate area to attract waste collection, recycling, and/or coprocessing companies since once at least 1 ton of waste is stockpiled, its market value is economically interesting. This chapter describes about the quantity of solid wastes generated in an operating ecological park and is carried out to promote among visitors and workers the habit of separating and commercializing solid wastes as an environmental education alternative.

10.1.1 STUDY AREA

In the State of Veracruz, it is planned to build the Tuzandepetl ecological park, located in the municipality of Ixhuatlán del Sureste, bordering to the north with the city of Nanchital, to the east with lands belonging to Ixhuatlán, to the South with the federal highway 180 Villahermosa—Veracruz, section Coatzacoalcos—Córdoba and to the west with the Coatzacoalcos River. The towns of Barragantitlan and Benito Canales belonging to the municipality of Ixhuatlán del Sureste are among the geographically nearest social actors. Las Águilas, Paraíso, Ixhuatlán, and Nanchital, the municipal seat, are further away (see Fig.10.1).

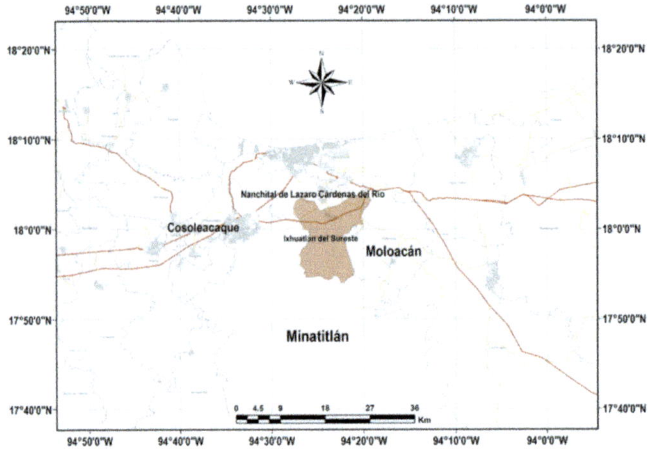

FIGURE 10.1 Municipal information system. Municipal booklets. Ixhuatlán del sureste.

Ixhuatlán del Sureste (Veracruz) is located in the southwestern part of the State of Veracruz of Ignacio de la Llave, bordering to the north with Coatzacoalcos; to the east and south with Moloacan and to the southeast with Minatitlán. It is located at about 320 km from the capital of the state (see Fig.10.2). Ixhuatlán del Sureste, so called since 1959, is the municipal seat administering 57 places among which only Ixhuatlán is a urban area.

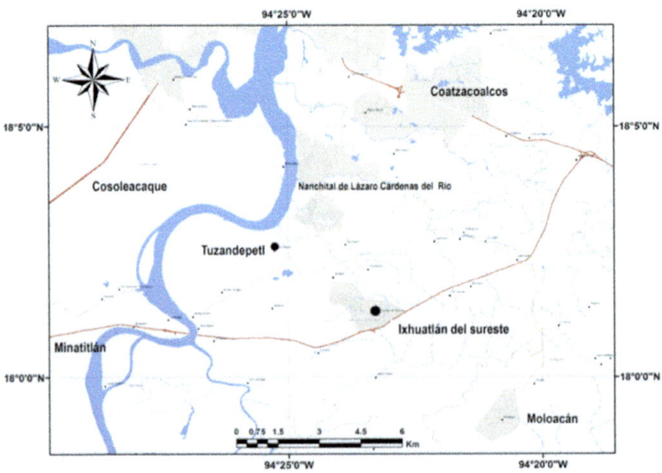

FIGURE 10.2 Regional map showing the Tuzandepetl park polygons as well as the location of the main towns having a direct and potential influence on the protected natural area.

10.2 METHODOLOGY

The methodology was divided into three phases:

10.2.1 FIRST PHASE

A comprehensive desk study was performed, including standards and laws related to the topic. The Laws, Regulations, and Standards related to the sustainable buildings of reception and storage centers and compositing plants were analyzed.

10.2.2 SECOND PHASE

In order to study the generation, separation and quantification of solid wastes, standards related to the studies of urban solid wastes generation were considered, as detailed under the lowercase roman numerals i–iv.

i) First, the Mexican Standard NMX-AA-061-1985 was taken as basis for determining the generation of solid wastes; however, adaptations were made to take into account the quantity of solid wastes generated in parks with a particular data stratification.

   In order to obtain the value of the per capita generation of solid wastes in kg/person/day corresponding to the date on which they were generated, the weight of the solid wastes was divided by the number of persons (eq 10.1), according to (NMX-AA-061-1985).

$$\text{Per Gen} = \frac{\text{Collected wastes in kg}}{\text{Number of persons}} \tag{10.1}$$

ii) The quartering method according to Mexican Standard NMX-AA-15-1985 was used.

   Because the production of solid wastes in this study was less than or equal to 50 kg/day, the whole sample was included in the analysis and not only part of it as provided for in the standard.

iii) The solid wastes found were classified according to the provisions of standard NMX-AA-22-1985 By-products Selection and Quantification.

- A record was kept for each classification of solid wastes found as well as the quantity expressed in kilogram and the percentage with regard to the total analyzed sample.
- A daily report was produced with regard to the classified materials and their quantity.

iv) In order to obtain the "in situ" volumetric weight, Mexican Standard NMX-AA-19-1985 was used.

The total weight of the solid wastes was determined according to the technical parameters. A 200-L drum, as shown in Figure 10.3, was used for determination purposes.

FIGURE 10.3 200-L iron drum.

As illustrated in Figure 10.4, the empty container was weighed and the value obtained was taken as the container tare. The container was filled with the daily solid wastes, placing them perfectly within the container, without squeezing the wastes when putting them in the drum in order not to modify the density data. The container was dropped three times on the floor from a height of 10 cm and solid wastes were added again up to the top.

FIGURE 10.4 Sequence for obtaining the "in situ" volumetric weight.

Finally, the average weight of the inorganic wastes was obtained (eq 10.2), calculating then the in situ volumetric weight of the wastes. This calculation was made every day throughout the study and the average volumetric weight of the inorganic wastes for 1 week was obtained using eq 10.3. The volumetric weight of the organic wastes was calculated using the data obtained from the FAO techniques.

Finally, the weight of the wastes was obtained through the difference between the tare and the weight of the container containing the solid wastes; then the "in situ" volumetric weight of the wastes was calculated.

$$P_{averege} = \sum_{i=\text{saturday}}^{\text{saturday}} \text{day}_i = Peso_{\text{saturday}} + Peso_{\text{sunday}} + \cdots P\text{day}_i \quad (10.2)$$

$$P_V = \frac{\text{Averege weekly weight (kg)}}{\text{Container volume } (\text{m}^3)} \quad (10.3)$$

where:

P_V = Average volumetric weight of the solid wastes (kg/m³).
P = Average weight of the wastes (gross weight minus tare) (kg).
V = Volume of the container (m³).

In order to obtain the volume of each solid waste, the weight of each of the solid wastes is divided by the previously obtained average volumetric weight (eq 10.4).

$$V = \frac{P(\text{kg})}{P_V\left(\frac{\text{kg}}{\text{m}^3}\right)} \quad (10.4)$$

wherein:

V = Solid wastes volume (m^3)
P_V = Average volumetric weight (kg/m^3)
P = Weight of solid wastes (kg)

The percentage by weight of each of the by-products was calculated through eq 10.5:

$$P_S = \frac{G_1}{G} \times 100 \quad (10.5)$$

wherein:

P_S = Percentage of the considered by-product.
G_I = Weight of the considered by-product, in kg; excluding the weight of the bag.
G = Total sample weight.

10.2.3 THIRD PHASE

Taking again as reference, the generation study of the ecological park Jaguaroundi, the space necessary for the temporary storage of the potentially valuable solid wastes was calculated, and a 3-month projection of the storage of the valuable inorganic wastes was made.

With regard to inorganic wastes, only the following specific areas were implemented: metal area, PET and plastic area, paper and tetra pack area, and area for other materials, the heaviest being PET and plastic (218 kg). A 3-m height was considered for the reception and storage center because more wastes can be accumulated and placed in vertically stacked containers.

The reception and storage center will be managed by a group of people actively working on the project. They will perform the functions of administration and planning of the activities of the center.

Any economic benefit resulting from the sales of recyclables will be destined to material expenses of the reception and storage center and environmental education. The budget will be handled transparently, a log of expenses and incomes will be kept, and receipts and invoices will be recorded. The following wastes have a certain economic value and can be collected:

- PET,
- Other plastics except Unicel,
- Aluminum can,
- Iron,
- Glass bottles or vials separated in transparent and color glass,
- Newspapers, paper,
- Cardboard,
- Used cell phones and their all accessories,
- Ink and toner cartridges,
- Used vegetable oil,
- Books, magazines, notebooks, and other materials that can be reused or be useful for someone else.

Once the materials are collected, they will be temporarily stored in 1.17-m^3 containers with plastic lid, in 1.5-m^3 plastic super sacks, and in 50-L drums (see some examples in Fig. 10.5). Storage containers with casters make transportation and movement easier.

Although the place is covered, it is advisable to use lidded containers for storing nonplastic materials such as cardboard, paper, newspapers, among others; in the case of plastics, super sacks will be used covering them with a tarp for protection against sun, air, and water.

On the other hand, in sustainable buildings, the treatment of organic wastes must be taken into account; composting is an excellent option.

The ecological park Tuzandepetl is a highly favored place because of its geographical conditions; moreover, its large size permits the construction of ecological buildings.

For the design of the composting plant, the quantity of organic wastes was determined in the generation study of the Jaguaroundi Park. Based on these data, the number, size, and volume of the needed piles were calculated. For this purpose, the volume of a mount obtained in 1 week from the collection of organic wastes was considered. Moreover, the surface necessary for the working yard was calculated, taking into account that the time requested for pile-processing completion and compost readiness is 3 months on average.

FIGURE 10.5 Containers appropriate for the temporary storage of solid wastes in the reception and storage center.

Based on the calculations of the workshop-composting techniques,[1] for obtaining a compost pile and considering with the weights obtained in the generation of organic material study, the size of a mound formed in 1 week was obtained and then the size of a pile formed in 3 months was computed.

10.3 RESULTS

Based on the desk study, it was determined that Veracruz has an Environmental Protection State Law, an Ecological Balance and Environmental Protection State Law, a Law for the Prevention and Integrated Management of Urban Solid and Special Wastes for the state of Veracruz of Ignacio de la Llave (Table 10.1). According to the laws, an adequate management of the wastes and their disposal is requested, the ideal situation being the integration of reception and storage centers and adequate composting plants within the management programs.

TABLE 10.1 Comparison of the Mexican Laws.

Law	Regulation
LGEEPA	[a]Reduction of the effects of the generation, storage, management, and treatment of nonhazardous waste on the environment.
Legislation of the Federal District	[a]It is the responsibility of any person; either natural or legal (Article 24, Solid Waste Law).
	[a]It shall the performed by the individuals at the request of the Secretariat (Article 71, Regulations of the Solid Waste Law).
Legislation of the State of Veracruz	[a]The participation of the generators is essential.
	[a]It is the responsibility of the owner to reduce the effects on the environment. (Article 173, State Law on environmental protection).

[a]LPGIRSUME: Law for the Prevention and Integrated Management of Urban Solid and Special Wastes for the State of Veracruz.

The legislation of the State of Veracruz is on a par with the federal legislation. For this reason, we recommend that the guidelines regarding the design and operation of the ecological park should adhere to the national legislation. As far as waste handling is concerned, it is recommended to follow the guidelines of the Comisión Nacional de Áreas Naturales Protegidas (National Commission of Protected Natural Areas) because they offer a wider scope than the federal, state, and municipal legislations.

As previously mentioned, it is necessary to implement an integral waste management program including separation, temporary storage, transportation, and adequate final disposal. These concepts will determine, together with other factors, the spatial and functional design of the installations. Moreover, strategically located waste-collection facilities must be supplied for tourists and other users. The purpose of said facilities and the environmental improvements generated by their use must be clearly indicated.

Furthermore, it is important to implement appropriate technologies and methods for treating organic and solid wastes. Because it is an ecological park, it has large wood and green areas and thus the material obtained from pruning, mowing, or maintaining said areas will be abundant, potentially permitting its use as compost. With regard to wastes not susceptible to be composted, it is advisable to reuse materials such as PET, paper, cardboard, glass, and tetra pack, or collect them for recycling purposes.

10.3.1 GENERATION OF SOLID WASTES IN AN OPERATING ECOLOGICAL PARK

First, a diagnostic of the type and quantity of solid wastes (trash) generated "in situ" was conducted. The Jaguaroundi Ecological Park was chosen for this purpose. This park is in operation, and employs 28 persons working from Monday to Sunday, normally from 9:00 to 17:00 h, 4 people being in charge of the cleaning services, while 2 persons are responsible for gardening.

The results of by-product classification are shown in Table 10.2. It can be seen that 22 different types of wastes were identified, mostly garden wastes (488.64 kg), food wastes (22.10 kg), and PET (7.75 kg).

As established in the methodology, in order to calculate the per capita generation of solid wastes in kg/person/day corresponding to the date on which they were generated, the total weight amounting to 549.88 kg was divided by the number of employees and visitors of the park (359), according to NMX-AA-061-1985.

Calculation algorithm:

$$\text{Per Gen} = \frac{549.85\text{ kg}}{359\text{ number of persons}} = 1.53 \frac{\text{kg}}{\text{persons/day}}$$

Based on the results of the generation study, gardening and food wastes were subtracted from the total waste weight (eq 10.6) and the result was divided by the same number of people (eq 10.7) in order to obtain the per capita generation of 0.109 kg/person/day without gardening and food wastes.

$$549.85\text{ kg} - 488.6\text{ kg} - 22.1\text{ kg} = 39.15\text{ kg} \tag{10.6}$$

$$\text{Per Gen} = 39.15\text{ kg} / 359\text{ persons} = 0.109\text{ kg} / \text{person} / \text{day} \tag{10.7}$$

Then, the average volumetric weight was calculated taking into account the daily generation percentage (Fig.10.6). The sum of the daily weight was divided by 7 days. This weight was divided by the volume of the container, in this case, a 0.2 m³ bin. Using this result, the size of the reception and storage center was determined, while in the case of organic wastes (food and gardening), an independent calculation of its volumetric weight was made in order to determine the size of the mounds of the composting pile.

TABLE 10.2 Results of the Study of Waste Generation at the Jaguaroundi Ecological Park.

Solid wastes	Weight (kg)	Percentage by weight (%)
Plastic bags	3.4	0.62
Cardboard	1.96	0.36
Wrappings	0.85	0.15
Aluminum cans	0.9	0.16
Wood	5.5	1.00
Metal	0.9	0.16
Others	2.9	0.53
Brown paper	1.27	0.23
Toilet paper	3.1	0.56
White paper	0.17	0.03
Aluminum foil	0.15	0.03
Newspaper	0.01	0.00
PET	7.75	1.41
Light plastic	2.35	0.43
Rigid plastic	3.2	0.58
Unicel	0.58	0.11
Food wastes	22.1	4.02
Gardening wastes	488.6	88.86
Tetra pack	2.85	0.52
Cloth	0.01	0.00
Transparent glass	1.22	0.22
Leaflets	0.08	0.01
Total	549.85	100%

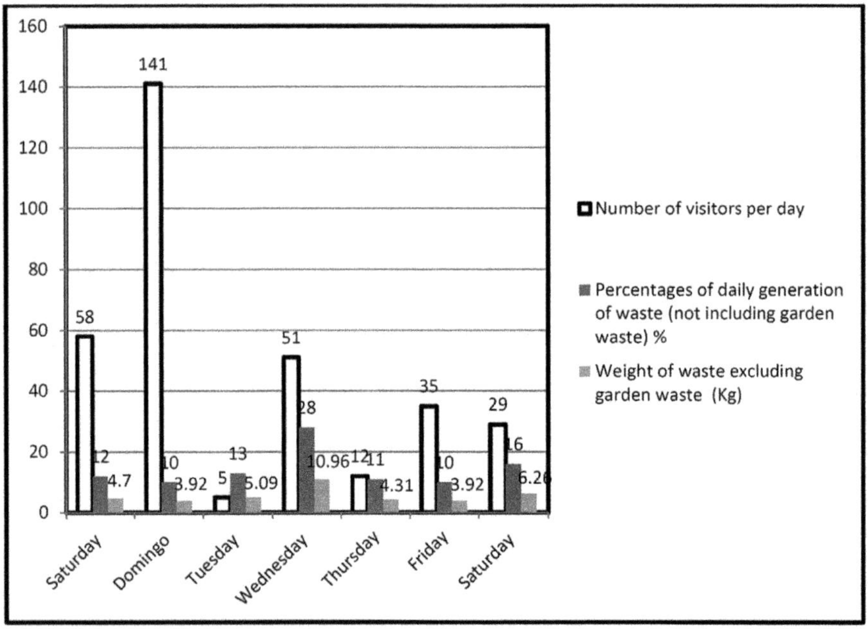

FIGURE 10.6 Daily percentage and total waste weight without gardening wastes.

The average weekly weight (without including gardening and food wastes) is calculated as follows:

$$P_{average} = \frac{(4.7 + 3.92 + 10.96 + 4.31 + 3.92 + 6.26)\ \text{kg}}{7\ \text{days}} = 5.6\frac{\text{kg}}{\text{day}}$$

The volumetric weight obtained without including gardening waste:

$$\text{Volumetric weight} = \frac{\text{Average weekly}\ (\text{kg})}{\text{Container volume}\left(\text{m}^3\right)} = \frac{5.6\ \text{kg}}{0.2\ \text{m}^2} = \frac{\text{kg}}{\text{m}^2}$$

10.3.2 OTHER TYPES OF WASTE CALCULATED AS ABOVE

The reception and storage centers have their own characteristics because they are designed to meet the specific needs of each project. The weight and volume of each type of waste have to be considered because they vary greatly from project to project. For this purpose, in order to design a reception and storage center having the appropriate size, the weight of each type of waste has to be converted into volume taking its density into account.

The advisable infrastructure is as follows:

(1) gabled or shed-style roof, for enhancing rainwater discharge, (2) maximum use of natural lighting, (3) surrounded by a perimeter fence and barbed wire, among others, (4) concrete slab or gravel, (5) partitions for separating the wastes that will be collected and stored (plastic, metal, glass, paper, and cardboard), (6) the materials should not be exposed to weathering such as rain, sun, and wind, except plastics (7) they must be in a safe place to prevent theft, and (8) they must be orderly, dry, and clean to check harmful fauna.

Safety measures: (1) use of extinguishers and (2) away from ignition sources.

10.3.3 CALCULATION, DESIGN OF COMPOST PILES, AND SIZING OF THE RECEPTION AND STORAGE CENTER

The weight of each type of waste generated per periods of 1 and 3 month and 1 year (see Table 10.3) was determined. Concurrently, the volume of each type of waste per periods of 1 and 3 month and 1 year was defined in order to find out the time needed to fill the reception and storage center and to size up the processed quantity. Accurate volume estimates may help prevent the saturation of the reception and storage center.

Once the volumetric weight is obtained, the volume is obtained from eq 10.4, the weight of the wastes obtained in 1 and 3 month and 1 year is divided by the volumetric weight of 28 kg/m³. With the 3-month volume for each waste, the requested size was determined; finally the size of the reception and storage center was obtained. Food and gardening wastes are not considered for the size of the reception and storage center.

PET and plastic weight = 218 kg

$$P_v = \frac{P}{V} \text{ obtaining the volume } V = \frac{P}{P_v}$$

$$\text{Substituting PET and plastic weight } v = \frac{218\text{ kg}}{28\text{ kg}/\text{m}^3}$$

TABLE 10.3 Waste Generation by Weight (kg) and Type, in 1 and 3 Months and 1 Year.

Solid wastes	Weight (kg)	Percentage by weight (%)	Wastes accumulated in 1 month (kg)	Wastes accumulated in 3 months (kg)	Wastes accumulated in 1 year (kg)
Plastic bags	3.4	0.62	13.6	40.8	163.2
Cardboard	1.96	0.36	7.84	23.52	94.08
Wrappings	0.85	0.15	3.4	10.2	40.8
Aluminum cans	0.9	0.16	3.6	10.8	43.2
Wood	5.5	1.00	22	66	264
Metal	0.9	0.16	3.6	10.8	43.2
Others	2.9	0.53	11.6	34.8	139.2
Brown paper	1.27	0.23	5.08	15.24	60.96
Toilet paper	3.1	0.56	12.4	37.2	148.8
White paper	0.17	0.03	0.68	2.04	8.16
Aluminum foil	0.15	0.03	0.6	1.8	7.2
Newspaper	0.01	0.00	0.04	0.12	0.48
PET	7.75	1.41	31	93	372
Light plastic	2.35	0.43	9.4	28.2	112.8
Rigid plastic	3.2	0.58	12.8	38.4	153.6
Unicel	0.58	0.11	2.32	6.96	27.84
Food wastes	22.1	4.02	88.4	265.2	1060.8
Gardening wastes	488.6	88.86	1954.4	5863.2	23,452.8
Tetra pack	2.85	0.52	11.4	34.2	136.8
Cloth	0.01	0.00	0.04	0.12	0.48
Transparent glass	1.22	0.22	4.88	14.64	58.56
Leaflets	0.08	0.01	0.32	0.96	3.84
Total	549.85	100%	2199.4	6598.2	26,392.8

Based on the volume obtained, the appropriate size of the PET and plastic container is calculated. A 1.5-m-wide and 2-m-long base is proposed, with an appropriate height of 3 m. Only wastes still having an economic value are considered and will have specific areas, such as: metal, PET and plastic, paper and tetra pack.

These areas were considered because they are the ones corresponding to the greatest waste generation; in the case of the metal area, the weights of aluminum cans, metal, and aluminum foil (23.4 kg) shall have to be taken into account. The PET and plastics weight is about 218 kg, the weight of paper and tetra pack is about 113 kg and it is estimated that the weight of other wastes reaches 116 kg.

Thus, the reception and storage center was divided in four areas considering a volume of about 7.8 m³ per area, as shown in Figure 10.7a–d. The site should be easily accessible, have water supply, and be covered in order to avoid rainwater penetration and excess sun. The area should also be ventilated and located at some distance from places where staff or visitors operate, such as the garden, courtyard, car park, roof, etc., because undesirable odors are likely to be generated during the training process.

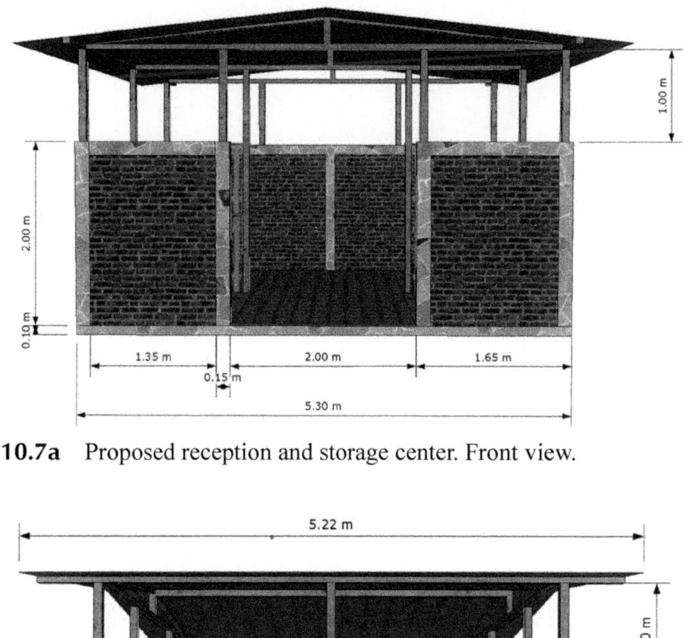

FIGURE 10.7a Proposed reception and storage center. Front view.

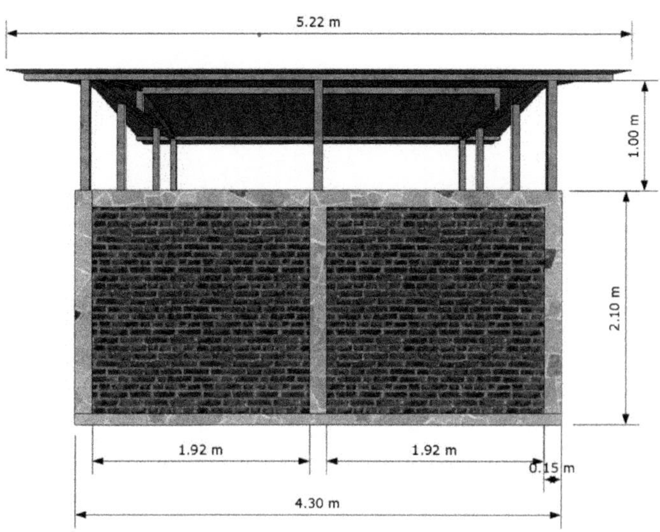

FIGURE 10.7b Proposed reception and storage center. Side view.

FIGURE 10.7c Proposed reception and storage center. Space for wastes.

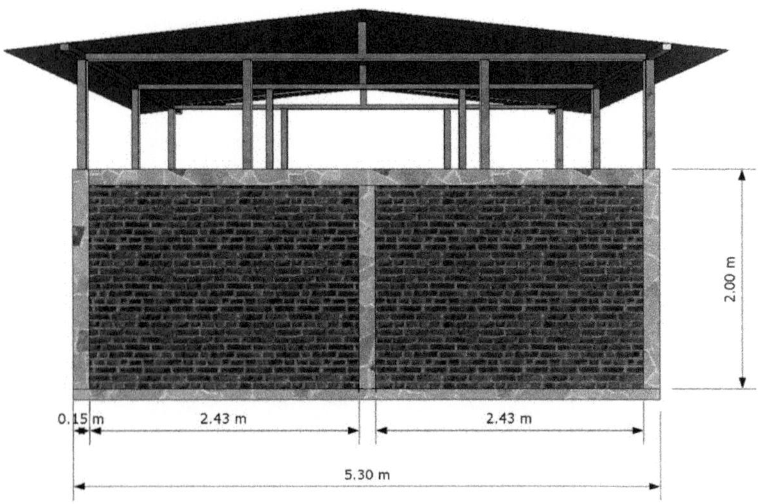

FIGURE 10.7d Proposed reception and storage center. Rear view.

To design the composting plant, we calculated the daily volume of wastes it will receive, dividing the expected quantity of wastes by their volumetric weights. In total, 10% was added to this volume to create a buffer space.

The calculation was made based on the results obtained from the Jaguaroundi Park that are shown in Table 10.3. The weights of gardening and food wastes were added giving a total of 510.7 kg per week. With information from the Taller—Técnicas de Compostaje (Workshop—Composting Techniques),[1] compost density was estimated at 250 kg/m^3, and the volume was obtained dividing 510.7 kg by 250 kg/m^3 resulting thus in 2.043 m^3, plus 10%, that is, 0.2043 m^3, adding up to 2.247 m^3. Using eq 10.8, the length of the mound was obtained.

Example of calculation algorithm

$$\text{Volume} = \pi \times h \times \text{Width} \times \frac{\text{Length}}{2} \tag{10.8}$$

Substituting the recommended volume, height, and width, the length was obtained.

$$2.247 \text{ m}^3 = \pi \times 1.2 \times 1.5 \times \frac{\text{Length}}{2} 6$$

Length = 0.79 m

Thus, the dimensions of the mound generated per week are:

Length = 0.79 ~ 0.8 m
Width = 1.5 m
Height = 1.2 m

The results of the sizing of the mounds and compost pile using the method developed by the FAO are shown in Figures 10.8 and 10.9.

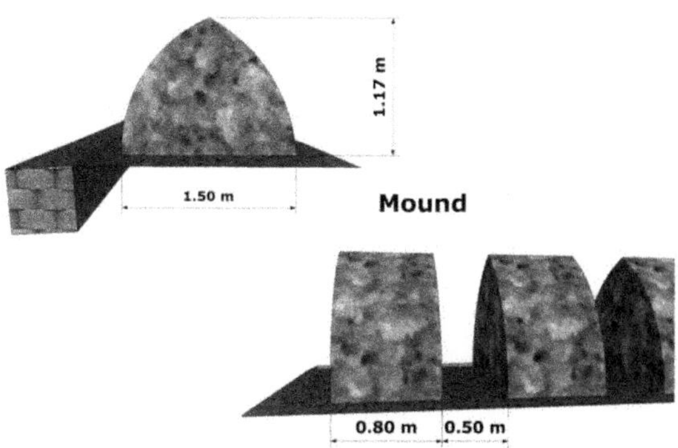

FIGURE 10.8 Compost mount generated in 1 week.

A lateral space (kerb) was designed (height: 0.5 m; width: 0.7 m; and volume: 5.6 m³) (Fig.10.9) in order to collect leachates during the compost maturation process. As it is a small plant, calculation made for a sanitary landfill was not made.

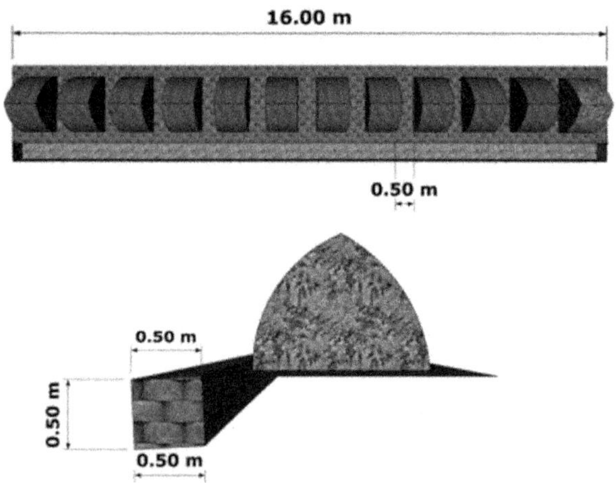

FIGURE 10.9 Pile formed in 3 months and specifications.

Figure 10.10 shows the mound and other implements (tubes) recommended for preventing the accumulation of excess biogas and promote natural aeration.

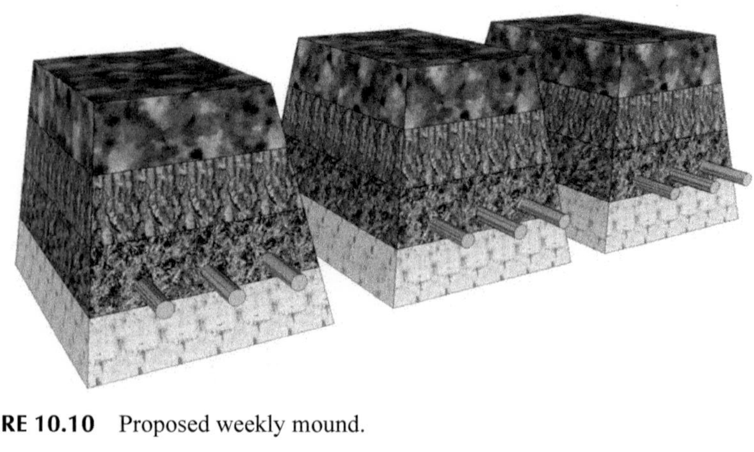

FIGURE 10.10 Proposed weekly mound.

The composting installations, besides the composting yard must have the following areas: (a) surveillance booth; (b) input reception; (c) trituration and mixture formation; and (d) maturing and storage (see Fig.10.11).

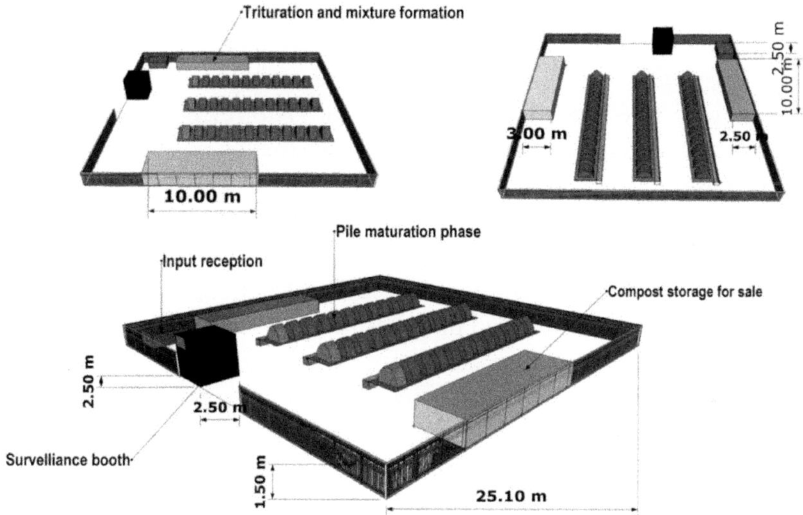

FIGURE 10.11 Proposed composting plant.

In order to ensure environmental protection and the health of the nearby population, actions must be implemented to prevent the propagation of pests and the mitigation of disagreeable odors.

The plot of land where composting is conducted must have a space for loading and unloading inputs, treating and storing finished product (compost), as well as possible rejected products, ensuring that these activities are performed within the plant.

The composting plant must have a control and surveillance booth. Sanitary installations must be available for the personnel working in the composts plants, according to the principles of safety and hygiene of the Federal Labor Law for preventing labor risks and damages to the worker.

The composting process is taken from the Environmental Standard Bill for the Federal District PROY-NADF-020-AMB-T-2011. There are several process types for the elaboration of compost such as: piles with mechanical turning, static piles with forced or passive ventilation, vertical or horizontal flow reactors, mounds, rows, containers, trenches, among others. In all of them, temperature, aeration, moisture, and initial mixture must be controlled. The selected process will be valid, if and only if, it fulfills the environmental safety and sanitary conditions of the Standard Bill.

Other factors that influence the composting process and must be taken into account are:

- *Composting materials*: The materials entering the composting plant must be free from inorganic material. The rejected materials and products must be sent to recycling, whenever possible, or to final disposal.
- *Material classification*: The received materials must be classified and stored separately till they are used.
- *Volume reduction*: The whole organic fraction, except grass and leaves, must be triturated or submitted to volume reduction, before being incorporated into the composting process.
- *Carbon-to-nitrogen ratio*: It is advisable that the materials subjected to composting be combined in such a way as to initiate with a carbon-to-nitrogen ratio (C/N) comprised between 25:1 and 40:1, the optimum ratio being 30:1.
- *Initial humidity*: The material mixture must be humidified till an initial value ranging from 50% to 60% is reached. The resulting mixture must be homogeneous.
- *Process humidity*: During composting, the mixture humidity must range from 40% to 70%. It should not be above 70% in order to prevent the runoff of fermented liquids and the formation of anaerobic conditions that could generate unpleasant odors.
- *Temperature*: During composting, the temperature must be recorded. The temperature reached by the material during composting is an indicator that the process is being appropriately conducted. Recommended temperature–time relationships to ensure the safety of the final product are established in Table 10.4.

TABLE 10.4 Temperature–Time Relationships to Ensure the Safety of the Final Product.

Average temperature	Time
55°C	During 2 weeks
60°C	During 1 week
Up to 65°C	During 3 days

The appropriate pH throughout the composting process must range from 4 to 9.

- *Aeration*: Whatever the composting method used, an appropriate aeration process must be conducted to prevent the formation of anaerobic conditions inside the mixture.
- *Logging*: All the processes conducted during compost production, as well as the key parameters must be identified and recorded in numbered logs. These logs must be made available to the authorities for revision and audit. During the first 2 weeks of the composting process, a daily recording of the temperature and humidity must be conducted. It is advisable to continue with the daily recording at least till the third week after the beginning of the process and then at least twice a week.
- *Process termination and storage*: When the control parameters established in the Environmental Standard Bill PROY-NADF-020-AMB T-2011 indicate that the composting process has concluded, the compost must be sieved for distribution or commercialization, taking into account and fulfilling the compost types specifications. In order to reach a higher quality grade according to the established categories, the final product will be eventually stored if it requires an addition maturation period. The materials that have not completed their degradation must be reincorporated to the beginning of the process.
- *Packaging*: Having met the specifications of any of the four compost types established in the Environmental Standard BillPROY-NADF-020-AMB T-2011, the product will be packed, either in sacks or bags of different materials.
- *Effluent handling*: Composting plants must have a plan for handling the effluents produced before, during, or after the composting process. Said liquid fraction must be collected and reincorporated into any of the process phases or treated for stabilization and then incorporation into preparations for vegetal nutrition or discharge. Only during the first composting phase, the effluents produced by the organic fraction of the solid wastes before initiating the composting must be collected and used for material watering before reaching the maximum temperature.

Please, find hereinafter a list of the minimum necessary equipment for a composting plant:

- Gross crushing mill.
- Fine crushing mill.
- Gross mill feeding hopper
- Pick-up van.
- Elevator belt conveyor.
- Drying equipment, possibly plot of land.
- Mechanical mixer.
- Warehouse.

Minimum staff:

- One compost technician administrator.
- One laborer for reception and yards.
- A laborer for fine milling.
- A guard.

The reception and storage center does not require a sophisticated infrastructure. It is recommended that 50% of the excess materials from other works such as walls, floors, roofs, and finishings be used. It has to be considered that once the wastes are separated and adequately handled, it is necessary to keep them in conditions that maintain their reutilization value, preventing as much as possible soaking, combination, or impregnation with other wastes.

For the reception and storage center, the requested technology and personal required for the process of recyclable wastes separation must be available. There are two types of selection plants: the plants that receive mixed wastes and the plants that receive separated wastes. It is important to state that the more complete the waste separation, the higher the efficiency of the plants will be because more wastes can be recycled.[17] It must be stated that the design of the reception and storage center is for a park and thus minimum equipment and staff are required because a smaller quantity of inorganic wastes will be handled.

Once the Tuzandepetl ecological park is operating, the pruning and mowing periods as well as the quantity of food wastes, shall have to be reviewed, for composting purposes. The drought periods shall have to be monitored in order to increase or reduce the compost maturation area. The drought periods influence the production of a pile and must be taken into account for determining the most appropriate distributions of the compost pile area.

10.3.4 BENEFITS

The benefits obtained from the separation of municipal wastes and, generally, of any type of wastes are numerous because they help society in various aspects. Organic wastes promote the agglomeration of bacteria that, in turn, cause putrefaction processes that affect the environment, generate odors, pests, and different types of diseases. Their separation contributes to the generation of alternative fuels, and makes energy expense for the production of new household materials unnecessary. Once wastes are separated, a recycling process may start that permits the manufacturing of new products, making it unnecessary to use new natural resources. When the inorganic wastes are separated and gathered, they can be sold for manufacturing of new products, permitting thus the generation of income; moreover, the separation process generates employment.

The benefits that can be derived from the operation of a composting plant and a reception and storage center in an ecological park are:

a) Environmental benefits:

- Safe elimination of organic wastes.
- Exploitation of leachates in the composting process.
- Exploitation of recyclable wastes.
- Reduction of ecosystem deterioration.
- Prevention of waste deposit in dumps.
- Reduction of greenhouse effect emissions.

b) Economic benefits:

- Savings in infrastructure costs, because local materials or materials requiring little maintenance will be used for the reception and storage center and the composting plant.
- Direct and indirect jobs are created.
- With the sales of compost and recyclable wastes, resources can be generated for the maintenance of these centers.

In order to have a more precise idea of the economic benefits generated by such a center, an analysis of the recycling market was made. In Figure 10.12, the wastes currently collected by waste collection companies are shown. For this pie chart (Fig. 10.12), the data obtained from interviews conducted at six reception and storage centers in the Federal District and two reception and storage centers in the city of Veracruz were averaged.

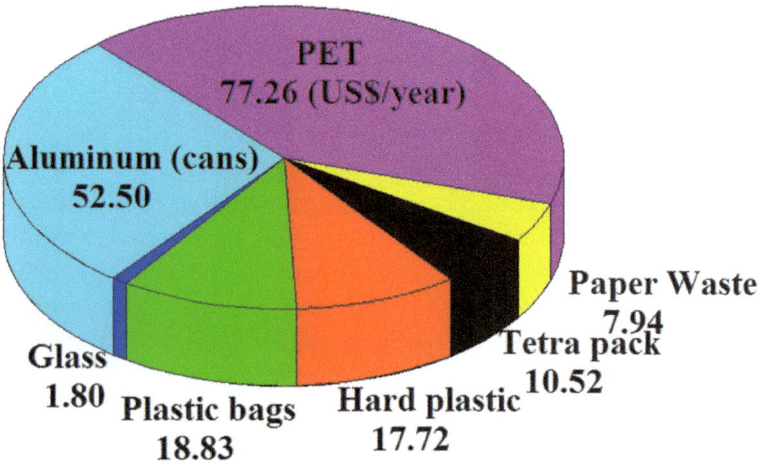

FIGURE 10.12 Estimated profits generated by the sales of recyclable wastes.

In most of the cases, recycling companies do not accept materials such as tetra pack, batteries, and Unicel, because their handling is complicated and the market is small; in other cases, they accept them but do not pay for them. Tetra pack has just started being recycled in Mexico and so there is a little market for it. Unicel presents many inconveniences for its collection and recycling. This is also the case of batteries that are difficult to handle because of the potential risk they represent. In many reception and storage centers, plastic bags are not accepted either, because this type of material is characterized by large volume and little weight, and is thus little profitable for the collecting companies.[15]

It must be taken into account that the compost generated from organic wastes can also be marketed. The compost market prices obtained from Internet pages vary from US$ 0.12 to US$ 0.31

per kg. It is not possible to obtain a precise estimate of compost generation because the generated quantity varies to a great extent depending on the applied treatments, but supposing that 30 kg of compost are obtained from 100 kg of organic wastes, and three piles are formed per year, a maximum income of US$ 1697 is obtained.

c) Social benefits:

- Contribution to the education and awareness of the population through the diffusion of information, the teaching of classes and counseling.
- Contribution to public health through the prevention of disease propagation.
- Generation of jobs.

10.4 CONCLUSIONS

Despite the fact that Mexico has developed a set of standards for the treatment of solid wastes, no specific legal provisions are available for the operation, construction, and design of composting plants or reception and storage centers, and this situation occurs not only in cities and sustainable buildings but also in ecological parks.

The design of the reception and storage center and composting plant was based on the results obtained from the generation study of an operating ecological park. The size of the reception and storage center obtained in the case in question was: length: 5.3 m; width: 4.3 m; and height:3 m. Inside, it was divided into four areas of 1.5 m × 2 m with a 2-m-high wall for keeping various recyclable wastes. The design of the composting plant was based on the FAO recommendations, being a technique that can be easily applied in this case. The composting plant consisted of mounds (length: 1.5 m; width: 0.8 m; and height: 1.2 m), and piles (length: 16 m and width: 2.6 m) representing a total area of 625 m^2.

Waste commercialization generates economic income, especially organic wastes because their volume and price is higher, while the income obtained from the sale of inorganic wastes may seem little. However, this can be used as an incentive for the cleaning staff in charge of conducting their collection.

The basic equipment needed for the reception and storage center is: 200-L drums and mega-bags, whereas for the composting plant, grass-crushing mill, fine-crushing mill, feeding hopper, drying equipment, possibly plot of land, and mechanical mixer are required. For both centers, a pick-up van is required for transporting the material to the composting center.

KEYWORDS

- **reception storage center**
- **ecological park**
- **wastes generation**
- **composting plant**
- **sustainable park**

REFERENCES

1. FAO, Organización de las Naciones Unidas para la agricultura y la alimentación. (Food and Agriculture Organization of the United Nations), Paraguay. Composting Techniques 2012. http://www.rlc.fao.org/fileadmin/content/events/taller_tcp-par-3303/compost.pdf (accessed July 18, 2013).
2. INE, Instituto Nacional de Ecología, Experiencias de la producción de composta en México. (National Institute of Ecology, Experiences in Production of Compost in Mexico) http://www2.ine.gob.mx/publicaciones/libros/499/experiencias.html#top (accessed April 14, 2013).
3. INEGI, Estadística básica sobre medio ambiente datos de Veracruz de Ignacio de la llave, 2013 (Basic Statistics on Environmental Data from Veracruz de Ignacio de la llave, 2013). http://www.inegi.org.mx/inegi/contenidos/espanol/prensa/Boletines/Boletin/Counicados/ speciales/2013/Abril/comunica22.pdf (accessed April 18, 2013).
4. INEGI, Estadística básica sobre medio ambiente, 2013 (Basic Statistics on Environment, 2013). http://www.inegi.org.mx/inegi/contenidos/espanol/prensa/Boletines/Boletin/Counicados/Especiales/2013/abril/comunica2.pdf (accessed June 25, 2013).
5. Informe de la situación del medio ambiente en México, edición 2012 (Report on the Environment Situation in Mexico, Edition 2012). http://app1.semarnat.gob.mx/dgeia/informe_12/07_residuos/cap7_1.html (accessed April 16, 2014).
6. Ley General del Equilibrio Ecológico y la Protección al Ambiente (General Law on Ecological Balance and Environmental Protection), Official Gazette of the

Federation, Mexico 1988, pp 1–7. http://www.diputados.gob.mx/LeyesBiblio/pdf/148.pdf (accessed Sept 12, 2012).
7. Ley Estatal del Equilibrio Ecológico y la Protección al Ambiente (State Law on Ecological Balance and Environmental Protection), Law Number 76, Mexico, 1990, pp 1–9. http://sistemas.cgever.gob.mx/2003/Normatividad_Linea/constitucion_codigos_y_leyes/LEY%20ESTATAL%20DEL%20EQUILIBRIO%20ECOLOGICO%20Y%20LA%20PROTECCION%20AL%20AMBIENTE.pdf (accessed Sept 13, 2012).
8. Ley Federal Del Trabajo (Federal Labor Law), Official Gazette of the Federation, 1970. http://www.diputados.gob.mx/LeyesBiblio/pdf/125.pdf (accessed Sept 26, 2012)
9. Norma Mexicana NMX-AA-061-1985 Protección al Ambiente—Contaminación del Suelo-Residuos Sólidos Municipales—Determinación de la Generación (Mexican Standard NMX-AA-061-1985 Environmental Protection—Soil Contamination-Municipal Solid Wastes—Determination of Generation). http://legismex.mty.itesm.mx/normas/aa/aa061.pdf (accessed Jan 25, 2013).
10. NMX-AA-015-1985. Protección al Ambiente—Contaminación del Suelo. Residuos Sólidos Municipales–Muestreo-Método de Cuarteo (Environmental Protection—Soil Contamination-Municipal Solid Wastes—Sampling-Quartering Method). http://legismex.mty.itesm.mx/normas/aa/aa015.pdf (accessed Jan 25, 2013).
11. NMX-AA-019-1985. Protección al Ambiente-Contaminación del Suelo. Residuos Sólidos Municipales-Peso Volumétrico "in situ" (Environmental Protection—Soil Contamination-Municipal Solid Wastes-"In Situ" Volumetric Weight). http://legismex.mty.itesm.mx/normas/aa/aa019.pdf (accessed Jan 25, 2013).
12. NMX-AA-022-1985. Protección al ambiente-Contaminación del Suelo Residuos Sólidos Municipales-Selección y Cuantificación de subproductos. (Environmental Protection-Soil Contamination-Municipal Solid Wastes-Selection and Quantification of Byproducts). http://legismex.mty.itesm.mx/normas/aa/aa022.pdf (accessed Jan 25, 2013).
13. PlandeOrdenamientoTerritorial(POT),2012(TerritorialOrganizationPlan,POT;2012). http://www.metrocuadrado.com/m2-content/cms-content/glosario/ARTICULO-WEB-GLOSARIO_M2-2033425.html (accessed March 1, 2012).
14. PROY-NADF-020-AMB T-2011. Proyecto de Norma Ambientalpara el Distrito Federal, 2012 (Environmental Standard Bill for the Federal District).(accessed April 17, 2014)
15. Rojas Valencia, Ma. Neftalí, 2012. Estudios técnicos para definir el desarrollo y funcionamiento del Parque Ecológico Tuzandepetl (Technical Studies for Defining the Development and Functioning of the Tuzandepetl Ecological Park). Institute of Engineering, UNAM. (accessed May 18, 2012).
16. Stephan, O. E. (coord.).In *Segundo Seminario Internacional de Investigadores de Xochimilco*, (*Second International Seminar of Xochimilco Researchers*), Vols. 1 and 2; the Asociación Internacional de Investigadores de Xochimilco, Ed.; Mexico 1999. (accessed Oct 31, 2013).
17. Tchobanoglous, G. *Integrated Solid Waste Management*. McGraw-Hill: US, **1993**; Vol. 1, pp 221–260. (accessed Aug 2, 2013).

CHAPTER 11

REMOVAL OF MERCURY (II) USING IMMOBILIZED PAPAIN: EXPERIMENT, MODELING, AND OPTIMIZATION

APARUPA BHATTACHARYYA[1], SUSMITA DUTTA[1], and SRABANTI BASU[2*]

[1]*Department of Chemical Engineering, National Institute of Technology Durgapur 713209, Durgapur, India*

[2]*Department of Biotechnology, Heritage Institute of Technology, Kolkata 700107, India*

[*]*Corresponding author. E-mail: srabanti_b@yahoo.co.uk*

CONTENTS

ABSTRACT

Papain is used to modify the surface of a low-cost adsorbent, charred citrus fruit peel (CCFP) for adsorptive removal of mercury (II). The immobilization condition of papain on CCFP is optimized using response surface methodology considering initial concentration of papain, pH, and weight of CCFP as independent factors and specific enzymatic activity as response. The immobilized sample obtained at optimum condition is termed as charred citrus fruit peel immobilized papain (CCFPIP). Comparative study shows that commercial activated carbon, CCFP, and CCFPIP can remove 21.55%, 21.37%, and 99.92% mercury (II), respectively, when they have been used individually to treat synthetic wastewater containing 5 mg/L mercury (II) under identical condition. The removal process of mercury (II) by CCFPIP has been optimized using RSM, considering initial concentration of mercury (II), weight of CCFPIP, pH, and temperature as input factors and percentage removal of mercury (II) as response. Theoretically predicted percentage removal (83.76%) matches well with the experimental observation (85%) at optimum condition. CCFPIP is equally able to treat industrial wastewater. Finally, recovery study with spent CCFPIP shows better (66.68%) recovery at lower pH (4) than at higher pH (7 and 9).

11.1 INTRODUCTION

Occurrence of mercury (II) in wastes and surface water is becoming a great concern today. Due to its nonbiodegradability and persistency, mercury (II) is bioaccumulated in the ecosystem causing severe environmental pollution problem.[1,2] Mercury (II) is introduced in the environment through the release of untreated or improperly treated industrial effluents from different industries such as chloralkali plant, paper mills, oil refinery, paint, electrical, and electronic industry, pesticides in the surface water.[2–5] Proper treatment of industrial waste for removal of mercury (II) is essential as it affects human health at very low concentration.[6] Though bulk techniques such as simple filtration or precipitation may reduce a considerable amount of heavy metals from wastewater, they are unable to reduce their concentration beyond the standard limit as prescribed by different environmental agencies and hence a polishing step is required to be employed.[7] Conventional polishing methods such as ion exchange or chelation with crown ethers have several disadvantages such as weak binding characteristics, lack of selectivity in metal binding, slow release kinetics, nonrecovery of metals, handling of toxic chemicals, etc.[7] Therefore, a novel cost-effective, environment friendly polishing technology having opportunity for metal recovery is in search.

Papain, a cysteine protease, has the metal-binding capacity due to the presence of sulfhydryl groups in its active site.[8] Thus, papain, immobilized on suitable matrix can efficiently be used for metal removal. The method is primarily based on the augmentation of the active sites for adsorption in porous adsorbent. Researches show that immobilization of papain on commercial activated carbon (CAC) can efficiently be used to remove mercury (II) from its simulated solution.[9] This surveillance opens up a new avenue to increase the adsorption capacity of activated carbon but the high initial cost of CAC restricts its use as a solid matrix for enzyme immobilization. This leads the present investigators to search for a low-cost alternative to CAC which can be used as a matrix for enzyme immobilization. Though a number of studies have been made to utilize various low-cost adsorbents prepared from different nonconventional industrial or agricultural waste materials for abatement of number of pollutants from wastewater,[10–23] no investigation was made to explore the possibility of usage of any of these low-cost adsorbents as a matrix for enzyme immobilization for removal of heavy metals. Therefore in the present study, a novel low-cost adsorbent, charred citrus fruit peel (CCFP), prepared by carbonizing citrus fruit peel,[14] has been chosen as the solid matrix for immobilization of enzyme instead of CAC. The immobilization condition is optimized using response surface methodology (RSM) considering initial concentration of papain in solution, weight of CCFP, and pH as input factors and specific enzymatic activity (SEA) of immobilized papain sample as response. The purpose of RSM is to find out the optimum experimental condition of the system or to establish a region that suits the operating principles.[24, 25] The adsorbent obtained at optimum condition is termed as charred citrus fruit peel immobilized papain (CCFPIP) and used for removal of mercury (II) from simulated and real wastewater. To judge its effectiveness in mercury (II) removal, a comparative study has been carried out with its native form, that is, CCFP and with CAC under identical conditions. Results reveal that CCFPIP is much more efficient than others.

11.2 EXPERIMENTAL

Details of the experimentations along with materials used have been described in this section. All the experiments have been performed thrice to check the repeatability and reproducibility of the results and the arithmetic mean has been taken. All materials unless otherwise stated are of AR grades.

11.2.1 PREPARATION OF CCFPIP

CCFP has been prepared following the protocol of Dutta et al. (2011) by carbonizing citrus fruit peel at optimum condition, namely, weight ratio of peel to activating agent: 3, temperature: 524.86°C, and time of carbonization: 0.75 h.[14] Papain (SRL) was immobilized on CCFP by physical adsorption method. The immobilization condition was optimized using RSM. Definite amount of papain was dissolved in distilled water to prepare papain solution of desired concentration. Specific amount of CCFP was then added to this papain solution. The solution was stirred continuously using magnetic stirrer (Remi Equipments Pvt. Ltd., Model No. 1MLH) in a batch contactor for 60 min. For immobilization, weight of CCFP (0.3 g–1.0 g), initial concentration of papain (20 g/L–50 g/L), and pH (5–9) were varied in significant manner. The solid was then separated by vacuum filtration and dried at ambient temperature. *SEA* of immobilized sample was determined spectrophotometrically (UV–VIS–NIR Spectrophotometer, U4100, HITACHI) using casein (Hammarsten quality, SRL) as substrate following standard protocol.[26]

RSM was employed to determine the optimum condition for immobilization. Design-Expert Software (8.0.4) is used for this purpose. RSM is a collection of mathematical and statistical techniques useful for developing, improving, and optimizing processes.[27] It can also be used to evaluate the relative significance of several affecting factors even in the presence of complex interactions.[27] The optimization process through RSM is consisting of three key steps such as designing of experiments, generating a mathematical model for prediction of response and determination of coefficient (R^2) of that model, and checking the competence of that model.[22] RSM assists to enumerate the relationship between responses and the input parameters by fitting the experimental data to empirical correlation.

$$Y' = f\left(x'_1, x'_2, \ldots, x'_n\right) \tag{11.1}$$

where, Y' is the response of the process and x'_i s are the input factors.

A standard RSM design called central composite design (CCD) method has been used to optimize the immobilization condition. For three input factors ($n = 3$), the total no of experiments (N) required is

$$N = 2^n + 2n + n_c = 2^3 + 2 \times 3 + 6 = 20 \tag{11.2}$$

where $n_{c=}$ number of centre points repeated.

The experimental sequence was randomized in order to minimize the effects of the uncontrolled factors.[22, 27] The independent parameters and the dependent output response were modeled and optimized using analysis of variance (ANOVA) to estimate the statistical parameters. Weight of CCFP, initial concentration of papain, and pH were considered as independent factors and their maximum (+1) and minimum (−1) levels were fixed at 0.3 and 1.0 g, 20.0 and 50.0 g/L and 5 and 9, respectively. SEA of immobilized papain sample was considered as dependent output response. Statistical design of experimentation as prepared by Design-Expert Software is presented in Table 11.1. The immobilized papain sample, prepared at optimum condition as suggested by the software, was termed as CCFPIP and used for further studies on mercury (II) removal.

TABLE 11.1 Experimental Design for Optimization of Immobilization of Papain on CCFP.

Run	A_1:pH	B_1:initial concentration of papain (g/L)	C_1:weight of CCFP (g)	R_1:SEA (g peptide formed/g papain*h)
1	5	50	1	0.05912
2	9	50	1	0.077
3	7	35	0.65	0.195867
4	7	9.77	0.65	0.06996
5	3.64	35	0.65	0.02676
6	5	50	0.30	0.05423
7	7	35	0.06	0.08558
8	7	35	0.65	0.195867
9	9	20	1	0.0442
10	9	50	0.30	0.057501

TABLE 11.1 *(Continued)*

Run	A_1:pH	B_1:initial concentration of papain (g/L)	C_1:weight of CCFP (g)	R_1:SEA (g peptide formed/g papain*h)
11	9	20	0.30	0.05471
12	7	35	0.65	0.195867
13	5	20	1	0.04442
14	7	60.23	0.65	0.147538
15	7	35	0.65	0.195867
16	10.36	35	0.65	0.016213
17	7	35	0.65	0.195867
18	7	35	0.65	0.195867
19	7	35	1.24	0.139693
20	5	20	0.30	0.055569

11.2.2 CHARACTERIZATION OF CCFPIP

11.2.2.1 SEM AND ENERGY DISPERSIVE X-RAY SPECTROSCOPY ANALYSES

To see the changes in surface characteristics, SEM analysis was performed. For preparation of SCCFPIP, 0.3 g of CCFPIP was contacted with 10 mg/L mercury (II) solution at pH 8 and incubated for 30 min at 37.5°C. Solid sample was then separated from mercury (II) solution and termed as spent charred citrus fruit peel (SCCFPIP). SCCFPIP was then treated with buffer of pH 4 to recover mercury (II) from SCCFPIP. Three samples, namely, CCFPIP, SCCFPIP, and CCFPIP after recovery of mercury were analyzed using scanning electron microscope (SEM; JSM 6700F, JEOL, Japan). All the samples were mounted on brass stubs individually using double-sided adhesive tape. SEM photographs were taken at the required magnification at room temperature. The working distance of 25 mm was maintained and acceleration voltage used was 15 kV, with the secondary electron image (SEI) as a detector. Energy dispersive X-ray spectroscopy (EDS) was performed for those three samples as well.

11.2.2.2 DETERMINATION OF PH AND TEMPERATURE OPTIMA AND PH AND TEMPERATURE STABILITY

Temperature and pH optima were determined following standard procedure to know at which temperature and pH the immobilized enzyme works best.[26] To determine temperature optima, 0.03 g of CCFPIP was taken in several glass tubes. Specific concentration of casein (10 g/L) was used as substrate and ethylene diamine tetra acetic acid (EDTA)-cysteine reagent as activating agent. The solution pH was maintained at 8 using 0.1 M Tris-HCl buffer and incubated for 20 min at different temperature (35–70°C). During incubation period, reaction mixture was stirred continuously to make the process reaction-rate controlled instead of diffusion-rate controlled. The action of enzyme was terminated using trichloro acetic acid (TCA) (MERCK, 100 g/L). The solid was then separated from solution by centrifugation and *SEA* was measured by taking absorbance of the solution at 280 nm. Similarly pH optimum was measured by determining *SEA* at different pH (2–11) conditions at constant temperature (35°C).

For assessment of temperature stability, 0.03 g of CCFPIP was incubated at different temperature (30–70°C) at constant pH (7) for an hour. Solid was then separated from the solution by filtration and dried at room temperature. SEA of dried sample was then determined spectrophotometrically following standard protocol as described above.[26] Likewise, pH stability was also determined by incubating CCFPIP at different pH (2–11) at 35°C for an hour. The separated solid was then checked for SEA according to the procedure described above.

11.2.2.3 DETERMINATION OF KINETIC PARAMETERS

Kinetics of protein hydrolysis by CCFPIP was performed to find out the kinetic parameters such as maximum forward velocity of the enzymatic reaction (V_{max}), rate constant for enzymatic reaction (k_2) and Michaelis–Menten constant (K_m). Velocity (SEA) of the reaction was measured by varying the amount of casein at pH 7 keeping the weight of CCFPIP constant. To reduce the external mass transfer resistance for bulk diffusion, the solution was stirred during reaction using magnetic stirrer.

11.2.2.4 DETERMINATION OF SHELF LIFE

Once CCFPIP was prepared, it was stored in refrigerator at 4°C. Determination of "shelf life" is important as it indicates the efficiency of immobilized enzyme during an extended period. Thus, to determine "shelf life," SEA of stored CCFPIP has been measured at regular intervals for 6 month.

11.2.3 REMOVAL OF MERCURY (II)

11.2.3.1 COMPARATIVE STUDY OF MERCURY (II) REMOVAL USING CAC, CCFP, AND CCFPIP

Synthetic solution of mercury (II) was prepared by dissolving definite amount of mercuric chloride (MERCK) in water. A comparative study was then carried out to compare the potential of CAC, CCFP, and CCFPIP in the removal of mercury (II) from simulated wastewater. The mercury (II) solution with initial concentration of 5 mg/L was contacted with same amount of CAC, CCFP, and CCFPIP separately for 30 min in a batch reactor at pH 7 and at 35°C. The adsorbents were separated and the residual concentrations of mercury (II) present in the solution were estimated by atomic absorption spectrophotometer (AAS) (Varian AA240, Graphite Tube Atomizer GTA 120, Flame mode AA240, Germany).

11.2.3.2 KINETIC STUDY OF REMOVAL OF MERCURY (II)

An extensive study on kinetics of mercury (II) removal has been carried out in a batch contactor. Four parameters, namely, initial concentration of mercury (II) (1.76–14.7 mg/L), weight of CCFPIP (0.03–0.1 g), pH (4–9) of solution, and incubation temperature (30–45°C), have been varied to see the individual effect of the parameters on removal of mercury (II) with time. During every reaction, 30 mL of mercury (II) solution was contacted with specific amount of CCFPIP for 30 min.. The solution was stirred vigorously to eliminate the external mass transfer resistance due to the bulk diffusion of mercury (II) to the surface of CCFPIP. Samples were collected at a particular interval of time. Solids were separated from the samples by filtration and remaining mercury (II) in solution was estimated by AAS.

11.2.3.3 EQUILIBRIUM STUDY OF REMOVAL OF MERCURY (II)

Different concentrations (1.76–14.7 mg/L) of mercury (II) solutions were prepared by diluting stock solution with distilled water. In each reaction tube, 0.5 g of CCFPIP was added and stirred for 60 min in magnetic stirrer. Equilibrium studies were carried out at 35°C. Reaction solutions were then centrifuged and the clear liquid was analyzed by AAS for residual mercury (II).

11.2.3.4 OPTIMIZATION OF REMOVAL OF MERCURY (II) USING CCFPIP FROM ITS AQUEOUS SOLUTION

11.2.3.4.1 Experimental

Though kinetic study on removal of mercury (II) gives the information of variation of percentage removal with time as a function of different parameters, it fails to assess the optimum condition of mercury (II) removal in the present parameter space. Thus, another investigation was made to find out optimum condition for mercury (II) removal and RSM was used as a tool for such purpose. The most popular software, Design-Expert (8.0.4), was used to design a set of experimental condition considering four parameters as independent numeric factors such as initial concentration of mercury (II) (10–25 mg/L), weight of CCFPIP (0.3–0.8 g), pH (5–9), and temperature (30–45°C), and percentage of removal was taken as dependent response. Experiments were performed at experimental condition designed by the software using the procedure stated in the Section 2.3.2. All the experiments were carried out taking 30 mL of mercury (II) solution. In kinetic study, it is seen that saturation has been achieved within first 15 min of operation. Therefore, to reduce the number of experiments, time was not included as independent factor and total run time of 30 min was maintained in each case.

11.2.3.4.2 Design of Experiment

A standard RSM design called CCD method was used. For four input factors ($n = 4$), the total no of experiments (N) required for optimization is

$$n = 2^n + 2n + n_c = 24 + 2 \times 4 + 6 = 30$$

The maximum (+1) and the minimum level (−1) of four numeric factors, namely, initial concentration of mercury (II), weight of CCFPIP, pH, and temperature were 10 and 25 mg/L, 0.3 and 0.8 g, 5 and 9, and 30°C and 45°C, respectively. The experiments were performed according to the condition as designed by the software. The reaction broth was stirred continuously to eliminate mass transfer resistance. Solid was then separated from slurry and residual concentration of mercury (II) in solution was analyzed by atomic absorption spectroscopy and percentage removal of mercury (II) was calculated. According to the software, the optimum response can be obtained when initial concentration of mercury (II), weight of CCFPIP, pH, and temperature are 10 mg/L, 0.3 g, 8.57, and 37.5°C, respectively. Spent adsorbent prepared at optimum condition was termed as SCCFPIP and further studies on recovery of mercury (II) were performed using this SCCFPIP.

11.2.3.5 REMOVAL OF MERCURY (II) FROM INDUSTRIAL EFFLUENT

An attempt has also been made to remove mercury (II) from industrial effluent collected from local chloralkali plant. The wastewater was characterized in terms of pH, color, odor, and turbidity. The suspended solids present in the wastewater were removed initially by gravity settling and then by filtration. It was then checked for initial concentration of mercury (II) by AAS. The wastewater was then treated with definite amount of CCFPIP for 20 min at 37.5°C at its own pH with continuous stirring. Finally, the residual concentration of mercury (II) was measured.

11.2.4 RECOVERY OF MERCURY (II) FROM SCCFPIP

To recover mercury (II) from SCCFPIP, 0.1 g of SCCFPIP was incubated at 100 mL of buffer solution at different pH (4–9) for 30 min and at 35°C temperature with constant stirring. After 30 min, the solutions were filtered and checked for mercury (II) concentration by AAS.

11.3 RESULTS AND DISCUSSIONS

11.3.1 PREPARATION OF CCFPIP

Papain, a cysteine protease, has been immobilized on CCFP by physical adsorption method to enhance the active site of CCFP for adsorption of mercury (II). Immobilization condition has been optimized using RSM considering three parameters, namely, weight of CCFP, pH of solution, and initial concentration of papain as input parameters and SEA of immobilized enzyme sample as response. RSM has been used for this purpose. Experimental conditions as designed by the software along with the responses are shown in Table 11.1. As suggested by the software square root transformation has been chosen and quadratic process order has been selected to analyze data. R^2, $R^2_{\mathrm{Pr}ed}$, and R^2_{adj} have been found to be 0.9725, 0.7910, and 0.9478, respectively.

$$\begin{aligned} Sqrt(R_1) &= -1.24523 + 0.37306 \times A_1 + 0.013138 \times B_1 + 0.37079 \times C_1 + \\ &1.81728E-004 \times A_1 \times B_1 + 5.13016E-003 \times A_1 \times C_1 + 2.3E-003 \times B_1 \times C_1 \\ &-0.02739 \times A_1^2 - 2.06032E-004 \times B_1^2 - 0.3529 \times C_1^2 \end{aligned} \quad (11.3)$$

where R_1 = SEA [(g peptide formed)/(g papain × h)], A_1= pH of solution, B_1 = concentration of papain (g/L), and C_1 = weight of CCFP (g).

The conjugate effects of different input parameters on response function, that is, SEA of immobilized samples are illustrated in Figure 11.1a and b. Figure 11.1a shows the combined effect of initial concentration of papain and pH on SEA of immobilized papain sample at constant weight of CCFP (0.65 g). It is seen from the figure that SEA initially increases and then decreases when initial concentration of papain changes from 20 to 50 g/L at constant pH 5. This may be due to the fact that at constant amount of CCFP, papain at low concentration cannot saturate the given amount of CCFP. Thus, increase in initial concentration of papain favors immobilization process and thereby SEA is enhanced. At constant initial concentration of 20 g/L, SEA initially increases and then decreases when pH changes from 5 to 7 and 7 to 9, respectively. This is because of the fact that in other pH values than neutral, structure of papain has been changed so much that immobilization is not being favored. Thus, maximum SEA is obtained at neutral pH. Finally, it can be said that initial concentration of papain and pH both have profound effect on enzyme immobilization at constant weight of CCFP.

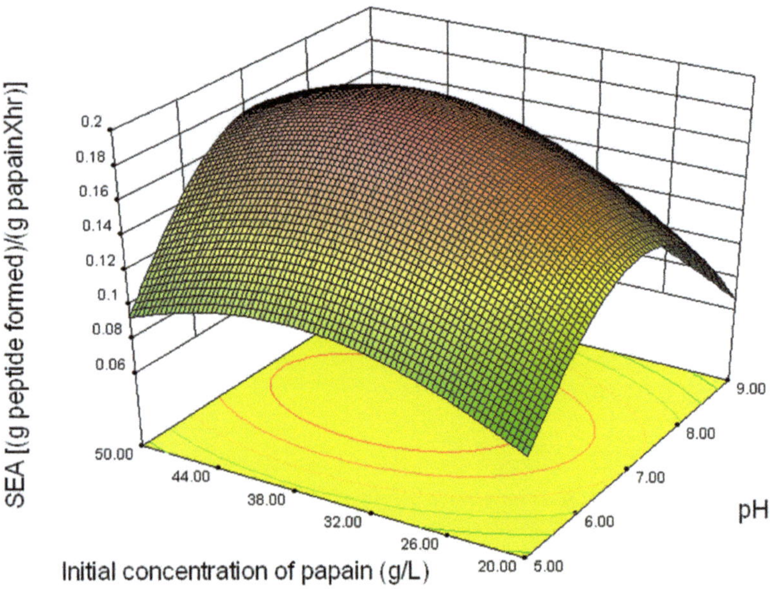

FIGURE 11.1a Combined effect of initial concentration of papain and pH on *SEA* of immobilized papain sample.

Figure 11.1b represents the combined effect of weight of CCFP and pH on SEA of immobilized papain sample at constant initial concentration of papain (35 g/L). SEA increases with an increase in pH from 5 to 7 and then decreases with further increase in pH at constant weight of CCFP at 0.3 g. Again, SEA increases little with increase in weight of CCFP up to a certain point and then decreases with further increase in weight of CCFP at constant pH 5. Thus, it can be said that change in weight of CCFP does not have any significant influence on SEA value at constant pH. This may be due to the fact that the present amount of papain is not sufficient enough to saturate CCFP.

Finally, to make the process cost effective, the following criteria have been fixed during optimization: initial concentration of papain: minimize, pH (7) and weight of CCFP: minimize, and response: maximize. The optimum condition for immobilization has been found as follows: weight of CCFP: 0.33 g, pH: 7, and initial concentration of papain: 20

g/L, respectively. The sample obtained at optimum condition is termed as CCFPIP and is used for further investigations on mercury (II) removal.

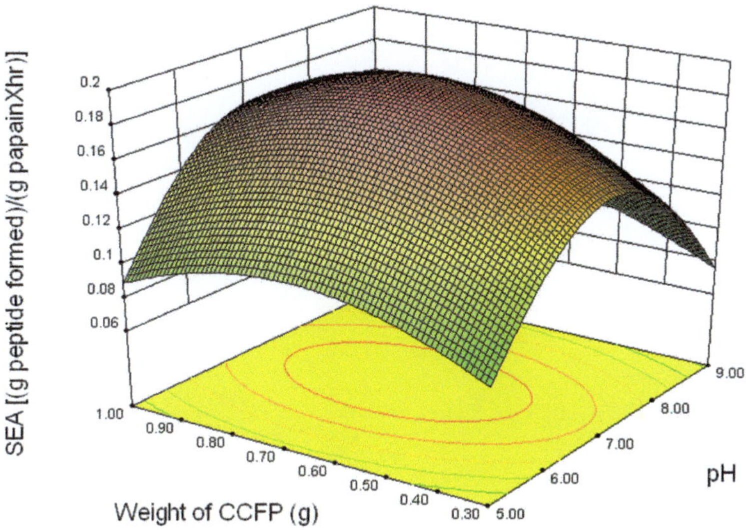

FIGURE 11.1b Combined effect of weight of *CCFP* and pH on *SEA* of immobilized papain sample.

11.3.2 CHARACTERIZATION OF CCFPIP

11.3.2.1 SEM AND EDS ANALYSES

Figure 11.2a–c represent SEM images of CCFPIP, SCCFPIP, and SCCFPIP after recovery of mercury from SCCFPIP, respectively. These images help to compare the changes in surface morphology upon binding of mercury (II) with CCFPIP (Fig. 11.2b) and after recovery of mercury (II) from SCCFPIP (Fig. 11.2c) with its native form CCFPIP (Fig. 11.2a). The surface has become smoother after binding of mercury (II) (Fig. 11.2b) but recovery has made the surface rough again (Fig. 11.2c). EDS study confirms the binding of mercury (II) with CCFPIP (Fig. 11.2d). Table 11.2 depicts the mass percentages of mercury (II) in CCFP, CCFPIP, SCCFPIP, and SCCFPIP after recovery of mercury. This shows that mass percentages of mercury (II) in CCFP and CCFPIP are very less, whereas the percentage in SCCFPIP is 72.66 which reconfirms the binding of mercury (II) with

CCFPIP. Finally, SCCFPIP on treatment with buffer losses its bound mercury and thus, mass percentage of mercury (II) becomes negligible.

FIGURE 11.2a SEM image of CCFPIP.

FIGURE 11.2b SEM image of SCCFPIP.

FIGURE 11.2c SEM image of SCCFPIP after recovery of mercury (II).

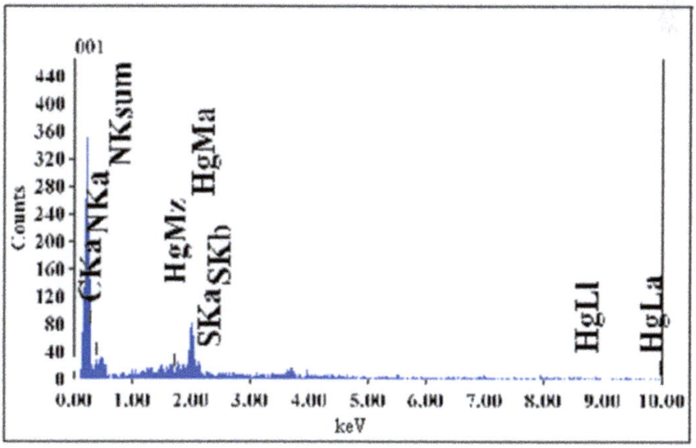

FIGURE 11.2d EDS spectra of SCCFPIP.

TABLE 11.2 Mass Percentages of Mercury (II) in Different Samples.

Sample	Mass percentages of mercury
CCFP	14.09
CCFPIP	19.98
SCCFPIP	72.66
SCCFPIP after recovery	–

11.3.2.2 DETERMINATION OF PH AND TEMPERATURE OPTIMA AND PH AND TEMPERATURE STABILITY

The maximum activity of CCFPIP has been found at pH 9 (0.5712). Activities of CCFPIP at pH 7 (0.5458) and 4 (0.4870) are almost similar to that at pH 9. At pH 2, the SEA of CCFPIP is very low (0.172). Though the maximum activity of CCFPIP has been found at pH 9, to make the process cost-effective all the reactions have been carried out at pH 7. CCFPIP has shown its maximum activity at 70°C (0.8034). A significant decrease in SEA (0.5458 and 0.6411) values has been observed at 35°C and 50°C, respectively, than that at 70°C. This may be due to the lower diffusion rate of substrate at lower temperature. Finally it can be stated that CCFPIP is active over a wide range of pH (2–11) and temperature (35–70°C).

11.3.2.3 DETERMINATION OF KINETIC PARAMETERS

Kinetic study of protein hydrolysis has been performed to determine the kinetic parameters such as V_{max}, k_2, K_m. The kinetic parameters are calculated by Lineweaver–Burk method. Table 11.3 represents the kinetic parameters calculated both for free papain and CCFPIP. The loading of enzyme on CCFP has been determined using Kjeldahl method. The amount of papain immobilized on CCFPIP has been found to be 0.784 mg/g of CCFPIP. The kinetic parameter V_{max} indicates the maximum velocity of product formation. The other parameter K_m gives the amount of substrate required to reach half the maximum velocity signifying affinity of the enzyme toward its substrate. The rate constant k_2 represents rate of breakdown of enzyme–substrate complex to product. All the kinetic parameters have been affected by immobilization. After immobilization of papain on CCFP, V_{max} value is decreased which indicates that the rate of product formation has been decreased. K_m value is increased which suggests that the affinity of the enzyme toward its substrate is decreased. Lower k_2 value corresponds to lower rate of breakdown of enzyme–substrate complex to product.

TABLE 11.3 Values of Kinetic Parameters.

Enzyme	Lineweaver–Burk equation				Eadie–Hofstee equation			
	V_{max}	K_m	K_2	R^2	V_{max}	K_m	K_2	R^2
Free papain	0.89	0.44	35.90	0.99	0.89	0.43	35.62	0.94
CCFPIP	0.351	0.966	8.199	0.993	0.032	0.734	7.48	0.752

11.3.2.4 DETERMINATION OF SHELF LIFE

Shelf life has been determined by checking the SEA of CCFPIP at regular interval. CCFPIP has been found active over 6 month. Enzymatic activity has not been altered significantly in this time period.

11.3.3 REMOVAL OF MERCURY (II)

11.3.3.1 COMPARATIVE STUDY OF MERCURY (II) REMOVAL USING CAC, CCFP, AND CCFPIP

To compare the efficacy of removal of mercury (II) between CAC, CCFP, and CCFPIP vis-à-vis to establish the usefulness of the present study, an investigation has been made with these adsorbents individually to remove mercury (II) from simulated solution under identical condition. The study reveals that the efficacy of percentage removal of mercury (II) by CCFPIP (99.92%) is much higher than CAC (21.55%) and CCFP (21.38%) (Fig. 11.3a). The result reveals that surface modification of *CCFP* with papain has enhanced the capacity of mercury (II) binding from aqueous solution and thus, usefulness of the present study is preliminary established.

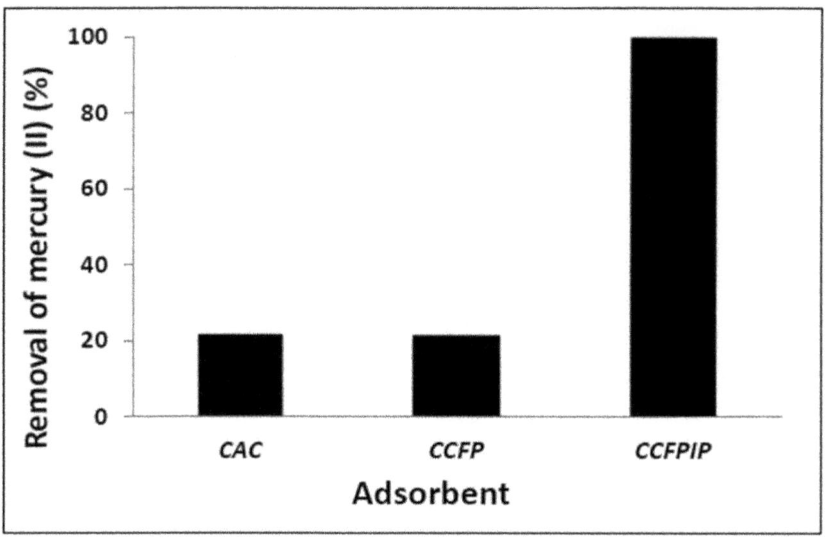

FIGURE 11.3a Percentage removal of mercury (II) by CAC, CCFP, and CCFPIP.

11.3.3.2 KINETIC STUDY OF REMOVAL OF MERCURY (II)

Kinetic study on removal of mercury (II) using CCFPIP has been performed by varying four parameters, namely, initial concentration of mercury (II) (1.76–14.7 mg/L), pH (4–9), weight (0.3–1 g), and temperature (30–45°C) in a judicial manner. Figure 11.3b shows percentage removal of mercury (II) at different initial concentrations of mercury (II) when weight of CCFPIP, pH, and temperature have been maintained constant at 0.5 g, 7, and 35°C, respectively. The figure shows that within 10 min, saturation level has been achieved for all initial concentration of mercury (II). When initial concentration is comparatively lower (1.47, 5, and 7 mg/L) saturation level has been reached within first 2 min. However for higher concentrations, it takes 10 min to get saturated. This may be due to the fact that when concentration is relatively lower (1.76–7 mg/L), mercury (II) can easily bind with the immobilized enzyme.

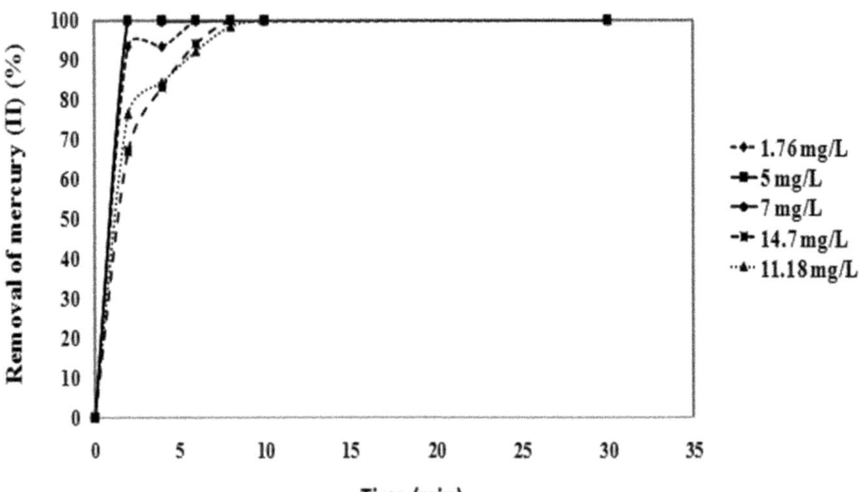

FIGURE 11.3b Removal of mercury (II) with time considering initial concentration of mercury (II) as parameter when weight of CCFPIP = 0.5 g, pH = 7, and temperature = 35°C.

Variations of percentage removal of mercury (II) at different weight of CCFPIP with time have been shown in Figure 11.3c. Initial concentration of mercury (II), pH, and temperature have been kept constant at 5 mg/L, 7, and 35°C, respectively. From the Figure 11.3c, it is clear that percentage removal of mercury (II) with 0.3 g of adsorbent is much lower (60%) than that with 0.5 g (90.6%) and 1 g of adsorbent (90.6%). This may be because of the fact that less amount of CCFPIP contains less amount of enzyme which is not sufficient to remove mercury (II) to the desired level, whereas 0.5 g and 1 g of CCFPIP contains more amount of immobilized enzyme, which can easily remove mercury (II) to the desired level. Here also, the major part of removal is achieved within 10 min. In total, 90.6% mercury (II) has been removed when mercury (II) solution having initial concentration of 5 mg/L has been contacted with 0.5 g CCFPIP at pH 7 and 35°C. The same removal has been achieved with 1 g CCFPIP.

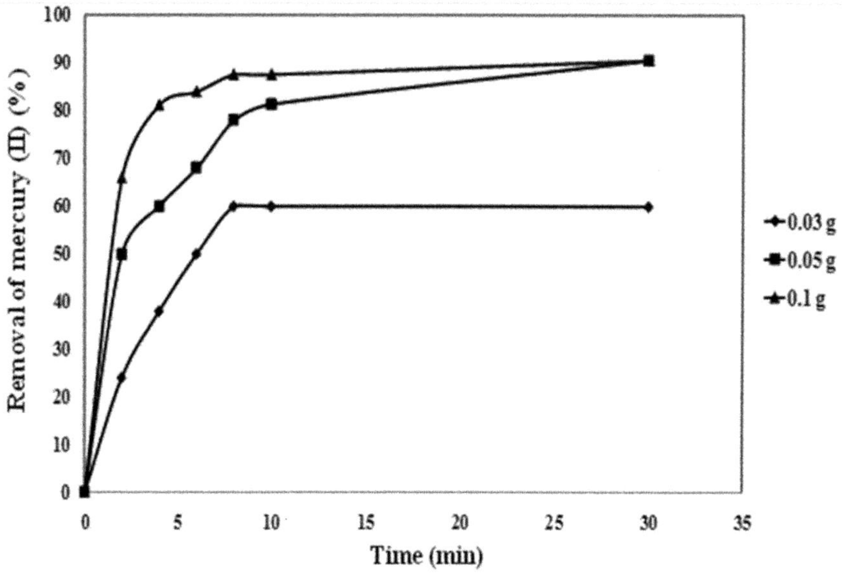

FIGURE 11.3c Removal of mercury (II) with time considering weight of CCFPIP as parameter when concentration of mercury (II) = 5 mg/L, pH = 7, and temperature = 35°C.

Effects of pH on the variation of percentage removal of mercury (II) with time are shown in Figure 11.3d. pH has been varied from 4 to 9. Weight of CCFPIP and temperature are kept constant at 0.5 g and 35°C, respectively. At pH 4, 68.7% removal has been obtained, whereas at pH 7 and 9 percentage removal is almost similar (99.9% and 99.8%, respectively). Being a cysteine protease, papain has sulfhydryl groups at its active sites which gets deproptonated at higher pH forming E^- molecule. This E^- molecule is more susceptible to bind with mercury (II) than its original undissociated form. This favors binding of metal ions with deprotonated enzyme molecules at higher pH. Thus, higher pH favors removal of mercury (II) which is in conformity with the observation made by Dutta et al. 2009.[9] At lower pH, due to the presence of unionized enzyme, binding of mercury (II) with the enzyme is not favored.

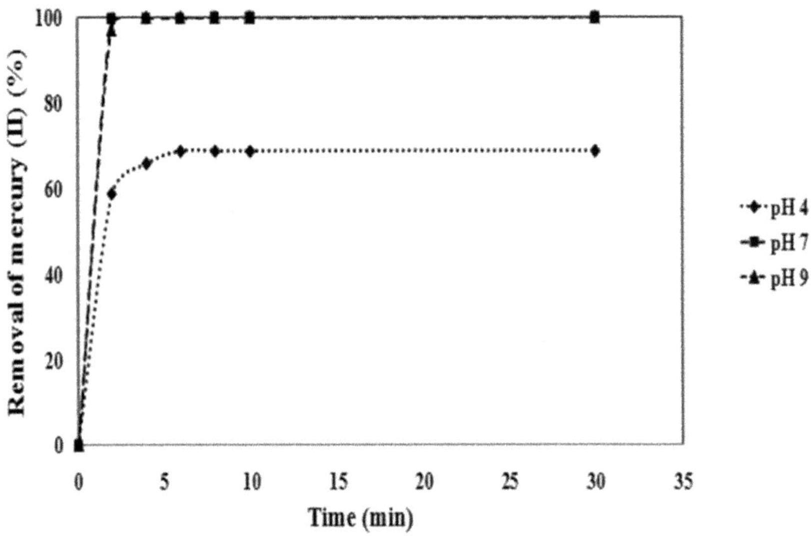

FIGURE 11.3d Removal of mercury (II) at different pH when concentration of mercury (II) = 5 mg/L, weight of CCFPIP = 0.5 g, and temperature = 35°C.

When temperature has been varied from 30°C to 45°C keeping initial concentration of mercury (II), weight of CCFPIP and pH at 17.65 mg/L, 0.5 g, and 7, respectively, 99.9% removal have been observed in each case (figure not shown). Thus, it is clear that temperature has no effect on percentage removal of mercury (II) in the temperature range studied. This may lie in the fact that in this temperature range, CCFPIP can easily remove mercury (II) to the desired level. Change of temperature beyond this range may affect the removal process.

Kinetic data of removal of mercury (II) have been fitted to different kinetic models. Morris–Weber model suggests that a plot of the square root of time versus the uptake will be linear if intraparticle diffusion is involved in the sorption process.[28] If this line passes through the origin, then intraparticle diffusion would be the rate-controlling step of the sorption process.[28] The equation for Morris–Weber model is as follows:

$$q_t = k_M t^{1/2} \tag{11.4}$$

Lagergren first-order kinetic model recommends a first-order kinetic rate based on surface reaction.[28] This model can be described by the following equation.

$$\log(q_e - q_t) = \log q_e - \frac{k_L t}{2.303} \tag{11.5}$$

Pseudo-second-order model (PSOM) assumes that the sorption process is of pseudo-second-order and that the rate-limiting step is of chemisorption nature.[28] The mechanism may possibly involves valence forces by sharing or through the exchange of electrons between sorbent and sorbate. The equation for PSOM is

$$\frac{t}{q_t} = \frac{1}{k_P q_e^2} + \frac{t}{q_e} \tag{11.6}$$

Values of kinetic parameters obtained by fitting the data to the above equations are shown in Table 11.4. From the Table 11.4, it is seen that experimental data fit most satisfactorily to PSOM. Therefore, the process is reaction-rate controlled rather than mass transfer controlled.

11.3.3.3 EQUILIBRIUM STUDY

Equilibrium study has been performed at 35°C. Both Langmuir and Freundlich adsorption isotherm models are used to analyze the equilibrium data. Langmuir model has been found to explain the equilibrium data more aptly than the other model as evidenced from the values of determination of coefficient (R^2). The values of the constants of these two models are shown in Table 11.5.

TABLE 11.4 Kinetic Parameters.

Kinetic Model		Parameter: initial concentration (mg/L)					Parameter: weight of CCFPIP (g)			Parameter: pH			Parameter: temperature (°C)		
		1.76	**5**	**7**	**11.18**	**14.7**	**0.3**	**0.5**	**1**	**4**	**7**	**9**	**30**	**37**	**45**
Moris–Weber Model	k_M	0.039	0.118	0.165	0.235	0.348	0.760	0.673	0.360	0.002	0.622	0.016	0.417	0.118	0.416
	R^2	0.875	0.622	0.625	0.406	0.621	0.538	0.476	0.129	0.731	0.118	0.645	0.630	0.622	0.641
Lagergren Model	k_L	0.432	0.642	0.055	0.096	0.071	0.016	0.009	0.207	0.845	0.642	1.807	0.014	0.642	0.921
	R^2	0.197	0.625	0.003	0.046	0.034	0.105	0.018	0.442	0.974	0.625	1.0	0.0	0.625	0.882
Pseudo-second-order model (PSOM)	k_P	31.25	1000	303.03	3.521	2.320	0.169	0.257	1.04	1666.6	1000	1818.2	222.2	1000	62.5
	R^2	0.998	1.0	1.0	0.999	0.998	0.952	0.990	0.995	0.999	1.0	1.0	1.0	1.0	1.0

TABLE 11.5 Adsorption Isotherm Parameters.

Adsorption isotherm model	Temperature 35°C	
	constants	R^2
Langmuir model	$q_{max} = 2.99$	0.932
	$K_L = 67.11$	
Freundlich model	$K_F = 2.02$	0.563
	$n_F = 3.51$	

11.3.3.4 OPTIMIZATION OF REMOVAL OF MERCURY (II) USING CCFPIP FROM ITS AQUEOUS SOLUTION

While kinetic study is able to give variation in percentage removal with time, it is unable to choose optimum condition in the parameter space. Therefore, optimization study on removal of mercury (II) is necessary from practical point of view. The condition for removal of mercury (II) has been optimized using RSM. Percentage removal of mercury (II) at each experimental condition as specified by Design-Expert software is determined and shown in Table 11.6. As the ratio of maximum (95.98) to minimum response (44.48) is 2.1578, much below than 10, no transformation is required. Regression analysis has been performed to fit the response. As suggested by the software, quadratic model is selected to analyze the data. The final regression equation in terms of coded factors used in representing statistical model is shown below:

$$\begin{aligned} R_2 &= +92.48 - 4.32\times A_2 + 9.69\times B_2 + 9.48\times C_2 + 1.27\times D_2 \\ &-0.084\times A_2\times B_2 + 1.14\times A_2\times C_2 - 0.44\times A_2\times D_2 - 1.69\times B_2\times C_2 + \\ &0.58\times B_2\times D_2 - 0.35\times C_2\times D_2 - 1.66 A_2^2 - 5.68\times B_2^2 - 6.37\times C_2^2 \\ &-1.02\times D_2^2 \end{aligned} \quad (11.7)$$

where, R_2 = percentage removal of mercury (II) (%), A_2 = initial concentration of mercury (II) (mg/L), B_2 = weight of *CCFPIP* (g), C_2 = pH and D_2 = temperature (°C). The high values of statistical parameters, namely, R_2 = 0.9763, R^2_{Pred} = 0.8632, and R^2_{adj} = 0.9541 represent the good fit of data with the model equation.

Analysis by ANOVA reveals that first three parameters, namely, initial concentration of mercury (II), weight of CCFPIP, and pH have significant effect on percentage removal of mercury (II). The quadratic effects of these parameters are also found to be significant.

TABLE 11.6 Experimental Design for Optimization of Removal of Mercury (II) Using CCFPIP.

Run	Factor 1 A_2: IC (mg/L)	Factor 2 B_2: wt CCFPIP g	Factor 3 C_2:pH	Factor 4 D_2: temp (°C)	R_2: % Removal
1	10.00	0.80	9.00	30.00	95.49
2	17.50	0.55	3.00	37.50	47.50
3	25.00	0.80	5.00	30.00	69.50
4	10.00	0.30	5.00	45.00	64.97
5	10.00	0.30	5.00	30.00	61.01
6	17.50	0.55	7.00	37.50	92.48
7	25.00	0.80	5.00	45.00	74.98
8	25.00	0.30	5.00	45.00	51.97
9	17.50	0.55	11.00	37.50	87.29
10	25.00	0.80	9.00	45.00	91.61
11	17.50	0.55	7.00	37.50	92.48
12	10.00	0.80	5.00	45.00	85.92
13	17.50	0.55	7.00	37.50	92.48
14	10.00	0.80	5.00	30.00	81.94
15	17.50	0.55	7.00	52.50	93.10
16	17.50	0.55	7.00	22.50	84.50
17	25.00	0.30	9.00	45.00	76.97
18	17.50	1.05	7.00	37.50	95.89
19	32.50	0.55	7.00	37.50	76.50
20	25.00	0.30	5.00	30.00	55.97
21	2.50	0.55	7.00	37.50	95.98
22	10.00	0.80	9.00	45.00	95.70
23	17.50	0.55	7.00	37.50	92.48
24	17.50	0.05	7.00	37.50	44.48
25	25.00	0.30	9.00	30.00	76.97
26	17.50	0.55	7.00	37.50	92.48
27	25.00	0.80	9.00	30.00	89.99
28	10.00	0.30	9.00	30.00	82.177
29	10.00	0.30	9.00	45.00	84.76
30	17.50	0.55	7.00	37.00	92.48

Figure 11.4a represents three-dimensional response surfaces of the interactive effect of initial concentration of mercury (II) and weight of CCFPIP at constant values of pH (7) and temperature (37.5°C). It is evident from Figure 11.4a that percentage removal is greatly influenced by weight of CCFPIP. The percentage removal increases from 80% to 90% and from 71.6% to 90.6% when weight of CCFPIP increases from 0.3 to 0.8 g at constant initial concentration of 10 and 25 mg/L, respectively. This may be attributed to the availability of more amount of enzyme which binds mercury (II) from its aqueous solution. The decrease in percentage removal with increase in the initial concentration of mercury (II) at constant weight of adsorbent may be due to the saturation of adsorbent.

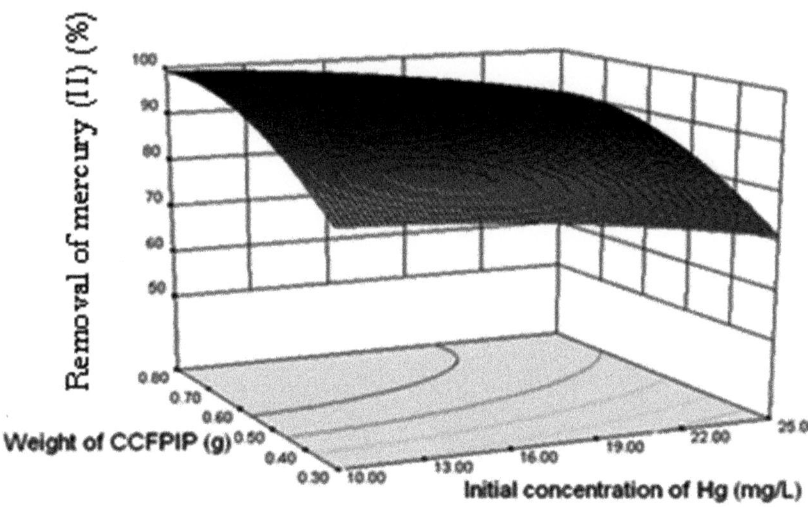

FIGURE 11.4a Combined effect of initial concentration of mercury (II) and weight of CCFPIP on percentage removal of mercury (II).

Figure 11.4b represents the combined effect of pH and weight of CCFPIP on the percentage removal of mercury (II) at constant initial concentration of mercury (II) (17.5 mg/L) and temperature (37.5°C). Figure reveals that both the pH and weight of CCFPIP have synergistic effect on percentage removal of mercury (II). The increase in percentage removal at high pH may be due to the fact that high pH favors protonation of enzyme molecules which, in turn, binds with mercury (II) ion easily. As a result, percentage removal of mercury (II) increases from 60.18 to

82.01 when pH increases from 5 to 9 at constant weight of CCFPIP. In addition, percentage removal increases from 60.18 to 82.5 when weight of CCFPIP increases from 0.3 to 0.8 g at constant pH. When pH and weight of CCFPIP are increasing, percentage removal reaches to the maximum value of 97.9.

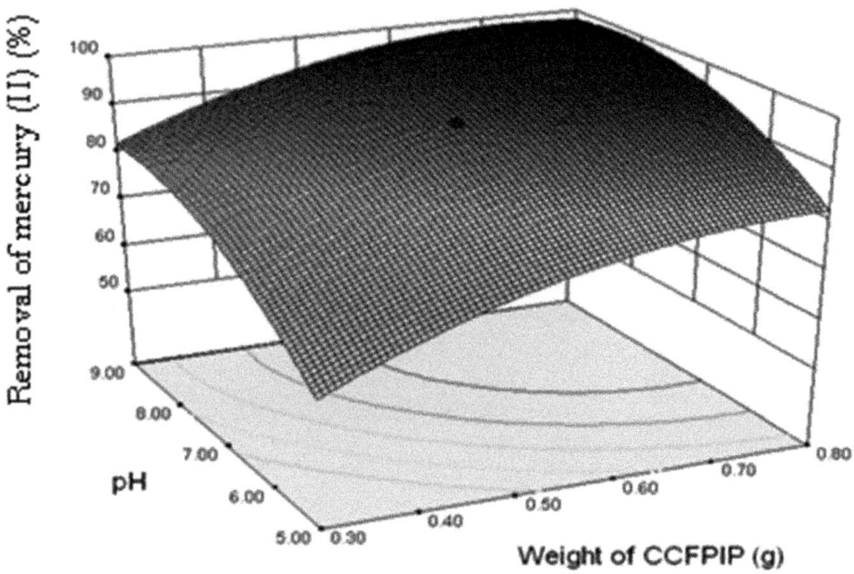

FIGURE 11.4b Combined effect of weight of CCFPIP and pH on percentage removal of mercury (II).

Figure 11.4c represents the interactive effect of temperature and pH on the percentage removal of mercury (II) from its aqueous solution at constant values of initial concentration (17.5 mg/L) and weight of CCFPIP (0.55 g). From Figure 11.4c, it is evident that pH has more pronounced effect on the percentage removal than temperature. It is seen that percentage removal increases from 74.16% to 77.22% and 93.88% to 98.68% when temperature increases from 30°C to 45°C at constant values of pH 5 and 9, respectively.

The removal process of mercury (II) using CCFPIP has been optimized by Design-Expert software. According to the software, the optimum response can be obtained when initial concentration of mercury (II), weight of CCFPIP, pH, and temperature are 10 mg/L, 0.3 g, 8.57, and 37.5°C, respectively. Theoretically predicted percentage removal (83.76%) matches well with the experimental observation (85%) obtained by performing the experiments at the aforesaid condition.

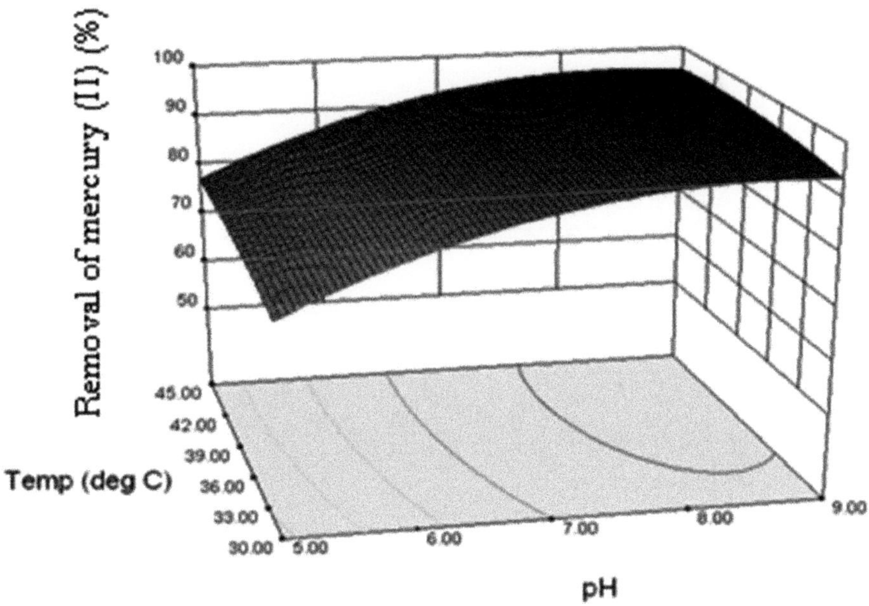

FIGURE 11.4c Combined effect of weight of pH and temperature on percentage removal of mercury (II).

11.3.3.5 REMOVAL OF MERCURY (II) FROM INDUSTRIAL EFFLUENT

Characteristics of industrial wastewater are given in Table 11.7. The initial concentration of mercury (II) in industrial solution has been found to be 3.8 mg/L. Figure 11.5 shows the percentage removal of mercury (II) with time. It is very clear from the figure that more than 90% removal has been achieved within first 5 min of reaction. This suggests again that the reaction is very fast. Maximum percent removal (98.15%) has been achieved after 30 min. In previous sections, it has been observed that CCFPIP can successfully remove mercury (II) from synthetic solution. The present experiment has proved that CCFPIP is very efficient in removal of mercury (II) from industrial effluent as well.

TABLE 11.7 Characteristics of Industrial Water.

Property	Value
pH	8
Color	Colorless
Odor	Odorless
Turbidity	400 NTU
Concentration of mercury (II)	3.8 mg/L

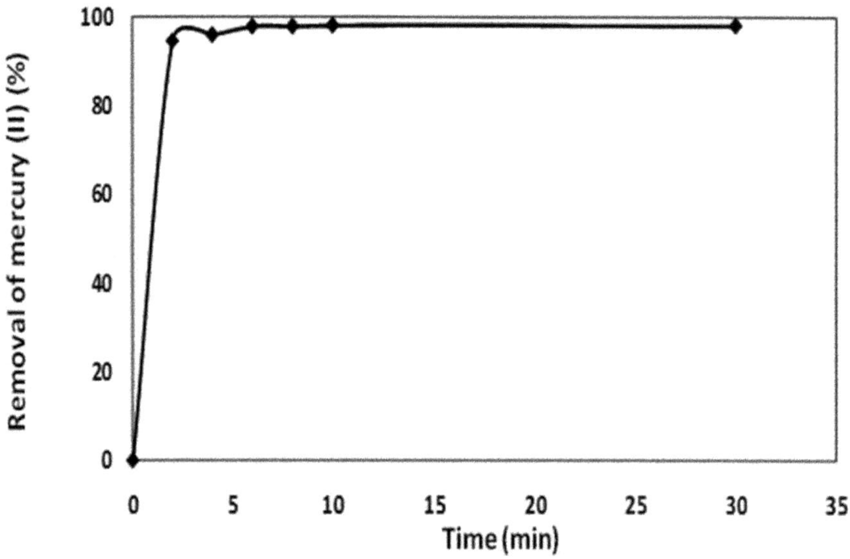

FIGURE 11.5 Percentage removal of mercury (II) with time from industrial solution when weight of CCFPIP, pH, and temperature was kept constant at 0.3 g, 7, and 35°C, respectively.

11.3.4 RECOVERY OF MERCURY (II) FROM SPENT CCFPIP

Recovery of mercury (II) from SCCFPIP has been performed in three different pH such as 4, 7, and 9. Maximum recovery has been obtained at pH 4 (66.67%). Recovery of mercury (II) is negligible at pH 7 and 9 which are 0.38% and 0.95%, respectively. This observation is in agreement with the previous study of Dutta et al. (2009), which supports that at low pH, papain has lost its mercury (II)-binding capacity and thereby recovery is more at low pH. At higher pH, more molecules of deprotonated enzyme are formed. Mercury (II) can bind with the deprotonated enzyme and mercury–enzyme complex is formed which reduces the chance of recovery of metal.

11.4 CONCLUSIONS

Papain is immobilized on CCFP, a low-cost matrix to augment its adsorption capacity. The immobilization condition has been optimized by RSM considering solution pH, weight of CCFP, and initial concentration of papain as independent parameters and SEA as dependent factor. The optimum immobilization condition in terms of initial concentration of papain, weight of CCFP, and pH is 20 g/L, 0.33 g, and 7, respectively. Papain immobilized at this condition has been termed as CCFPIP. EDS study of SCCFPIP shows that the adsorbent can successfully bind with mercury (II). Analysis of SEM image reveals that upon binding with mercury (II), the porous surface of CCFPIP becomes smooth. A comparative study of mercury (II) removal efficacy has then been performed with CAC, CCFP, and CCFPIP under identical condition. Results reveal that CCFPIP is much more efficient in removal of mercury (II). To observe the individual effects of four parameters, namely, initial concentration of mercury (II), weight of CCFPIP, pH, and temperature with time, kinetic study has been performed. Maximum removal has been achieved within first 10 min of operation, which suggests that the reaction is very fast. Data obtained in kinetic study have been fitted to three kinetic models among which PSOM is found to be best. Equilibrium study reveals that data follow Langmuir adsorption isotherm model. The removal process is optimized using RSM as well. Four parameters, namely, initial concentration of mercury (II), weight of CCFPIP, pH, and temperature are considered as numeric input factors, whereas percentage removal is considered as response. The optimum removal condition has been found as initial concentration of mercury (II): 10 mg/L, weight of CCFPIP: 0.3 g, pH: 8.57, and temperature: 37.5°C. The experimental observation (85%) at this condition matches well with the predicted value (83.76%). CCFPIP is equally efficient in removal of mercury (II) from industrial effluent as well. Recovery study shows that mercury (II) can be recovered and CCFPIP can be regenerated at low pH.

ACKNOWLEDGMENT

We acknowledge CSIR (Council of Scientific and Industrial Research, INDIA) for financial support.

KEYWORDS

- **papain**
- **mercury**
- **adsorption**
- **response surface methodology**
- **low-cost adsorbent**

REFERENCES

1. Rmalli, S. W. A.; Dahmani, A. A.; Abuein, M. M.; Gleza, A. A. Biosorption of Mercury from Aqueous Solutions by Powdered Leaves of Castor Tree (*Ricinus communis* L.). *J. Hazard. Mater.* **2008,** *152*, 955–963.
2. Starvin, A. M.; Rao, T. P. Removal and Recovery of Mercury (II) from Hazardous Wastes Using 1-(2-Thiazolylazo)-2-Naphthol Functionalized Activated Carbon as Solid Phase Extractant. *J. Hazard. Mater.* **2004,** *75*, 113–115.
3. Anirudhan, T. S.; Divya, L.; Ramachandran, M. Mercury (II) Removal from Aqueous Solutions and Wastewaters Using a Novel Cation Exchanger Derived from Coconut Coir Pith and Its Recovery. *J. Hazard. Mater.* **2008,** *157*, 620–627.
4. Donia, A. M.; Atia, A. A.; Heniesh, A. M. Efficient Removal of Hg (II) Using Magnetic Chelating Resin Derived from Copolymerization of Bisthiourea/Thiourea/Glutaraldehyde. *Sep. Purif. Technol.* **2008,** *60*, 46–53.
5. Krishnan, K. A.; Anirudhan, T. S. Removal of Mercury (II) from Aqueous Solutions and Chlor-alkali Industry Effluent by Steam Activated and Sulphurised Activated Carbons Prepared from Bagasse Pith: Kinetics and Equilibrium Studies. *J. Hazard. Mater.* **2002,** *92*, 161–183.
6. Namasivayam, C.; Kadirvelu, K. Uptake of Mercury (II) from Wastewater by Activated Carbon from an Unwanted Agricultural Solid By-product: Coirpith. *Carbon* **1999,** *37*, 79–84.
7. Malachowski, L.; Stair, J. L.; Holcombe, J. A. Immobilized Peptides/Amino Acids on Solid Supports for Metal Remediation. *Pure Appl. Chem.* **2004,** *76*, 777–783.

8. Light, A.; Frater, R.; Kimmel, J. R.; Smith, E. L. Current Status of the Structure of Papain: The Linear Sequence, Active Sulfhydryl Group and the Disulfide Bridges. *Biochemistry* **1964,** *52*, 1276–1283.
9. Dutta, S.; Bhattacharyya, A.; De, P.; Ray, P.; Basu, S. Removal of Mercury from its Aqueous Solution Using Charcoal Immobilized Papain (*CIP*). *J. Hazard. Mater.* **2009,** *172*, 888–896.
10. Agarwal, H.; Sharma, D.; Sindhu, S. K.; Tyagi, S.; Ikram, S. Removal of Mercury from Wastewater Use of Green Adsorbents—A Review. *EJEAF. Che.* **2010,** *9*, 1551–1558.
11. Azhar, S. S.; Liew, A. G.; Suhardy, D.; Hafiz, K. F.; Hatim, M. D. I. Dye Removal from Aqueous Solution by Using Adsorption on Treated Sugarcane Bagasse. *Am. J. Appl. Sci.* **2005,** *2*, 1499–1503.
12. Bhatnagar, A.; Sillanpää, M. Utilization of Agro-industrial and Municipal Waste Materials as Potential Adsorbents for Water Treatment—A Review. *Chem. Eng. J.* **2010,** *157*, 277–296.
13. Dutta, S.; Basu, J. K.; Ghar, R. N. Studies on Adsorption of p-nitrophenol on Charred Saw-dust. *Sep. Purif. Technol.* **2001,** *21*, 227–235.
14. Dutta, S.; Bhattacharyya, A.; Ganguly, A.; Gupta, S.; Basu, S. Application of Response Surface Methodology for Preparation of Low-cost Adsorbent from Citrus Fruit Peel and for Removal of Methylene Blue. *Desalination* **2011,** *275*, 26–36.
15. Sousa, F. W.; Oliveira, A. G.; Ribeiro, J. P.; Rosa, M. F.; Keukeleire, D.; Nascimento, R. F. Green Coconut Shells Applied as Adsorbent for Removal of Toxic Metal Ions Using Fixed-bed Column Technology. *J. Environ. Manage.* **2010,** *91*, 1634–1640.
16. El-Said, A. G.; Badawy, N. A.; Garamon, S. E. Adsorption of Cadmium (II) and Mercury (II) onto Natural Adsorbent Rice Husk Ash (RHA) from Aqueous Solutions: Study in Single and Binary System. *Am. J. Sci.* **2010,** *6*, 400–409.
17. El-Shafey, E. I. Removal of Zn (II) and Hg (II) from Aqueous Solution on a Carbonaceous Sorbent Chemically Prepared from Rice Husk. *J. Hazard. Mater.* **2010,** *175*, 319–327.
18. Ghorbani, M.; Lashkenari, M. S.; Eisazadeh, H. Application of Polyaniline Nanocomposite Coated on Rice Husk Ash for Removal of Hg (II) from Aqueous Media. *Synth. Met.* **2011,** *161*, 1430–1433.
19. Karacan, F.; Ozden, U.; Karacan, S. Optimization of Manufacturing Conditions for Activated Carbon from Turkish Lignite by Chemical Activation Using Response Surface Methodology. *Appl. Therm. Eng.* **2007,** *27*, 1212–1218.
20. Ngah, W. S. W.; Hanafiah, M. A. K. M. Removal of Heavy Metal Ions from Wastewater by Chemically Modified Plant Wastes as Adsorbents: A Review. *Bioresour. Technol.* **2008,** *99*, 3935–3948.
21. Ravikumar, K.; Krishnan, K. S.; Ramalingam, S.; Balu, K. Optimization of Process Variables by the Application of Response Surface Methodology for Dye Removal Using a Novel Adsorbent. *Dyes Pigm.* **2007,** *72*, 66–74.
22. Sahu, J. N.; Acharya, J.; Meikap, B. C. Response Surface Modeling and Optimization of Chromium (VI) Removal from Aqueous Solution Using Tamarind Wood Activated Carbon in Batch Process. *J. Hazard. Mater.* **2009,** *172*, 818–825.

23. Zvinowanda, C. M.; Okonkwo, J. O.; Shabalala, P. N.; Agyei, N. M. A Novel Adsorbent for Heavy Metal Remediation in Aqueous Environments. *Int. J. Environ. Sci. Tech.* **2009,** *6*, 425–434.
24. Gardea-Torresdey, J. L.; Tiemann, K. J.; Armendariz, V.; Bess-Oberto, L.; Chianelli, R. R.; Rios, J.; Parsond, J. G.; Gamez, G. Characterization of Cr(VI) Binding and Reduction to Cr(III) by the Agricultural by Products of Avena Monida (Oat) Biomass. *J. Hazard. Mater.* **2000,** *80*, 175–188.
25. Lalvani, S. B.; Wiltowski, T.; Hubner, A.; Weston, A.; Mandlich, N. Removal of Hexavalent Chromium and Metal Cations by a Selective and Novel Carbon Adsorbent. *Carbon* **1998,** *36*, 1219–1226.
26. Arson, R. Papain. *Methods in Enzymology*; Perlman, G. E.; Lorand L., Eds.; Academic Press: New York, 1970.
27. Ravikumar, K.; Ramalingam, S.; Krishnan, S.; Balu, K. Application of Response Surface Methodology to Optimize the Process Variables for Reactive Red and Acid Brown Dye Removal Using a Novel Adsorbent. *Dyes Pigm.* **2006,** *70*, 18–26.
28. Al-Asheh, S.; Banat, F.; Masad, A. Physical and Chemical Activation of Pyrolyzed Oil Shale Residue for the Adsorption of Phenol from Aqueous Solutions. *Environ. Geol.* **2003,** *44*, 333–342.

CHAPTER 12

STUDIES ON THE EFFECT OF COCONUT PITH ON THERMOPLASTIC VULCANIZATES BASED ON RECYCLED POLYPROPYLENE/RECLAIMED ETHYLENE PROPYLENE DIENE RUBBER COMPOSITES

ABITHA V. K.[1*], AJAY VASUDEO RANE[2], K. KANNY[2], and SABU THOMAS[1]

[1]*School of Chemical Sciences, Mahatma Gandhi University, Kerala, India*

[2]*Department of Mechanical Engineering, Durban University of Technology, Durban, South Africa*

[*]*Corresponding author. E-mail: abithavk@gmail.com*

CONTENTS

ABSTRACT

The concern for environment and sustainable growth has created more awareness among the researchers to develop composites based on recycled materials and materials from nature. Reuse and recycling extends the useful life of the raw material resources in instances where a market exists for the recycled products, and it is economical to carry out collection and reprocessing. Recycling and reuse of materials have been getting more interest nowadays. Ethylene propylene diene monomer (EPDM)/ polypropylene (PP) thermoplastic vulcanizates are most commonly used ones. In the current study, reclaimed EPDM (REPDM) and recycled PP (RPP) thermoplastic vulcanizates were prepared by melt blending method. REPDM and RPP has been melt blended in a Brabender plasticorder in different ratios and the blend with optimized results was selected, as well as RPP/REPDM coconut pith composites have been prepared and developed for cost-effective footwear applications. A comparative study of composites in presence and absence of maleic anhydride-grafted EPDM (MA-g-EPDM) was carried out and the effectiveness of the properties was determined. Coconut pith in different ratios was added into REPDM/ RPP blends along with MA-g-EPDM and other additives. Physicomechanical and thermal properties of the coconut pith REPDM/RPP composite have been carried out. The dispersion of coconut pith in the matrix has been carried out by scanning electron microscopy (SEM) analysis. It was found that the mechanical and thermal properties are increasing with the addition of compatibilizer as compared to the control one. SEM analysis showed that the compatibility between coconut pith RPP and REPDM has been improved by the addition of MA-g-EPDM as a compatibilizer and hence, can be used as a low-cost filler for footwear soling application.

12.1 INTRODUCTION

Population growth in large urban centers, social, and technological developments and unusual changes in human habits have led to greatly increasing amounts of solid wastes. Thus, nowadays, waste management is one of the most significant issues that modern society deals with. The suitability of polymers for a large number of applications and uses is a consequent result of their important properties, that is, polymers are lightweight, flexible, and versatile, offering many practical benefits to various uses, easy to process in any desirable shape, and available with various types of modified properties. They match optimum design with functional solutions; they are economic to produce in custom-made forms and are extremely durable.[1] Recently the importance of recycling waste materials has been increasing for all industries worldwide. Reuse and recycling extends the useful life of the raw material resources in instances where a market exists for the recycled products, and it is economical to carry out collection and reprocessing. The concern for environment and sustainable growth has created more awareness among the researchers to develop composites based on recycled polymeric materials and materials from nature.[2] Plastics materials are one of the major components of global municipal solid waste and as such it presents a promising raw material source for styrene-butadiene thermoplastic elastomers (TPE's). Using recycled material rather than virgin material provides an additional market for recycled materials, thereby helping to reduce the burden of waste disposal in landfills.[3] Today there is a growing importance in producing composite products from recycled materials. Although most research in this area has been concentrated on the use of homogenous polymer waste, heterogeneous material research has not received much attention. Using a postconsumer polymer to produce reinforced composites has economic and environmental importance.[4] During the recycling process, the material undergoes various operations that bring out several modifications in the molecular structure. As a matter of fact, the mechanical properties of the recycled products and their structural organization are quite different compared to those composed of virgin material. So the studies in recycled materials are important. Composites are combinations of two or more than two materials in which one of the materials is reinforcing phase (fibers, sheets, or particles) and the other is matrix phase (polymer, metal, or ceramic). Composite materials are usually classified by type of reinforcement, such as polymer composites, cement-, and metal–matrix composites.[5,6] Polymer–matrix composites are mostly commercially produced composites in which resin is used as matrix with different reinforcing materials. Polymer matrix can be classified as given in Figure 12.1:

Polymer (resin) is classified in two types: thermoplastics [polyethylene (PE), polypropylene (PP), polyether ether ketone (PEEK), polyvinyl chloride (PVC), polystyrene (PS), polyolefin, etc.] and thermosets (epoxy, polyester, and phenol formaldehyde resin, rubbers, etc.) which reinforce different types of fibers such as natural (plant, animal, and mineral) and man-made fiber for different applications. In metal matrix composites, metal is one of important part of element and other part may be metal, ceramic, or organic compounds. Cement matrix composites are made up of cement and with aggregate and basically used in building applications. Reinforcing constituents in composites, as the word indicates, provide the strength that makes the composite what it is. But they also serve certain additional purposes such as heat resistance or conduction, resistance to corrosion, and provide rigidity. Reinforcement can be made to perform all or one of these functions as per the requirements. Reinforcements for the composites can be fibers, fabrics particles, or whiskers. The classifications of reinforcement are as shown in Figure 12.2.

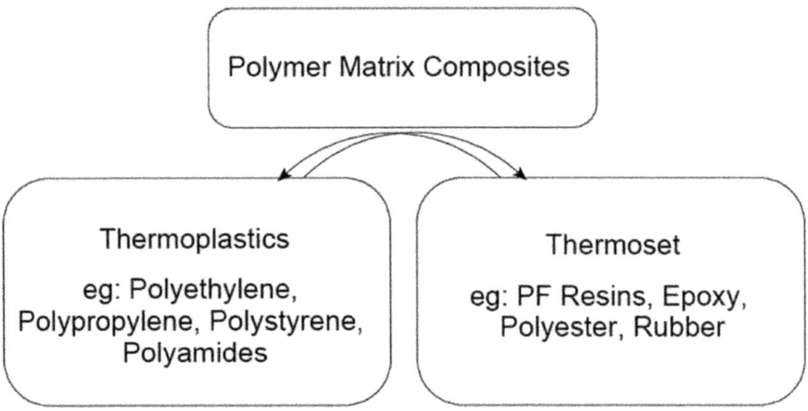

FIGURE 12.1 Classification of polymer matrices.

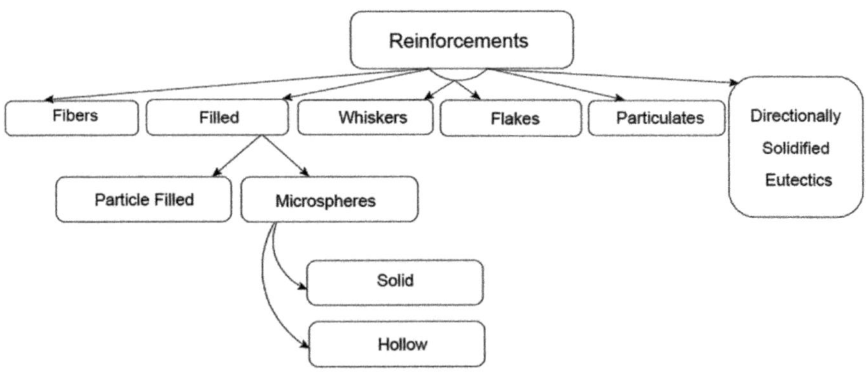

FIGURE 12.2 Classification of reinforcements.

Fibers are the important class of reinforcements, as they satisfy the desired conditions and transfer strength to the matrix constituent influencing and enhancing their properties as desired. Due to increase in population, natural resources are being exploited substantially as an alternative to synthetic materials. Due to this, the utilization of natural fibers for the reinforcement of the composites has received increasing attention. Natural fibers have many remarkable advantages over synthetic fibers. Nowadays, various types of natural fibers have been investigated by Utracki and Wilkie[7] for use in composites including flax, hemp, jute straw, wood, rice husk, wheat, barley, oats, rye, cane (sugar and bamboo), grass, reeds, kenaf, ramie, oil palm, sisal, coir, water hyacinth, pennywort, kapok, paper mulberry, banana fiber, pineapple leaf fiber, and papyrus.[7]

12.1.1 ADVANTAGES OF COMPOSITES

- High resistance to fatigue and corrosion degradation.
- High "strength or stiffness to weight" ratio. As enumerated above, weight savings are significant ranging from 25% to 45% of the weight of conventional metallic designs.
- Due to greater reliability, there are fewer inspections and structural repairs, directional tailoring capabilities to meet the design requirements. The fiber pattern can be laid in a manner that will tailor the structure to efficiently sustain the applied loads.
- Improved dent resistance is normally achieved. Composite panels do not sustain damage as easily as thin-gage sheet metals.
- High resistance to impact damage.
- Composites are dimensionally stable, that is, they have low thermal conductivity and low coefficient of thermal expansion. Composite materials can be tailored to comply with a broad range of thermal expansion design requirements and to minimize thermal stresses.
- Close tolerances can be achieved without machining.
- The improved weatherability of composites in a marine environment as well as their corrosion resistance and durability reduce the downtime for maintenance.

12.1.2 LIMITATIONS OF COMPOSITES

Some of the associated disadvantages of advanced composites are as follows:

- High cost of raw materials and fabrication.
- Composites are more brittle than wrought metals and thus are more easily damaged.
- Transverse properties may be weak.
- Matrix is weak, therefore, low toughness.
- Reuse and disposal may be difficult.
- Difficult to attach.
- Materials require refrigerated transport and storage and have limited shelf life.
- Hot curing is necessary in many cases requiring special tooling.
- Hot or cold curing takes time.
- Analysis is difficult.
- Matrix is subject to environmental degradation.

12.1.3 SCOPE

Coconut pith have gained much interest among technologists and scientist for applications in civil, military, industrial, space craft, and biomedical sectors. In the past two decades, growing interest for natural fibers composites has resulted in extensive research. The driving forces are (1) cost reduction, (2) weight reduction, and (3) marketing (application of renewable materials). Technical requirements were of less importance; hence, application remained limited to nonstructural parts for a long time. The reason for this is the traditional shortcomings of coconut pith composites, the low impact resistance and moist degradation. Recent research, however, showed that significant improvements in the properties of the blends of polyolefenic and thermoset rubber are observed if the coconut pith is used as a reinforcing material.

This benefits offered by using coconut pith in the blends are:

- Economic: lower costs on account of significantly reduced cycle times, energy savings.
- Technical: mechanical properties identical to those of traditional reinforcements, reduced tool wear and tear, high geometric stability of the manufactured parts, and good insulation characteristics.

- Environmental: renewable resource, easy to recycle, no material toxicity, reduced fossil fuels content, and CO_2-neutral materials.

The year 2009 has been assigned by the UN to be the international year of natural fibers. Natural fiber industries employ millions of people all over the world, especially in the developing countries. As the major nonfood commodity, natural fibers and their products are processed in many small and large industries and consumers all over the world profit from the provided products. Development of a sustainable global economy, which permits improving purchasing power and living standards without exhaustion of resources for future generations, requires a fundamental change in attitude. Competitive products based on renewable resources need to be developed that have high quality, show excellent technical performance, and harm the environment less than current products based on petrochemical materials.

The present research is concerned with the environmental impact of coconut pith composite production and industrial applications, highlights the need to comprehensively evaluate the environmental benefits that emanate from the use of coconut pith, as well as the possibility of utilizing the existing research results for promoting coconut pith.

12.1.4 OBJECTIVES

- Optimizing the blend ratio of recycled PP (RPP)/reclaimed EPDM (REPDM) on the basis of mechanical properties such as flexing, impact, abrasion resistance for low-cost footwear (sole) application.
- Also influence of maleic anhydride-grafted EPDM (MA-g-EPDM) on RPP/REPDM blend properties and properties of RPP/ REPDM without MA-g-EPDM are to be evaluated for the comparative study.
- Compatibilization effect due to MA-g-EPDM and coconut pith on mechanical, thermal, and morphological analysis of composites needs to be studied,
- Mixing coconut pith into the optimized blend ratio using different dosages and evaluating the properties.
- MA-g-EPDM was made using the reactive process and added as a compatibilizer in RPP/REPDM composites.

12.2 MATERIALS

RPP was obtained from Chethan Plastics, Mumbai. REPDM with 30% rubber hydrocarbon (RHC) for mixing was obtained from Gujarat Reclaim and Rubber Products Limited. Paraffinic oil was procured from Neelam Lubricants. Coconut pith was obtained from Rubber Park; Ernakulum. Dicumyl peroxide 40 (DCP 40) for dynamic vulcanization was obtained from BP Chemicals.

12.2.1 MACHINERIES

12.2.1.1 BRABENDER MIXER

Brabender is a mixer including a temperature control for blending of polymers flow under selected conditions of shear and temperature. The heart of the Brabender is a jacketed mixing chamber whose volume is approximately 300 cc for the model used. Mixing or shearing of the material in the mixing chamber is done by two horizontal rotors with protrusions. Mixer speed can vary from 0 to 120 min^{-1}. Mixer temperature can be controlled up to 3000°C. Once mixing conditions (rpm and temperature) are set, sufficient time was given for the temperature to attain the set value and become steady.

FIGURE 12.3 Brabender mixer.

12.2.1.2 COMPRESSION MOLDING

Molding is carried out in compression molding machine with water cooling facility and having two zones. Molding pressure is limited up to 50 kg/cm^2. Special features of thermoplastic molding compared to rubber molding:

- There is no curing process, only mold filling and cooling.
- No need of high pressure.
- Wax paper normally uses as mold releasing agent.
- Longer cycle time and includes preheating time, mould filling time, and cooling time.

FIGURE 12.4 Compression molding.

12.2.1.2.1 Optimized Conditions of Molding

The molding parameters such as molding temperature, molding pressure, mold filling time, and cooling time are optimized by trial and error method and are selected as follows:

Molding pressure: 50 kg\cm^2,
Mold filling time: 7 min,
Cool time: 25 min,

Total molding time (1 h) = preheating time (30 min) + mold filling time (7 min) + cooling time (20 min). Molding is carried out for tensile slabs (2-mm thickness) and flexing slabs (4-mm thickness).

12.2.2 TESTING EQUIPMENTS

12.2.2.1 TENSILE TESTING (ASTM D 638)

Tensile tests measure the force required to break a specimen and the extent to which the specimen stretches or elongates to that breaking point. The data are often used to specify material, to design parts to withstand application forces, and as a quality control check of materials. Tensile testing is based on ASTM D 638. The thickness of test specimens for hardness was about 6 mm.

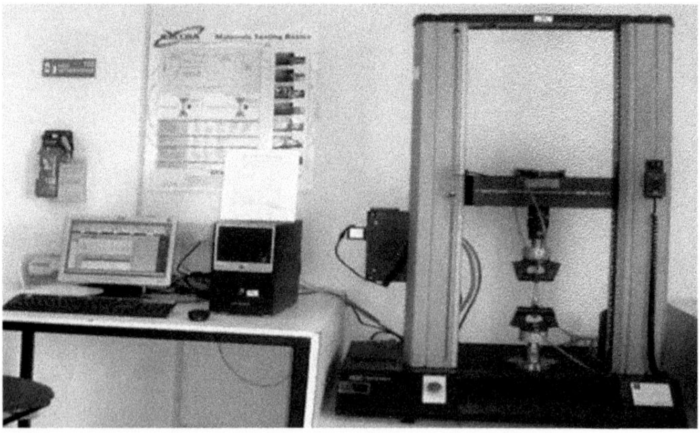

FIGURE 12.5 Tensile testing machine.

Place specimens in the grips of the universal testing machine at a specified gauge length and pull until failure. The testing speed is determined by the material specification. An extensometer can also be attached to test specimen to determine the elongation and modulus.

Data: The following calculations are the most common results given:

- Tensile strength (at yield and at break).
- Tensile modulus (this is stress at a given % of strain.
- Elongation at break.

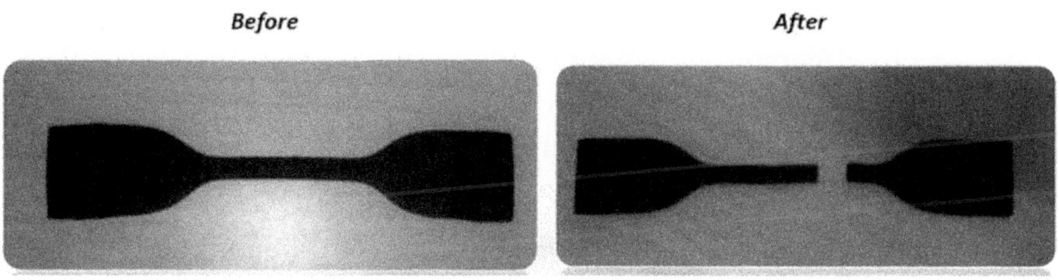

FIGURE 12.6 Specimens of tensile strength before and after.

12.2.2.2 TEAR STRENGTH (ASTM D 624)

Vulcanized rubber often fails in service due to the generation and propagation of a special type of rupture called a tear. This test method measures the resistance to tearing action. Tear strength may be influenced to a large degree by stress-induced anisotropy (mechanical fibering), stress distribution, strain rate, and test piece size. The results obtained in a tear strength test can only be regarded as a measure under the conditions of that particular test and may not have any direct relation to service performance. The significance of tear testing must be determined on an individual application or product performance basis. The sample is die cut into the desired size and test is carried out in a universal testing machine; the force required to tear the test sample is taken as tear strength.

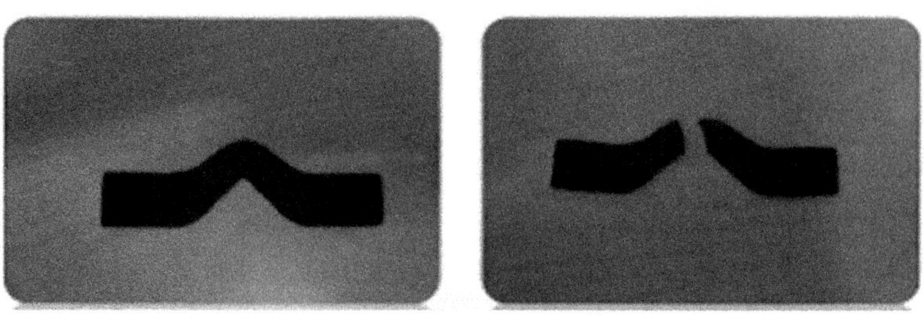

FIGURE 12.7 Specimens of tear strength before and after.

12.2.2.3 HARDNESS (ASTM D 2240)

Elastomeric materials are usually measured with either a Shore A Scale Durometer. These tests are designed for use with samples approximately 6 mm thick, and a surface area sufficient to permit at least three test points 5 mm apart, 13 mm from any edge. This test method is based on the penetration of a specific type of indenter when forced into the material under specified conditions. The indentation hardness is inversely related to the penetration and is dependent on the elastic modulus and viscoelastic behavior of the material. The geometry of the indentor and the applied

FIGURE 12.8 Durometer (Shore A).

force influence the measurements such that no simple relationship exists between the measurements obtained with one type of durometer and those obtained with another type of durometer or other instruments used for measuring hardness. This test method is an empirical test intended primarily for control purposes. No simple relationship exists between indentation hardness determined by this test method and any fundamental property of the material tested. Micro hardness testers, Shore "A" Micro (M) or IRHD Micro (M), are utilized to measure samples which are too small for testing with macro instruments. The Shore type instrument is a spring-loaded indentation device, in which, values are obtained as a function of the viscoelastic property of the material. The truncated cone indentor extends 0.098 in (2.5 mm) and is pressed onto the sample against an 822 g spring. Each 0.001 in of deflection of the indenter is shown as 1° Shore (A), therefore, the harder the material, the more the deflection and the higher the number. Materials reading below 10° and above 90° can be tested using Shore O and D scales, respectively.

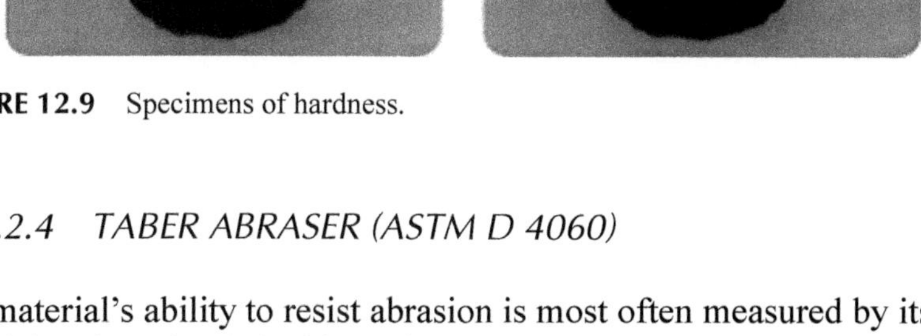

FIGURE 12.9 Specimens of hardness.

12.2.2.4 TABER ABRASER (ASTM D 4060)

The material's ability to resist abrasion is most often measured by its loss in weight when abraded with an abraser. The most widely accepted abraser in the industry is called the Taber abraser. A variety of wheels with varying degree of abrasiveness are available. The grade of "calibrase" wheel designated CS-17 with 1000-g load seems to produce satisfactory results with almost all plastics. For softer materials, less abrasive wheels with a smaller load on the wheels may be used. The test specimen is usually a 4-in diameter disc or a 4-in^2 plate having both surfaces substantially plane and parallel. A 1/2-in diameter hole is drilled in the center. Specimens are conditioned employing standard conditioning practices prior to testing. To commence testing, the test specimen is placed on a revolving turntable. Suitable abrading wheels are placed on the specimen under certain set dead weight loads. The turntable is started and an automatic counter records the number of revolutions. Most tests are carried out to at least 5000 revolutions. The specimens are weighed to the nearest milligram. The test results are reported as weight loss in milligrams per 1000 cycles. The grade of abrasive wheel along with amount of load at which the test was carried out is always reported along with results.

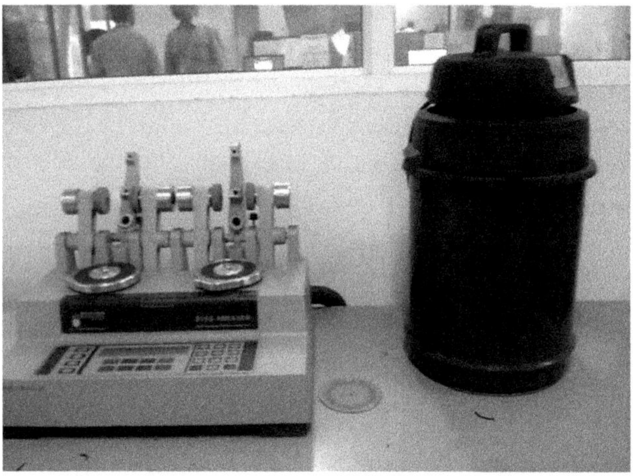

FIGURE 12.10 Taber abraser.

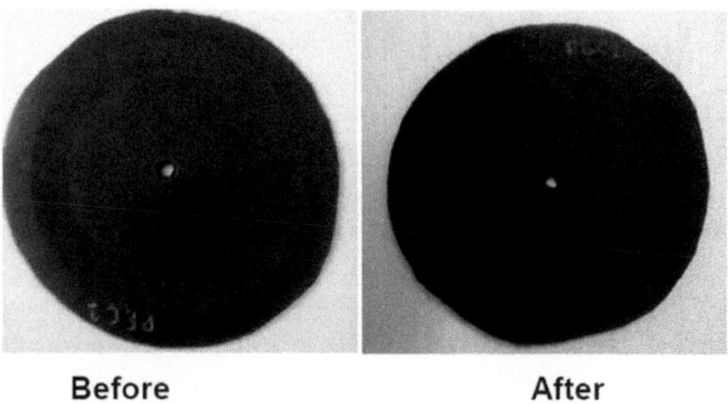

FIGURE 12.11 Specimens of abrasion (before and after).

Wear Index—Compute the wear index, I, of a test specimen as follows:

$$I = \frac{A - B}{C}$$

where:

A = weight of test specimen before abrasion, mg,
B = weight of test specimen after abrasion, mg, and
C = number of cycles of abrasion recorded.

Weight Loss—Compute weight loss, L, of the test specimen as follows:
$L = A - B$

where:
A = weight of test specimen before abrasion, mg, and
B = weight of test specimen after abrasion, mg.

12.2.2.5 ROSS FLEXING (ASTM D 1052)

The test gives an estimate of the ability of rubber vulcanizates to resist crack growth of a pierced specimen when subjected to bend flexing. The machine, as illustrated in Figure 12.12 below, allows one end of the test specimen to be clamped firmly to a holder arm while the pierced end is placed between two rollers that must permit a free bending movement of the test specimen during the test. During each cycle, the pierced area of the test specimen is bent freely over a 10-mm (0.4 in) diameter rod through a 90° angle. The machine shall operate at 1.7 ± 0.08 Hz (100 ± 5 cpm). At least two, preferably three, test specimens of each sample shall be tested simultaneously. The test specimens shall be 25 ± 1 mm (1.00 ± 0.05 in) in width, a minimum of 152 mm (6.0 in) in length, and 6.35 ± 0.03 mm (0.25 ± 0.01 in) in thickness, and shall be cut from a vulcanized sheet 6.35 ± 0.03 mm (0.25 ± 0.01 in) in thickness and of suitable dimensions or from finished articles by cutting and buffing. If obtained from a manufactured article, the piece of rubber shall be free of surface roughness and fabric layers.

Start the machine and record the number of cycles by the use of a counter. Make frequent observations, recording the number of cycles and the increase in cut length, measured to the nearest 0.5 mm for the purpose of determining the rate of increase in cut length. When observing cut growth, the holder arm shall be at an angle approximately 45° from the vertical. The test shall be continued until the cut length has increased 500%, that is, until the combined length of the cut and crack has increased to a total of 15.0 mm (0.60 in) or when 250 kc has been reached with slow cracking samples.

FIGURE 12.12 Ross flex tester.

12.2.2.6 IZOD IMPACT STRENGTH (ASTM D 256)

The objective of the Izod–Charpy impact test is to measure the relative susceptibility of a standard test specimen to the pendulum-type impact load. The results are expressed in terms of kinetic energy consumed by the pendulum in order to break the specimen. The energy required to break a standard specimen is actually the sum of energies needed to deform it, to initiate its fracture, and to propagate the fracture across it and the energy expended in tossing the broken ends of the specimen. This is called the "toss factor." The energy lost through the friction and vibration of the apparatus is minimal for all practical purposes and usually neglected. The specimen used in Izod test must be notched. The reason for notching the specimen is to provide a stress concentration area that promotes a brittle rather than a ductile failure. The specimen is held as a vertical cantilever beam and is broken by a single swing of the pendulum. The line of initial contact is at a fixed distance from the specimen clamp and from the centerline of the notch and on the same face as the notch. For a given system, greater stress concentration results in higher localized rates-of-strain. Since the effect of strain-rate on energy-to-break varies among materials, a measure of this effect may be obtained by testing specimens with different notch radii. In the Izod-type test, it has been demonstrated that the function, energy-to-break versus notch radius, is reasonably linear from a radius of 0.03 to 2.5 mm (0.001 to 0.100 in), provided that all specimens have the same type of break. The excess energy pendulum impact test indicates the energy to break standard test specimens of specified size under stipulated parameters of specimen mounting, notching, and pendulum velocity-at-impact.

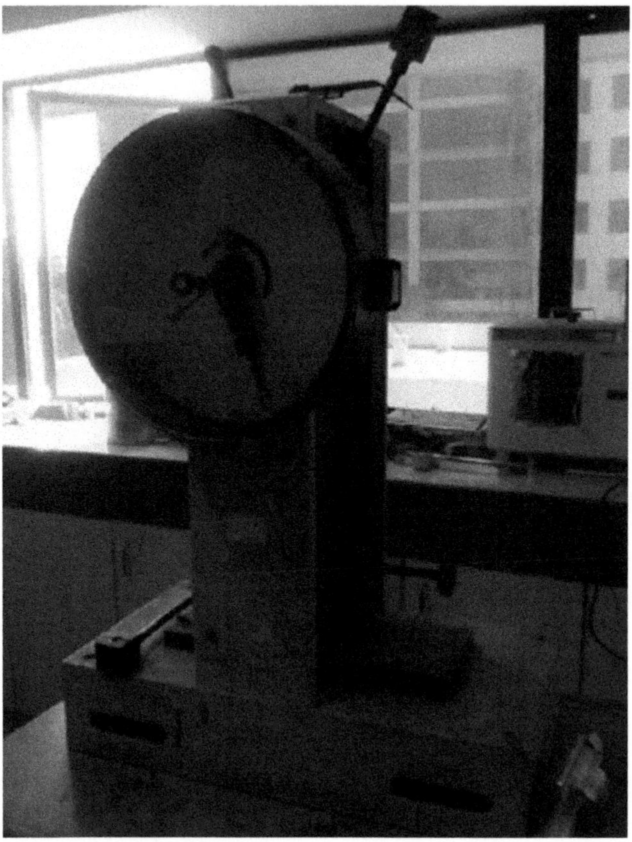

FIGURE 12.13 Izod impact tester.

12.2.2.7 THERMO GRAVIMETRIC ANALYSIS

Thermo gravimetric (TG) analysis provides determination of endotherms, exotherms, and weight loss on heating, cooling, and more. Materials analyzed by TGA include polymers, plastics, composites, laminates, adhesives, food, coatings, pharmaceuticals, organic materials, rubber, petroleum, chemicals, and explosives. TG analysis uses heat to force reactions and physical changes in materials. TGA provides quantitative measurement of mass change in materials associated with transition and thermal degradation. TGA records change in mass from dehydration, decomposition, and oxidation of a sample with time and temperature. Characteristic TG curves are given for specific materials and chemical compounds due to unique sequence from physicochemical reactions occurring over specific temperature ranges and heating rates. These unique characteristics are related to the molecular structure of the sample.

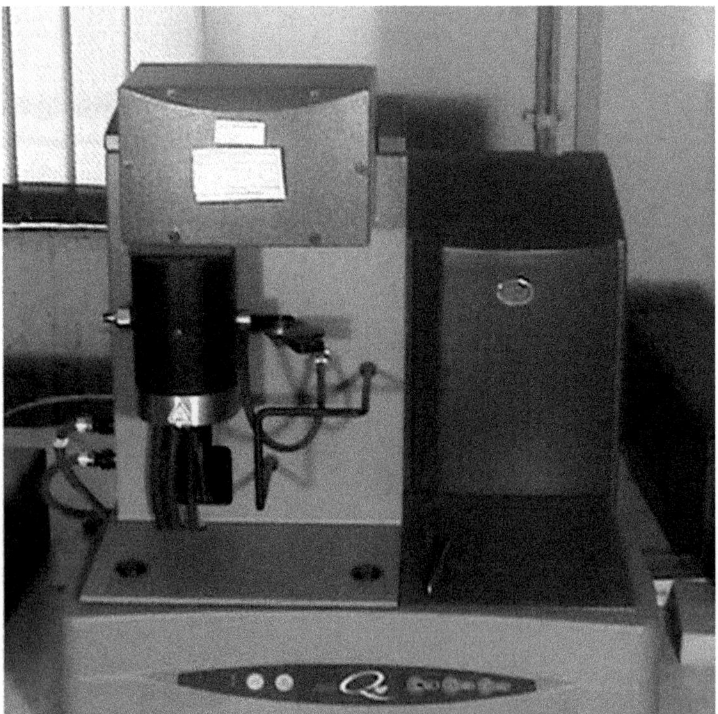

FIGURE 12.14 Thermal gravimetric analyzer.

12.2.2.8 SCANNING ELECTRON MICROSCOPY

The scanning electron microscopy (SEM) is an instrument that produces a largely magnified image by using electrons instead of light to form an image. A beam of electrons is produced at the top of the microscope by an electron gun. The electron beam follows a vertical path through the microscope, which is held within a vacuum. The beam travels through electromagnetic fields and lenses, which focus the beam down toward the sample.

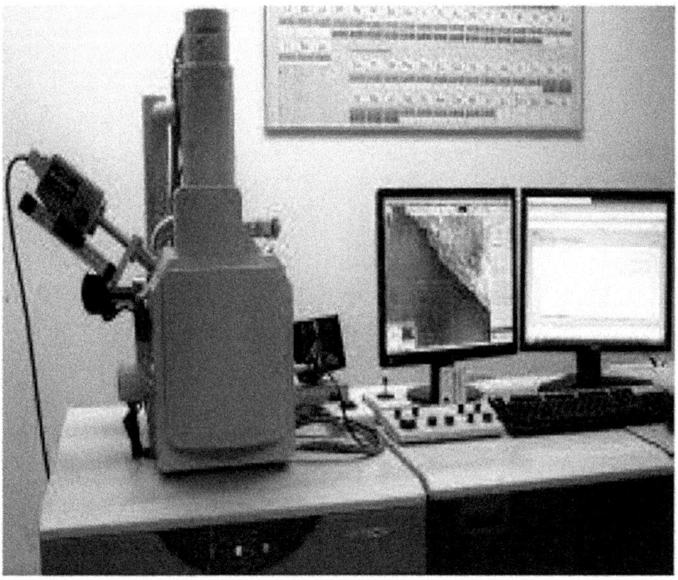

FIGURE 12.15 Scanning electron microscope.

12.3 METHODS

12.3.1 BLEND PREPARATION AND OPTIMIZATION

The materials used for preparation of blend are RPP, REPDM. The formulations for blend preparation are given in Table 12.1. The materials are melt mixed in Brabender plasticorder at 180°C and at a speed of 50 rpm for 10 min. After melt mixing, the molten mass was taken out from the laboratory mixer and while hot, passed through two-roll mixing mill to chill it and sheet it to about 2-mm thick. The sheet was then cut and press-molded for

7 min in a compression molding machine hydraulic press at 18°C, under specified pressure. Silicone wax paper was placed between the sheet and the press plates to avoid adhesiveness. The sheet was then cooled down to room temperature still under pressure. The test specimens were die-cut from the compression molded sheet and used for measuring mechanical properties after 24 h of conditioning at room temperature. The specimen for abrasion resistance was prepared from 6″ × 6″ sheet.

TABLE 12.1 Selection and Optimization of Recycled Polypropylene for Blend.

Recycled polypropylene	90	80	70	60
Reclaimed EPDM	10	20	30	40

On the basis of results obtained (Table 12.2) by flexing test, impact test, and abrasion resistance of different blend ratios, a graph was plotted against the blend ratios and the obtained values and optimum value of the blending polymers were determined. From the graph theory, it is clear that the optimum value is obtained at REPDM content at 35. Optimum values of the blending polymers were determined and 65:35 (RPP:REPDM) was finalized for the preparation of composites.

TABLE 12.2 Mechanical Properties of RPP/REPDM Blends.

Properties	90/10	80/20	70/30	60/40
Hardness (Shore A)	88	75	70	65
Tensile strength (kg/cm^2)	148	66	41	30
Tear strength (kg/cm)	60	27	16	15
Ross flex	25	58	80	90
Taber abrader (wt loss)	0.1	0.15	0.05	0.04
Taber abrader (wear index)	1×10^{-4}	1.5×10^{-4}	5×10^{-5}	4×10^{-5}
Izod impact (J/m)	60	68	110	170

12.3.2 PREPARATION OF RPP/REPDM COCONUT PITH COMPOSITES

After the optimization study, composite with 65:35 (REPDM:RPP) were prepared at different coconut pith content (0, 10, 20, and 30 phr). Paraffinic oil was used as processing oil and MA-g-EPDM was used as compatibilizer. DCP 40 was used for dynamic vulcanization. Also batches were made by without adding compatibilizer. Compression molding under specified pressure, preparation of test specimen were carried out in a similar manner as described in preparation of blends. Tables 12.3 and 12.4 give the recipe for the preparation of ternary composites without and with MA, respectively. After melt mixing, the molten mass was taken out from the laboratory mixer and while hot, passed through two-roll mixing mill to chill it and sheet it to about 2-mm thick. The sheet was then cut and press molded for 15 min in a compression molding machine hydraulic press at 180°C, under specified pressure. Wax sheet was placed between the sheet and the press plates to avoid adherence to mold surfaces. The sheet was then cooled down to room temperature still under pressure. The test specimens were die-cut from the compression molded sheet and used for measuring mechanical properties after 24 h of conditioning at room temperature.

TABLE 12.3 Recipe of Recycled PP/Reclaimed EPDM/Coconut Pith Composites (Without MA).

Ingredients	PEC00	PEC0	PEC1	PEC2	PEC3
Recycled PP	65	65	65	65	65
Reclaimed EPDM	35	35	35	35	35
MA-g-EPDM	0	0	0	0	0
Coconut pith	0	0	10	20	30
Paraffinic oil	10	10	10	10	10
DCP 40	1.2	1.2	1.2	1.2	1.2

TABLE 12.4 Recipe of Recycled PP/Reclaimed EPDM/Coconut Pith Composites (with MA).

Ingredients	PEC0	PECM0	PECM1	PECM2	PECM3
Recycled PP	65	65	65	65	65
Reclaimed EPDM	35	35	35	35	35
MA-g-EPDM	0	10	10	10	10
Coconut pith	0	0	10	20	30
Paraffinic oil	10	10	10	10	10
DCP 40	1.2	1.2	1.2	1.2	1.2

12.4 RESULTS AND DISCUSSIONS

12.4.1 COMPARATIVE STUDY OF RPP/REPDM COCONUT PITH COMPOSITES WITH AND WITHOUT MA

12.4.1.1 MECHANICAL PROPERTIES

The measurements of tensile properties, tensile strength at break as ASTM D 638 and tear strength ASTM D 624, of samples were carried out by an Instron 1185 tester according to standards, at a crosshead speed of 50 mm/min. Hardness testing was carried out using Shore A durometer, according to ASTM D 2240. The impact strength of the samples was carried out using Izod impact strength tester according to ASTM D 256 standards. Flexing cycles of samples were determined by Ross flex tester using ASTM D 1052 standards and abrasion properties as per ASTM D 4060.

12.4.1.2 HARDNESS

Hardness is the property that is defined as the resistance to indentation, the hardness of the RPP/REPDM coconut pith composites increases as the coconut pith content increases as compared to the blend ratio. After adding MA, there is an increase in hardness values, the increase in hardness is due to the interaction of coconut pith with the matrix, that is, the incorporation of coconut pith filler in the blend is proper and hence there is an increase in hardness due to the addition of filler, on contrary, if grafted EPDM rubber is used, there is also an increase in the hardness, as quantity of rubber required for incorporating more filler is available. With the addition of coconut pith, the hardness values are increased by three units and with the addition of MA-g-EPDM and coconut pith the hardness values are increased by eight units.

TABLE 12.5 Hardness Values (Shore A).

Coconut pith (phr)	Blend	0	10	20	30
Hardness (without maleic anhydride)	67	68	70	75	80
Hardness (with maleic anhydride)	67	70	75	80	82

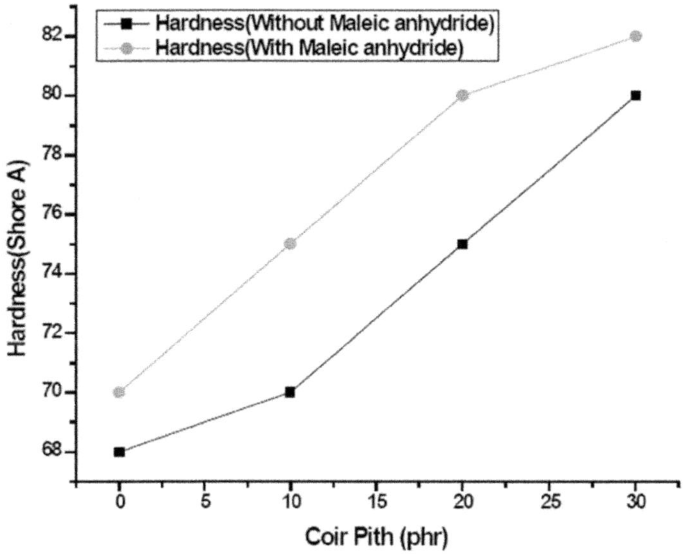

FIGURE 12.16 Comparative chart of hardness.

12.4.1.3 TENSILE STRENGTH

The tensile strength values are decreasing in both the cases. The tensile strength properties of composites with MA show an increasing trend as compared with and without compatibilizer. The decrease in tensile strength are because of the reason that large interphase voids exists between the blending polymers; the large voids are due to the filler in the reclaimed rubber, and herein we are again adding coconut pith as a cheap filler between the blending polymers and increasing the voids to a large dimensions, and hence, addition of coconut pith decreases the tensile strength values. The decrease in tensile strength values may be also due to the increased material stiffness from the interaction of coconut pith into the RPP/REPDM matrix.

TABLE 12.6 Tensile Strength Values (kg/cm^2).

Coconut pith (phr)	Blend	0	10	20	30
Tensile strength (without maleic anhydride)	32	28.2	26.8	26.6	28
Tensile strength (with maleic anhydride)	32	32	27	29	28.9

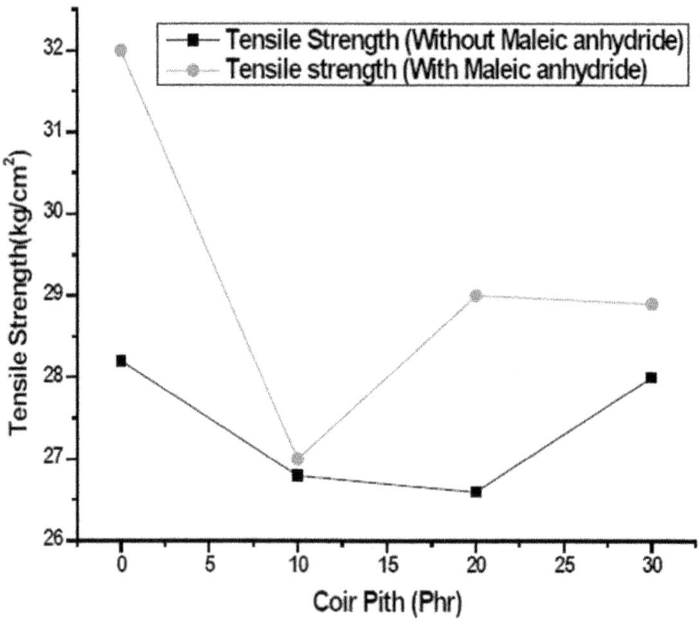

FIGURE 12.17 Comparative chart of tensile strength.

12.4.1.4 TEAR STRENGTH

Tear strength values are increasing with the addition of coconut pith which is advantageous for footwear applications, and it is decreasing after 20 phr. The surface area of coconut pith increases while the concentration increases; due to this, an increase in tear strength is observed. The composite with MA shows a slight improvement in tear properties as compared to the composites without MA. This is because the grafted EPDM improves the interaction of coconut pith with the matrix. The increases in tear strength properties are beneficial to the footwear applications. After 20 phr it shows a decreasing trend because of the poor dispersion of coconut pith in the matrix.

TABLE 12.7 Tear Strength Values (kg/cm).

Coconut pith (phr)	Blend	0	10	20	30
Tear strength (without maleic anhydride)	15	15	15.5	15.85	15
Tear strength (with maleic anhydride)	15	15.24	15.78	16	15.2

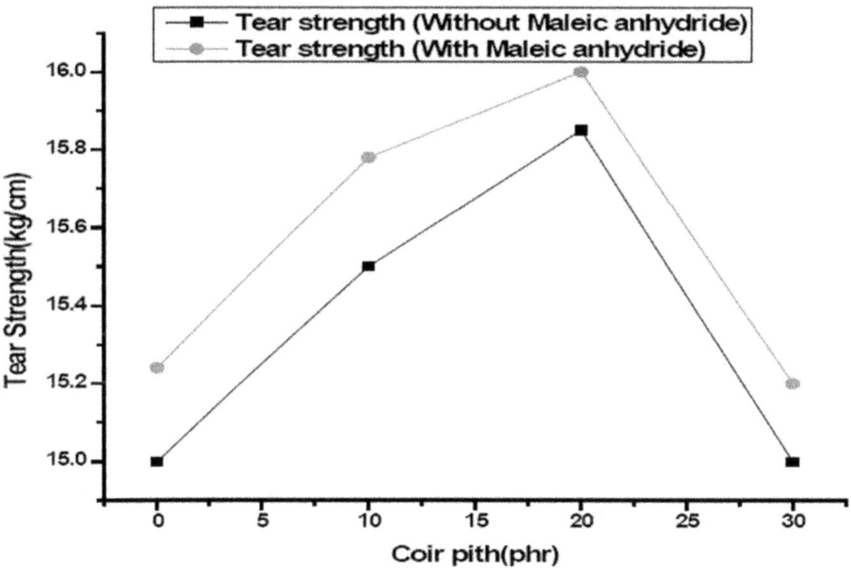

FIGURE 12.18 Comparative chart of tear strength values.

12.4.1.5 ROSS FLEXING

Flexing cycles for the failure of RPP/REPDM composites increased as the coconut pith content increases. After the addition of MA there is an increase in flexing properties as compared with composites without the addition of MA. The failures of samples were occurring either by crack initiation or by crack propagation. Addition of filler helps in interfering with the propagation step. A crack front approaches filler particles which have good adhesion to the matrix. The front of the crack is slowed by filler particles because of their interaction with the tip stress field. Cavitation and coalescence of voids is followed by the matrix breaking away from particles and the crack front progressing to the next obstacle, due to this mechanism, crack propagation slows and flexing cycles increases. At 30 phr coconut pith concentration, there is a decrease in flexing property because at this point, the coconut pith dispersion is poor; hence, instead of reinforcement it acts as a diluent. The improvement in flexing cycles for failure is beneficial for footwear application.

TABLE 12.8 Flexing Cycle's Counts.

Coconut pith (phr)	Blend	0	10	20	30
Flexing cycles (without maleic anhydride)	90	95	111	125	115
Flexing cycles (with maleic anhydride)	90	106	119	130	100

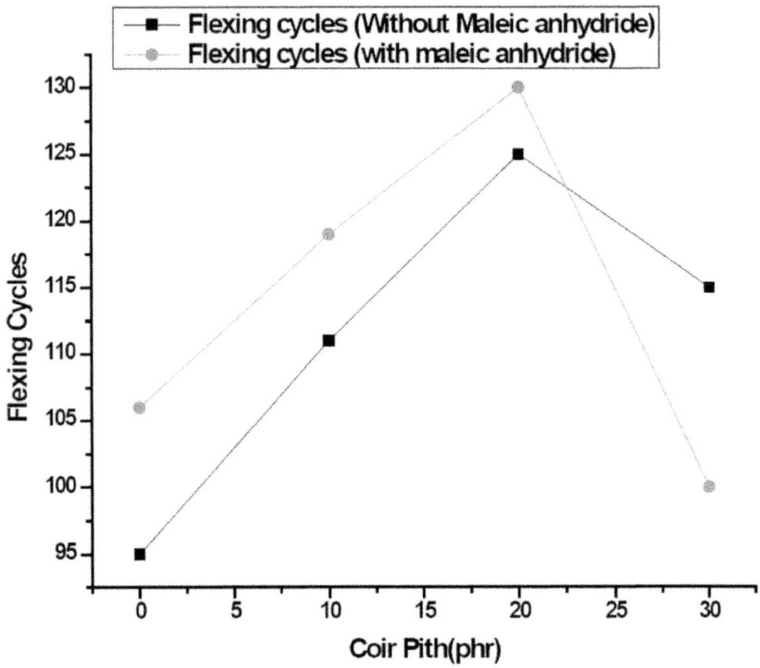

FIGURE 12.19 Comparative chart for counts of flexing cycles.

12.4.1.6 WEIGHT LOSS

As the coconut pith content increases, weight loss increases; hence the abrasion resistance decreases. After the addition of MA as a compatibilizer, there is an increment in weight loss and thus, decrease in abrasion resistance. The weight loss increase because of the poor interaction of coconut pith with the matrix; the rubber phase in the blend wears away more rather than the plastic phase. As compared to the blend properties, the weight loss is negligible. We can take 20 phr as an optimum filler loading.

TABLE 12.9 Abrasion Resistance (Weight Loss Method).

Coconut pith (Phr)	Blend	0	10	20	30
Weight loss (without maleic anhydride)	0.04	0.04	0.07	0.075	0.08
Weight loss (with maleic anhydride)	0.04	0.07	0.08	0.07	0.09

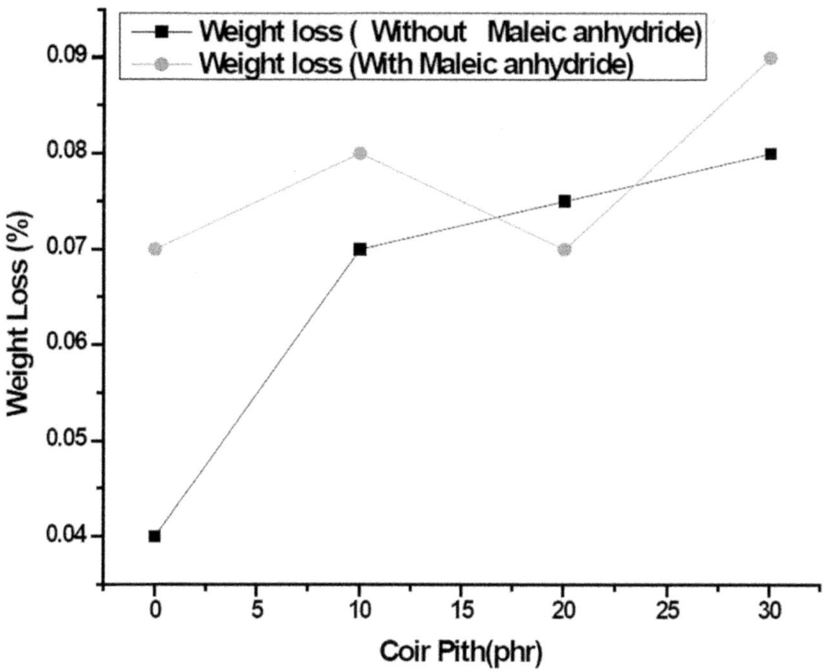

FIGURE 12.20 Comparative chart for weight loss during abrasion.

12.4.1.7 WEAR INDEX

Wear Index values are also important with the abrasion resistance. So, increase in wear index with the addition of coconut pith leads to decrease in abrasion resistance. It is mainly because in reclaimed rubber and recycled plastics, the filler content is more and if we add coconut pith again, fillers cannot incorporate into the matrix. But as comparing to the wear index of the blends, the wear index values are negligible, so we can make use of these composites in footwear applications. After 20 phr, there is an increase in values of wear index. So we can conclude that 20 phr of coconut pith is optimum loading.

TABLE 12.10 Abrasion Resistance (Wear Index Method).

Coconut pith (Phr)	Blend	0	10	20	30
Wear index(without maleic anhydride)	4×10^{-5}	4×10^{-5}	7×10^{-5}	7.5×10^{-5}	8×10^{-5}
Wear index (with maleic anhydride)	4×10^{-5}	7×10^{-5}	8×10^{-5}	7×10^{-5}	9×10^{-5}

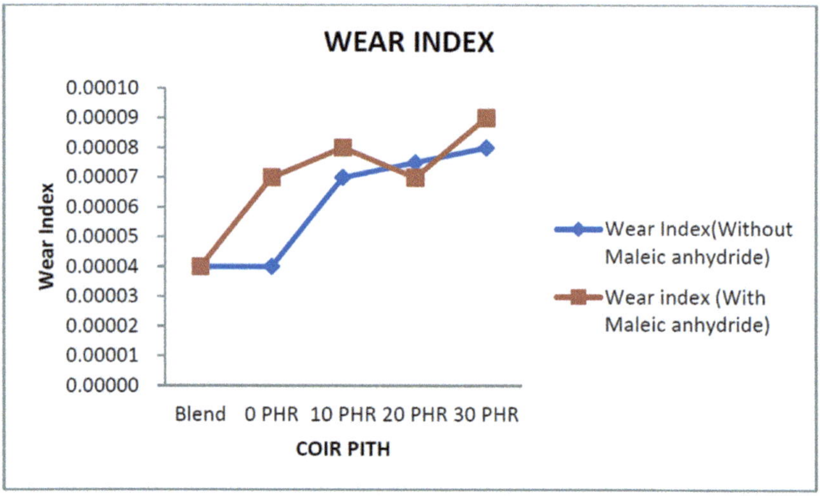

FIGURE 12.21 Comparative chart for wear index during abrasion.

12.4.1.8 IMPACT STRENGTH

The impact properties were increasing after the addition of coconut pith. After 20 phr of coconut pith, the impact values are decreased because of the dilution effect of coconut pith. The MA-g-EPDM acts as a compatibilizer between coconut pith and REPDM/RPP so that the impact property increases. There is a decrease in properties after 20 phr due to the dilution effect of coconut pith. Thus, we can conclude that 20 phr of coconut pith is the optimum loading for getting good impact properties. The increase in impact strength is due to the nucleation effect of coconut pith in the matrix. The size of the coconut pith particles is fine; being fine particle sized, there is proper dispersion although incorporation is difficult, and this proper dispersion also improves the impact strength of the composites.

TABLE 12.11 Impact Strength Values (J/m).

Coconut pith (Phr)	Blend	0	10	20	30
Impact strength (without maleic anhydride)	160	160	165	168	150
Impact strength (with maleic anhydride)	160	165	165	170	155

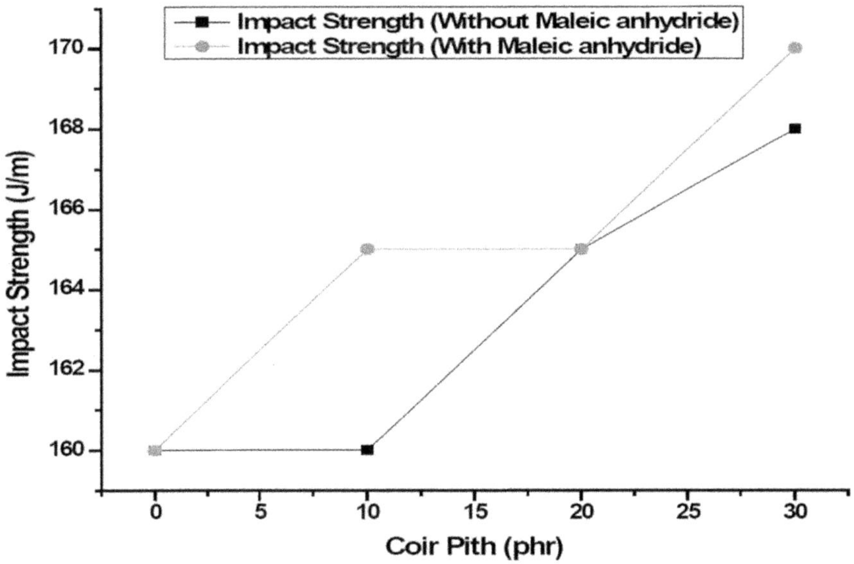

FIGURE 12.22 Comparative chart for impact values of coir composites.

12.4.1.9 THERMAL PROPERTIES

In order to investigate the influence on thermal stability of REPDM/RPP/coconut pith composites, TGA study was carried out. TGA was carried out in Perkin Elmer TGA 400 with a heating rate of 200°C/min. Figure 12.23 below shows the TGA thermographs of REPDM/RPP/coconut pith composites with variable dosage of coconut pith. Thermographs reveal that the onset of degradation of the composites shifts toward a higher temperature on increasing coconut pith concentration in the blend indicating higher thermal stability. After the addition of MA-g-EPDM as a compatibilizer to the REPDM/RPP, the degradation temperature shifts to higher as compared with REPDM/RPP/coconut pith composites.

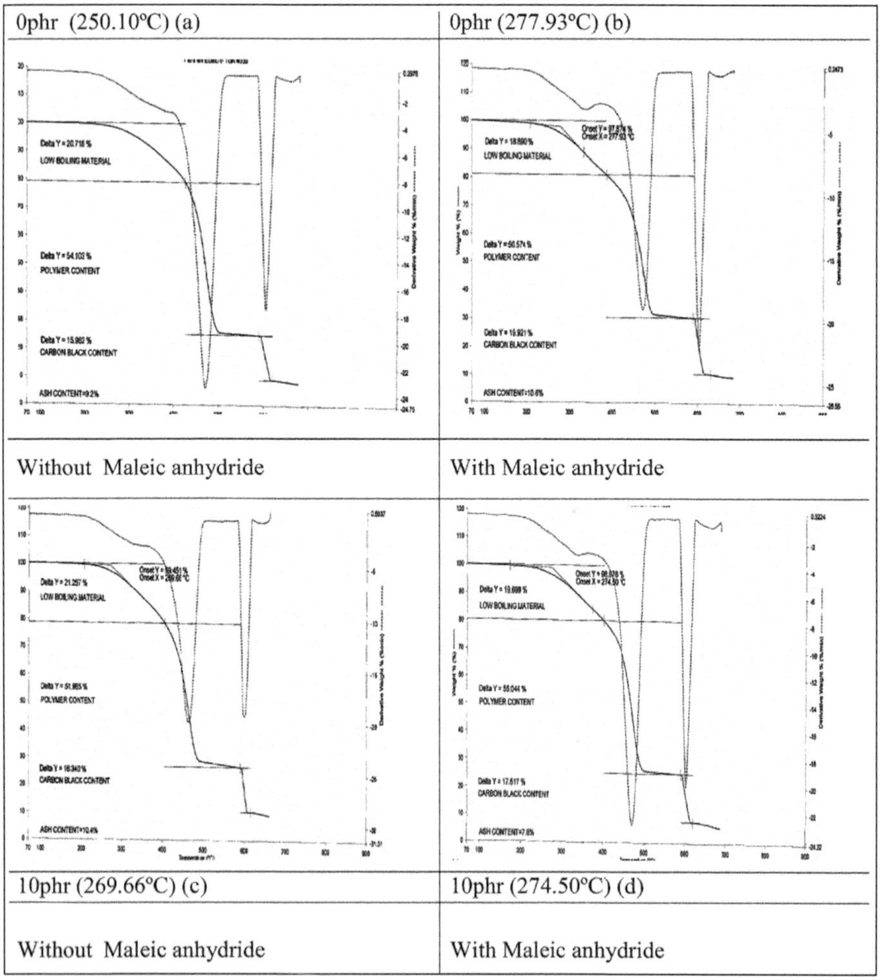

FIGURE 12.23 Graphical representation of 0 and 10 phr coconut pith for with and without MA systems (a, b, c, and d).

The increase in thermal stability of RPP/REPDM–coconut pith composites is due to the higher thermal stability of coconut pith that enhances the thermal properties of RPP/REPDM.

20 phr coconut pith(270.5°C)(e)

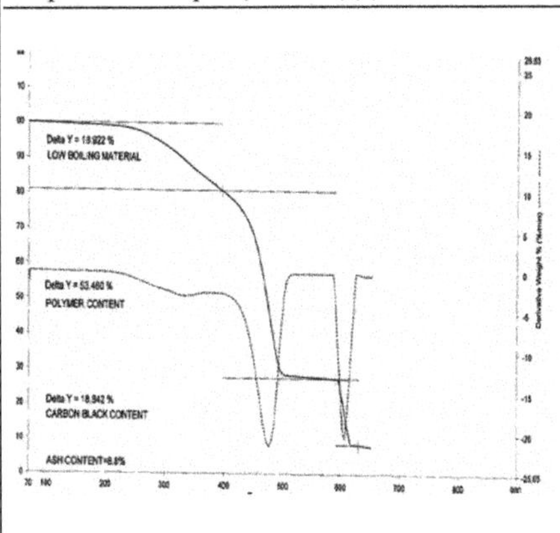

Without maleic anhydride

20 phr coconut pith (280.21°C)(f)

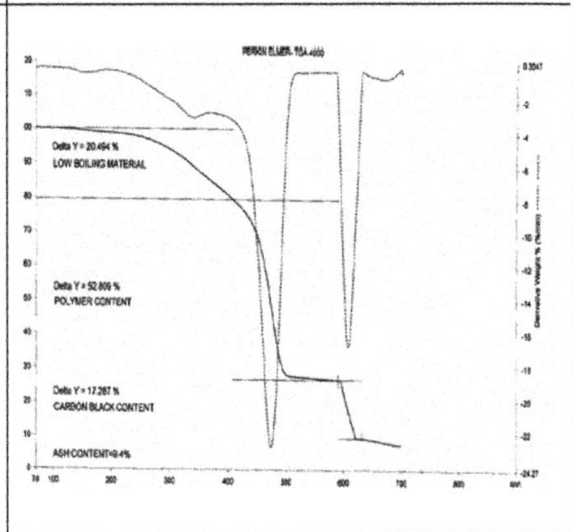

With maleic anhydride

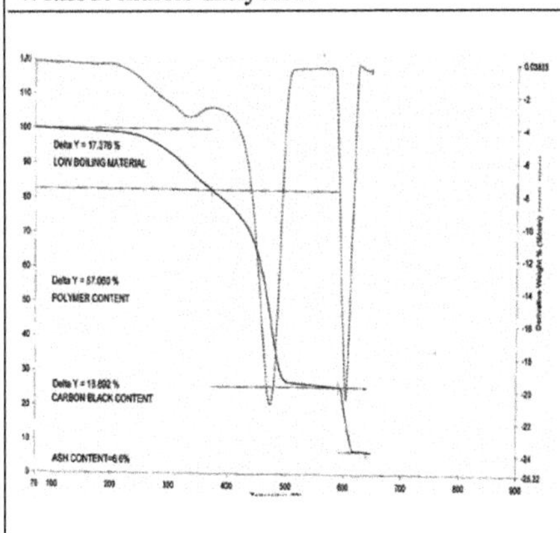

30 phr coconut pith (300.12°C) (g)

Without Maleic anhydride

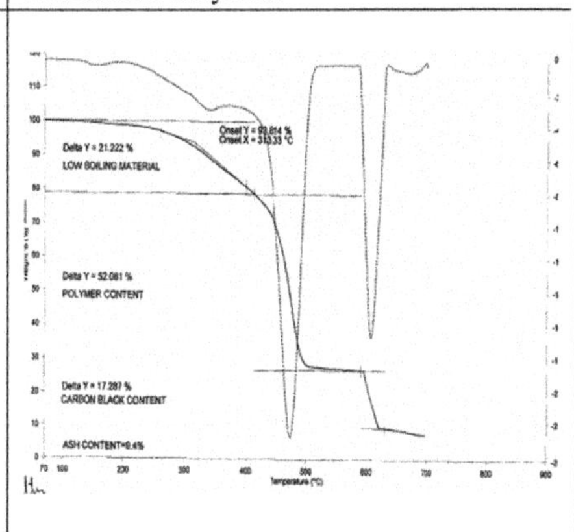

30 phr coconut pith (313.33°C) (h)

With Maleic anhydride

FIGURE 12.24 Graphical representation of 20 and 30 phr coconut pith for with and without MA systems (e, f, g, and h).

12.4.1.10 MORPHOLOGICAL ANALYSIS

Morphological analysis was carried out in order to study the matrix after the addition of coconut pith and effect of compatibilizer in the dispersion of coconut pith into the matrix. Phase morphology of the various blends was investigated by a digital SEM. The images were obtained at a tilt angle of 0° with an operating voltage of 20 kV.

We can see from the Figure 12.25 that the rubber particles are dispersed throughout the PP matrix in the form of aggregates and the size of rubber particles is less than 2 μm in the unfilled thermoplastic elastomer sample. It should be noted that after addition of coconut pith, the size of rubber particles increases and the voids increases. Initially the coconut pith cannot penetrate into the rubber phase but after adding curing agent DCP 40, the rubber phase becomes more polar. Therefore, it is possible that some coconut pith goes to REPDM phase before the curing cycle

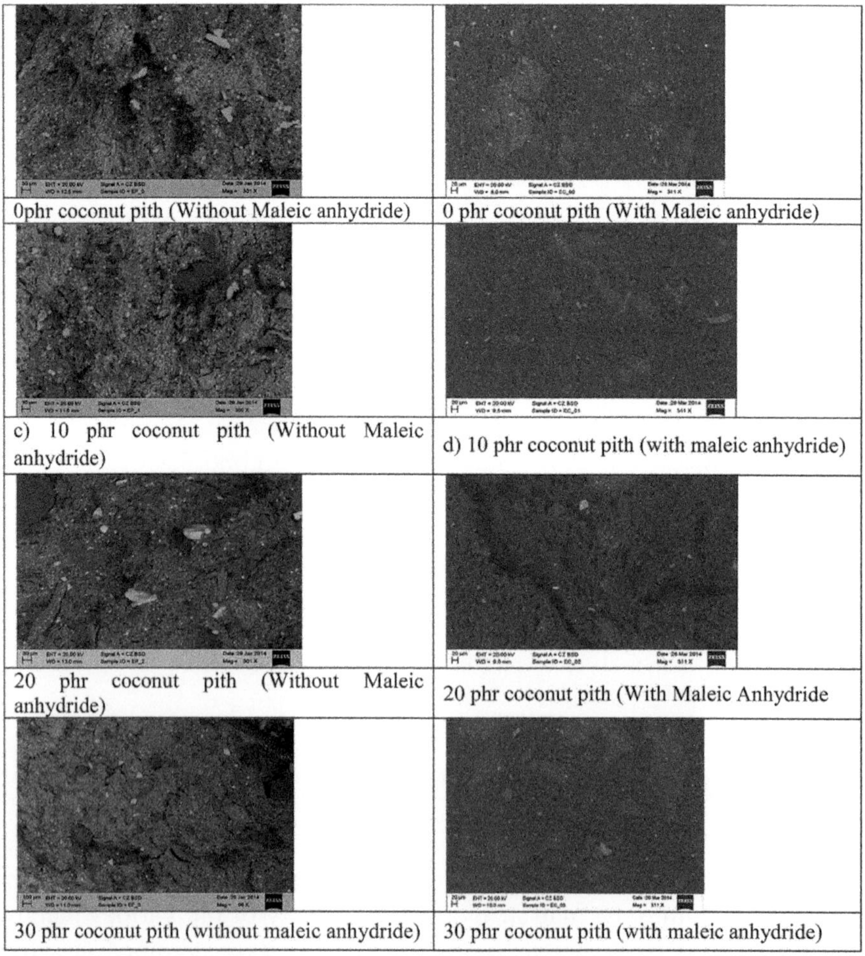

FIGURE 12.25 Micrographs of SEM for coir composites with and without maleic anhydride at 0, 10, 20, and 30 phr loading of coconut pith.

ends. Hence, there is change in the viscosity ratio between the two phases, and consequently, the size of the rubber phase increases. The interaction between coconut pith and RPP/REPDM is poor when there is no compatibilizer addition. After the compatibilizer addition, the interaction between the coconut pith and RPP increases due to these mechanical properties such as tensile, abrasion, flexing, and impact were increasing. At 30 phr concentration, the dispersion of coconut pith is less so the properties were decreasing.

12.5 CONCLUSIONS

The study was conducted to determine the RPP/REPDM coconut pith composites for low-cost sole in footwear application and we at IRMRA could achieve the same through the results obtained from various test conducted in the IRMRA laboratory. The preliminary studies on blends showed that 65:35 ratio of RPP/REPDM gives optimum properties based on mechanical properties such as tensile, abrasion, flexing, and impact; thus, the blend 65:35 (RPP:REPDM) was optimized and selected. Later on, the effect of coconut pith loading on 65/35 blend ratio was determined. The hardness, flexing cycles, and impact strength increases up to 20 phr of coconut pith loading and are considered to be optimum for the targeted application but thereafter at 30 phr, mechanical properties decreases due to formation of large voids due to agglomeration of coconut pith in the matrix. Abrasion resistance decreases as coconut pith loading increases but the values are comparable with the blend ratio. The presence of maleic anhydride increases the mechanical and thermal properties of RPP/REPDM coconut pith composites and hence MA-g-EPDM should be used while preparing the ternary-recycled composites. Morphological analysis through SEM revealed that there is an increase in interaction between coconut pith and REPDM/RPP after the addition of maleic anhydride, as addition of MA in the form of MA-g-EPDM decreases the interphase interaction and there is more compatibility in the blends formed. After 20 phr addition of coconut pith, there is a decrease in dispersion of coconut pith into the RPP/REPDM matrix which leads to detrimental effects on the properties of the composites and hence, 20 phr loading is optimum for 65:35 (RPP:REPDM) polymer matrix. Hence, I conclude that presence of coconut pith and MA as a grafting agent in RPP/ REPDM composites improves the properties which are applicable to footwear application.

KEYWORDS

- **recycled polypropylene**
- **reclaimed EPDM**
- **compatibilization**
- **coconut pith**
- **shoe sole**

REFERENCES

1. Raj, M. M.; Patel, H. V.; Raj, L. M.; Patel, N. K. Studies on Mechanical Properties of Recycled Polypropylene Blended with Virgin Polypropylene. *Int. J. Sci. Invent. Today* **2013,** *2*(3), 194–203.
2. Fukumori, K.; Matsushita, M. Material Recycling Technology of Crosslinked Rubber Waste. *Res. Develop. Rev. Toyota CRDL* **2003,** *38*(1), 39–47.
3. Cui, Y.; Lee, S.; Noruziaan, B.; Cheung, M.; Tao, J. Fabrication and Interfacial Modification of Wood/Recycled Plastic Composite Materials. *Compos. A: Appl. Sci. Manuf.* **2008,** *39*(4), 655–661.
4. Choudhury, A.; Kumar, S.; Adhikari, B. Recycled Milk Pouch and Virgin Low-density Polyethylene/Linear Low-density Polyethylene Based Coconut Fiber Composites. *J. Appl. Polym. Sci.* **2007,** *106*(2), 775–785.
5. Friedrich, K.; Fakirov, S.; Zhang, Z. *Polymer Composites: From Nano- to Macro-Scale*; Springer: New York, 2005.
6. Mazumdar, S. K. *Composites Manufacturing Materials, Product, and Process Engineering*; CRC Press, 2002.
7. Utracki, L. A., & Wilkie, C. A *Polymer Blends Handbook.* Dordrecht: Kluwer Academic Publishers. 2000; Vol. 1, p 2.

CHAPTER 13

CALCIUM LACTATE AS A PROMISING COAGULANT FOR THE PRETREATMENT OF LIGNIN-CONTAINING WASTEWATER

A. Y. ZAHRIM* and A. NASIMAH

Chemical Engineering Programme, Faculty of Engineering, Universiti Malaysia Sabah, Jalan UMS, 88400 Kota Kinabalu, Sabah, Malaysia

Corresponding author. E-mail: zahrim@ums.edu.my

CONTENTS

ABSTRACT

Lignin is the major colorant in the agro-based industry wastewater such as pulp and paper mill, palm oil mill, olive oil, etc. These industries are characterized by their water-intensive nature. This review is dedicated to explore the lignin and its removal from wastewater via coagulation/flocculation. Iron- and aluminum-based coagulants are widely used for this purpose. However, the application of aluminum- and iron-based coagulant may hinder the performance of biological posttreatment. Hence, there is interest in developing the use of calcium-based coagulants alone or hybrid calcium–other metals coagulant that might fully dismiss or minimize the adverse impact of the aluminum- and iron-based coagulants. On the whole, it looks for the possibility of calcium-based salt as a potential coagulant. Finally, we also discussed a nontoxic coagulant, that is, calcium lactate for the treatment of anaerobically digested palm oil mill effluent. It can be concluded that calcium lactate has potential for the treatment of lignin-containing wastewater.

13.1 INTRODUCTION

The development of agro-industries toward commercialization of their agricultural systems has improved the traditional farming and informal agro-industries in developing system such as India and Malaysia.[1] The daily operation of agro-based industries, such as pulp and paper mill, palm oil mill, textile, and dairy parlor, consume a considerable quantities of water that is derived from cleaning, washing, and disinfection stages.[2] The consumption of water from the process usually ends up as colored wastewater that can cause pollution harmful to the receiving environment.[3] The wastewater generated from agro-industries is known to be loaded with organic wastes such as oil and grease, proteins and sugars, resulting in high chemical oxygen demand (COD) and biological oxygen demand (BOD) in the wastewater. Besides that, the presence of plant components, that is, lignin, tannin,[2] as well as its biodegradation product, for example, melanoidin,[4] enhances the generation of colored wastewater. Untreated discharge of wastewater contains colored compounds in surface water that can lead to serious surface and groundwater contamination, besides causes disturbance of the aquatic biosphere due to reduction of sunlight penetration and depletion of dissolved oxygen.[5] The potential of colored compounds to release the carcinogenic amines makes it to be toxic and mutagenic.[2,6]

Over the past decades, several feasible techniques have been utilized for the treatment of palm oil mill wastewater (POMW); research has been mainly directed in the area of biodegradation technology. Biodegradation is found to be environment-friendly and cost-effective treatment in comparison to chemical processes. Anaerobic wastewater treatment is traditionally applied for treatment of highly concentrated organic matter wastewaters (i.e., BOD >1000 mg/L). The anaerobic treatment (AT) is one of the most well-known treatments that are being applied in Malaysia either in pond system or close digesting tank systems to treat highly concentrated POMW. This is because the anaerobic process has considerable advantages such as (1) it demands less energy, (2) sludge formation is minimal, and (3) anaerobic bacteria efficiently break down the organic substances to methane.[7] The pond system has been applied in Malaysia for POME treatment since 1982 for the stabilization of POMW.[8] However, the biological treatment is not guaranteed to always work properly. For a toxic wastewater such as from pulp and paper industry, the toxicity breakthroughs are reported to occasionally occurs.[9] Other than that, after AT, the color of the wastewater increases, which may be due to the organic material being converted into smaller chromophoric units rather than being mineralized.[10] For other applications such as water reuse, colored-treated water was identified as not suitable.[4] Therefore, it is necessary to develop innovative methods of treating colored wastewater effectively and economically. Normally, after AT of POMW, the wastewater undergoes aerobic treatment. Despite that, the sustainability issues have become very important in the wastewater treatment due to a concern of reducing the energy consumed from an aerobic degradation technology[11] and current aerobic treatment has failed to decolorize the wastewater. Therefore, replacing the current aerobic treatment with appropriate physicochemical process such as coagulation/flocculation could be a solution for lignin-containing wastewater.

This review will focus on the lignin as the main colorant and its treatment using coagulation/flocculation. Prior to that, the lignin structure and class is discussed.

13.2 MAIN COLORANT: LIGNIN

Lignin is classified as the second most abundant natural raw material and nature's most abundant aromatic (phenolic) polymer.[16] As shown in Figure 13.1, lignin is a natural polymeric product that is structured from an enzyme initiated dehydrogenative polymerization of the three primary precursors.[15,16] While, in another study, Essington[17] proposed the lignin structure as shown in Figure 13.2. Composition of lignin is different based on its classes (Table 13.1): softwood, hardwood, and grass lignin. Generally, lignin particles exist as negatively charged in the water.[5]

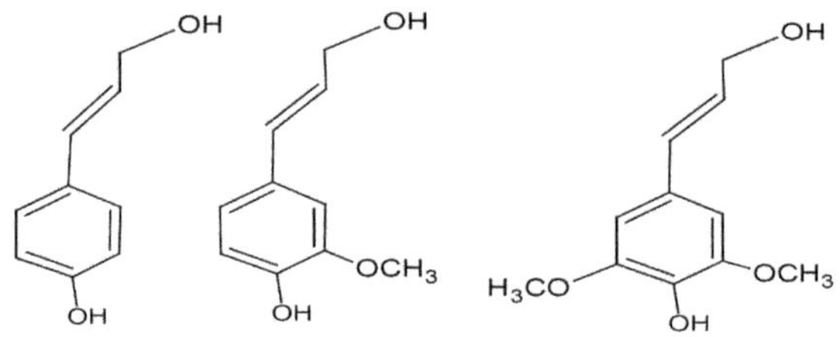

FIGURE 13.1 Lignin precursors. (Reprinted with permission from Chakar, F. S.; Ragauskas, A. J. Review of Current and Future Softwood Kraft Lignin Process Chemistry. *Ind. Crops Prod.* **2004,** *20*, 131–141. © 2004, Elsevier.)

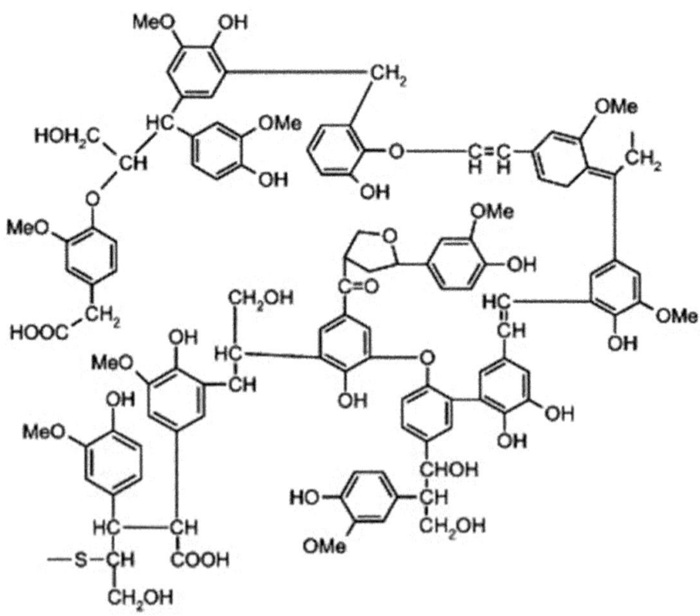

FIGURE 13.2 Lignin molecule. (Reprinted with permission from Norgren, M. and Edlund, H.. Lignin: Recent advances and emerging applications. *Curr. Opin. Colloid Interface Sci.*, **2014,** *19*, 409–416. © 2014, Elsevier.)

TABLE 13.1 Classes of Lignin.

Class of lignin	Other name	Unit	Reference
Soft wood	Guaiacyl Coniferous	Coniferyl alcohol	[14]
Hard wood	Dicotyledonous angiosperm	Coniferyl alcohol Sinapyl alcohol	
Grass lignin	Annual plant Monocotyledonous angiosperm	Coniferyl alcohol Sinapyl alcohol p-coumaryl alcohol	

13.3 COAGULATION/FLOCCULATION

The feasibility of treating effluent within a short period of time without involving a vast area of land by using coagulating and flocculating agents may offer a solution to the current treatment problems such in case of palm oil mill industry. Physicochemical treatment (coagulation–flocculation) was reported as an important step in the removal and settlement of colloidal, suspended particles and oily materials present in the wastewater.[18] The coagulation–flocculation process to be mainly applicable for a pretreatment in industry biological wastewater treatment plant are due to its simplicity, low cost, good removal efficiency, and easy on-site implementation.[11] Generally, coagulation is the destabilization of colloids by neutralizing the forces that keep them apart,[19] while flocculation is the process whereby, destabilized particles or particles formed as a consequence of destabilization are induced to come together, make contact, and thereby form larger agglomerates.[20] The destabilization of particles can occur only with an addition of coagulant(s). Coagulants are classified into three main categories: (1) inorganic-based coagulants, (2) organic-based flocculants, and (3) hybrid materials.[21] In general, inorganic coagulants consist of aluminum sulfate (alum), ferric chloride, and polyaluminum chloride (PAC). Every coagulant is effective in removing wide range of impurities from water by affecting the destabilization degree of the colloid particles differently.[22]

There are number of different mechanisms involved in a coagulation process that are very important in forming flocs which then could be easily settled and finally removed from the wastewater[23]. Since in the normal water condition lignin particle is negatively charged, the mechanisms of lignin removal could be as follows:[24–26]

- Chelation–precipitation/complex chemical reactions (inorganic-based coagulants).
- Electrostatic patch [organic-based flocculants, e.g., poly (diallyldimethylammonium chloride) and hybrid materials].
- Adsorption–precipitation (inorganic-based coagulants).
- Charge neutralization [inorganic-based coagulants, organic-based flocculants e.g., poly(diallyldimethylammonium chloride) and hybrid materials].
- Interparticle bridging [organic-based flocculants e.g., poly (diallyldimethylammonium chloride) and hybrid materials].
- Sweep coagulation (inorganic-based coagulants at $pH > 7$)

The organic matter removal and good floc characteristics in the coagulation process usually depend on the coagulant types, injection time, mixing rate, initial pH, and loading rate. These factors affected the flocs formation by[21]:

i. The rate of transport of coagulant or flocculant molecules to the lignin particles in the fluid,
ii. The rate of adsorption of coagulant/flocculant on the surface of lignin particles,
iii. The rate of aggregation of lignin particles having adsorbed coagulant/flocculant,
iv. The time scale needed for the coagulant/flocculant layer to reach equilibrium, and
v. The frequency of collisions of lignin particles with adsorbed particles to form flocs.

Based on Table 13.2, aluminum- and iron-based coagulants are widely used for the treatment of lignin-containing wastewater. These types of coagulant are traditionally and widely used as coagulants in wastewater treatment due to their proven effectiveness in removing organic substances.

Insufficient dosage of coagulant could deteriorate the coagulation performance. For example, in pulp and paper wastewater treatment, Rohella et al.[27] reported that aluminum chloride (10–20 mg/L) is unable to remove color efficiently. However, the application of cationic polyelectrolyte (0.2 mL/L) in same study shows that the removal of color is 82.58%.

Besides that, single coagulant may not be effective compared to mixed coagulants. Irfan et al.[28] has investigated the performance of different coagulants and flocculants such as alum, ferric chloride, aluminum chloride, ferrous sulfate, PAC, and cationic and anionic polyacrylamide polymers in individual form as well as in different combinations for pulp and paper wastewater treatment. They found that mixed coagulants were found to be more effective in reducing COD, total suspended solid (TSS), and color than the individual form. Combination of ferric chloride, PAC, and cationic polymer was excellent for reduction of 81% COD and 95% TSS, whereas combination of aluminum chloride, PAC, and anionic PAM was good in 88% color reduction.[28] This fact is actually a challenge for engineers and scientists to come out with a good formula for mixed coagulants and the optimization study of the coagulant mixture is vital. Many researchers have focus on hybrid coagulants for the treatment of lignin-containing wastewater. In another study, Ganjidoust et al.[29] studied the coagulation of pulp and paper wastewater by using hybrid aluminum sulfate–synthetic polymers. The result shows that the TOC and lignin removal is ~30% and ~80%, respectively. In another study of coagulation/flocculation of pulp and paper mill wastewater, Wang et al.[30] reported the use of aluminum chloride as coagulant and a modified natural polymer, polyacrylamide and poly(2-methacryloyloxyethyl) trimethyl ammonium chloride (starch-g-PAM-g-PDMC), as flocculant aids. They concluded that the coagulation/flocculation was able to reduce the turbidity up to 95.7% with water recovery as much 72.7%.[30]

In Malaysia, Tan et al.[18] have carried out investigation on various combinations of inorganic coagulants with commercial polymer SR316 as flocculant which would be useful in the pretreatment of POMW. The result of the study showed the reduction of turbidity and COD by 97% and 64%, respectively, through the combination of 10% w/w of ferric sulfate, 1% w/w of alum, and 1% of ammonium sulfate with very high molecular polyacrylamide SR316. However, the authors stated that study on the economics aspect on these combinations and the utilization of the recovered chemically treated sludge is required.[18]

Amuda and Amoo[31] combined an inorganic coagulant (ferric chloride) and nonionic polyacrylamide as flocculant to treat beverage industry wastewater. The results revealed at the optimal operating pH 9, 91%, 99%, and 97% removal of COD, total phosphorus (TP), and TSS, respectively, were achieved with the addition of 100 mg/L ferric chloride. In addition, the combination of coagulants resulted in the reduction of sludge volume of 60% of the amount produced, when coagulant was solely used for the treatment. Generally, organic polymers generate less sludge than inorganic salts, since they do not add weight or chemically combine with other ions in the water to form precipitate.[31]

Beside polymer, fly ash can be added to increase the performance of coagulation especially for color removal. In another study, Srivastava et al.[32] used PAC followed by adsorption with bagasse fly ash (BFA). They reported that at the optimal condition (pH 3; PAC dosage 300 mg/L), it can remove about 80% COD and 90% color.[32] The Teh et al.[33] demonstrated the performance of coagulation–flocculation of POME using a combination of alum and unmodified rice starches as coagulants. Higher TSS reduction with 88.4% removal was achieved. An addition of rice starches as a natural coagulant could potentially reduce the required alum dosage without sacrificing the efficiency of the process.[33]

13.3.1 ADVERSE EFFECT OF IRON- AND ALUMINUM-BASED COAGULANTS

Recently, the usage of the conventional inorganic coagulants has been questioned. These coagulants create hazardous activated sludge which contains residual aluminum which may cause side effect when discharged into the open water course.[23] As shown at Table 13.2, high valence inorganic coagulants such as aluminum sulfate and ferric chloride were being applied for lignin-containing wastewater. This is probably because as the valence electron increases, the metal ion concentration for coagulation decreases[34] (Fig. 13.3); and this affected the choice of coagulant due to the lower cost. It has been pointed out above that intake of large amount of aluminum salt may cause Alzheimer's disease.[23]

Besides that, these conventional coagulants have been shown to produce less biodegradable wastewater since the coagulation process also removes the amino acids, proteins, and long-chain fatty acids from wastewater.[35] It have been reported that the residual alum- and ferric-based coagulants with an initial concentration for both coagulants of 25 mg/L, will lead to inhibit the biological treatment process which is indicated by the reduction of microorganism respiration rate and low organic matter removal.[36] In addition, the predatory growth in biological wastewater treatment plant is also enhanced significantly.[36] The adverse effects of alum addition on

TABLE 13.2 Coagulants Used to Treat Lignin-containing Wastewater.

Industry	Metal	Polymer	Operating condition			Other removal (%)	Lignin removal			Reference
			pH	Dosage (mg/L)	Temperature (°C)		Before (mg/L)	After (mg/L)	Removal (%)	
Palm oil mill (400 mL diluted POME)	Ferric chloride	–	5	500	–	Turbidity—76%	–	–	–	[41]
Pulp and Paper mill	Aluminum sulfate	–	6	999.76	–	Turbidity—99.8 TSS—99.4 COD—91	–	–	–	[42]
Pulp and Paper mill	Aluminum sulfate	–	7	30	–	–	–	–	88	[43]
pulp and paper mill	Copper sulfate		5	5000	–	COD—74 Color—76	–	–	–	[40]
Pulp and Paper mill	Aluminum sulfate	–	–	–	–	–	–	–	80	[29]
Pulp and paper mill	Ferric chloride + Aluminum chloride	–	3	$FeCl_3$—799.97 $AlCl_3$—800.04	–	COD—18 TSS—49 Color—48	–	–	–	[28]
Kraft pulp mill (washing water)	Sodium	–	9	22989.80—114949.00	25	–	450—400	150—100	66.7–75.0	[34]
	Calcium	–	9	200.39—2805.46	25	–	450—400	150—100	66.7–75.0	
	Magnesium	–	11	680.54—729.15	25	–	450—400	100—50	77.8–87.5	
	Aluminum	–	9	134.91—1888.71	25	–	450—400	120—90	73.3–77.5	
Palm oil mill (AnPOME)	Calcium lactate	–	8–8.4	500.00	–	–	1173–1517	–	91	[44]

TABLE 13.2 *(Continued)*

Industry	Metal	Polymer	Operating condition			Other removal (%)	Lignin removal			Reference
			pH	Dosage (mg/L)	Temperature (°C)		Before (mg/L)	After (mg/L)	Removal (%)	
Wastewater from ECF birch pulp bleaching	–	Methacrylates copolymer	5.5	142	72	Wood extractive—92	–	–	–	[45]
Sugar mill	–	PAC	3	300	–	COD—80	–	–	–	[46]
Oily wastewater	–	Poly-zinc-silicate-sulfate (PZSS) + Anionic polyacrylamide (APAM)	2.0	Zn/Si ratio—1.00–1.50	Ambient	COD—supe-rior TSS—95 Turbidity—96.3	–	–	–	[47]
Paper and pulp mill	–	Polydadmac + Poly-acrylamide (PAM)	–	Polydadmac — 120 PAM—2.00	Ambient	COD—98 TSS—96.8	–	–	71.7	[48]
Pulp mill	–	Acrylamide + Starch + 2-methyarcyloyloxy-ethyl trimethyl ammo-nium chloride(DMC)	8.35	22.30 and 22.30	Ambient	Turbidity—95.7 Water recovery —72.7	–	–	83.4	[30]
Pulp mill	–	Chitosan + 2-methyar-cyloyloxyethyl trimeth-ylammonum chloride (DMC)	7.1	Chitosan—17.80 DMC—17.80	Ambient	Turbidity—99.4 COD—90.7 Water recovery—89.4	–	–	81.3	[49]
Palm oil mill		Chitosan	4	500	–	residue oil—>95 SS—>95%	–	–	–	[23]
Wood	–	Polyaluminum chloride	9.2	125.0	Ambient	COD—40.9 Color—83.8 LES—92.8	–	–	58.4	[50]

TABLE 13.2 *(Continued)*

Industry	Metal	Polymer	Operating condition			Other removal (%)	Lignin removal			Reference
			pH	Dosage (mg/L)	Temperature (°C)		Before (mg/L)	After (mg/L)	Removal (%)	
Pulp mill	–	Polyethylene	2	350	Ambient	COD—15	–	–	15	[51]
Pulp and Paper mill	–	Natural polymer (chitosan)	–	–	–	TOC—70	–	–	90.0	[29]
Pulp and Paper mill	–	Cationic polyacrylamide	7.3–8.3	5	–	Turbidity—95 COD—93	–	–	–	[52]
Pulp and Paper mill	Aluminum sulfate	Polyaluminium chloride (PACl)	6	PACl—500.00 Alum—1000.00	Ambient	COD—91 TSS—99.4	–	–	–	[53]
Palm oil mill	Ferric sulfate + aluminum sulfate + ammonium sulfate	polyacrylamide	3	Ferric sulphate—10 %w/w aluminum sulfate and ammonium sulfate—1%w/w polyacryl-amide—5 mL	–	Turbidity—97 COD—64	–	–	–	[18]
Pulp and Paper mill	Ferric chloride	poly aluminum chloride (PACl) + poly acryl-amide (PAM)	Ferric chlorid— 2 PAC—3 PAM—2	Ferric chloride and PACl—200.00 PAM—4.00	Ambient	COD—81 TSS—95 Color—Not efficient	–	–	–	[28]
Wastewater from beverage industrial	Ferric chloride	nonionic polyacrylamide	9	Ferric chlo-ride—100 nonionic poly-acrylamide—25	–	COD—91 TP—99 TSS—97	–	–	–	[31]

TABLE 13.2 *(Continued)*

Industry	Metal	Polymer	Operating condition			Other removal (%)	Lignin removal			Reference
			pH	Dosage (mg/L)	Temperature (°C)		Before (mg/L)	After (mg/L)	Removal (%)	
Pulp and Paper mill	Aluminum chloride + ferrous sulfate	Anionic PAM	Aluminum chloride and ferrous sulphate—6	Aluminum chloride—800.00 Ferrous sulfate—800.00 PAM—4.00	Ambient	COD—76 TSS—95 Color—95	–	–	–	[28]
Palm oil mill	Aluminum sulfate	Polyaluminum chloride (PACl)	4.5	Alum —8000.00 PACl—600.00	Ambient	COD—95 TSS—98 Residue oil and suspended solids—99	–	–	–	[54]
Palm oil mill	Ferric chloride	Anionic polymer—polyacrylamide	8	Ferric chloride—100 Anionic polymer-poly-acrylamide—100	–	Turbidity—~98	–	–	–	[55]
Palm oil mill	Aluminum sulfate + Activated Carbon	Cationic polyacrylamide	6	Alum—1700	Ambient	COD—85 TSS—99.9 BOD—86.3 Residue oil and suspended solids—95	–	–	–	[56]
Pulp and Paper mill	Aluminum sulfate	Synthetic polymer (HE, PEI, and PAM)	6	–	–	TOC—30	–	–	80	[57]
Palm oil mill	Ferric chloride	Cationic polymer	7–9	Ferric chloride—200–300 Cationic polymer—70–100	–	COD—(47–53) TS—(43–49) TSS—(92–94)	–	–	–	[41]

TABLE 13.2 *(Continued)*

Industry	Metal	Polymer	Operating condition			Other removal (%)	Lignin removal			Reference
			pH	Dosage (mg/L)	Temperature (°C)		Before (mg/L)	After (mg/L)	Removal (%)	
Pulp and Paper mill	Aluminum chloride	Modified natural polymer, starch-g-PAM-g-PDMC [polyacrylamide and poly (2-methacryloy-loxyeth-yl) trimethyl ammonium chloride]	8.35	Alum—871 mg/L, Flocculant dosage—22.3 mg/L	–	Turbidity—95.7 Water recovery efficiency—83.4	–	–	72.7	[30]
Kraft paper mill waste water	Aluminum chloride	Anionic Polyelectrolyte	–	Alum—300 mg/L Anionic poly-electrolyte—0.05 mg/L	–	Suspended solid—91.6 COD—97	–	–	66.7	[58]
Pulp and Paper mill	Aluminum chloride (alum)	Cationic Polyelectrolyte	–	Alum—20 mg/L Cationic poly-electrolyte—0.02 mg/L	–	Turbidity—96.26 COD—55.65	–	–	82.58	[27]

the treatment performance of membrane bioreactor has been reported by Zahid and El-Shafai.[37] The authors reported that the alum doses above 60 mg/L have toxic effect on the autotrophic bacteria with significant reduction in ammonia oxidation and hence the nitrogen removal.[37]

Furthermore, the addition of iron- and aluminum-based coagulants could turn treated wastewater into acidic condition[38,39] in which the neutralization step is required prior to biological treatment. In order to reduce the dosage of neutralizing agent, Kumar et al.[40] has applied copper sulfate (a less commercial coagulant) for the treatment of lignin wastewater. It is reported that the organic removal by the application of copper sulfate is less, compared to other alum-based coagulants.[40]

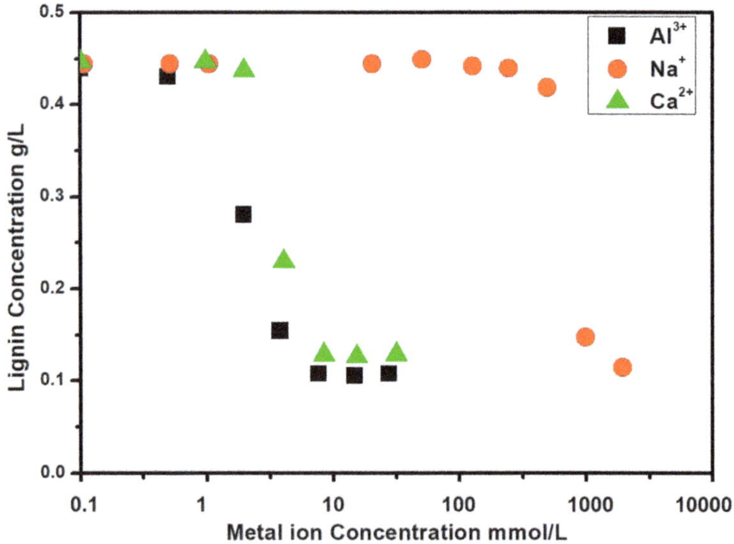

FIGURE 13.3 The influence of Al^{3+}, Ca^{2+}, and Na^{+} on spruce kraft lignin in solution of pH 9 and at 25°C.

13.4 POTENTIAL OF CALCIUM-BASED COAGULANT

Nowadays, it has become a necessity to develop a more efficient, environmental-friendly coagulant which has similar potential as conventional inorganic coagulants with an enhanced economic profile due to the high demand for organic matter removal such as lignin-containing wastewater.[41–49] Alternatively, among other coagulants, calcium-based coagulant had been proposed in several studies. There is interest in developing the use of calcium-based coagulants alone[50,59] or hybrid calcium–other metals coagulant[60,61] that might minimize or fully dismiss the adverse impact of the aluminum- and iron-based coagulants. Herewith, we provide some examples of studies for the wastewater treatment that have been using various calcium-based coagulants.

Aktas et al.[62] proposed lime pretreatment to reduce the polluting effect of olive mill wastewater (OMW). It is an applicable method in practice since lime can easily be purchased anywhere and it is cheaper than other chemicals such as aluminum sulfate, ferric chloride, magnesium sulfate, etc.[62] After lime treatment, COD values of the wastewater samples could be reduced by 41.5–46.2% in pressing and centrifugal methods, respectively. The average removal percentage of the other parameters are 29.3–46.9% for total solids, 41.2–53.2% for volatile solids, 74.4–37.0% for reduced sugar, 94.9–95.8% for oil–grease, 73.5–62.5% for polyphenols, 38.4–32.0% for volatile phenols, and 60.5–80.1% for nitrogenous compounds, respectively.[62] Other organic compounds that are responsible for the OMW color have been studied by Boukhoubza et al.[63] The application of 1% of lime concentration reduced COD, TSS, and polyphenols content in OMW to around 72%, 73%, and 60%, respectively.[63]

In another study, Ayeche[3] has conducted a coagulation–flocculation of dairy wastewater using the residual lime of National Algerian Industrial Gases Company (NIGC-Annaba). The addition of lime reduces the organic matter and increases the pH in range of 7–12. At optimum dose of 4000 mg/L of lime, COD, BOD_5, TSS, and TP achieved are 49%, 54%, 92%, and 83% of removal, respectively.[3]

13.5 CALCIUM LACTATE FOR POMW TREATMENT: A CASE STUDY

Calcium lactate is an eco-friendly coagulant and is a biodegradable compound. It can be obtained from agricultural waste by biotechnological methods, which involves fermentation of the substrates with lactic acid bacteria.[58] Calcium lactate is the calcium salt of lactic acid. The molecular formula of calcium lactate is $C_6H_{10}CaO_6$. The structure of calcium lactate is shown in Figure 13.4. The molecular mass is 218.22 g/mol and has a solubility of 3.4 g/100 g of water at 20°C.[64]

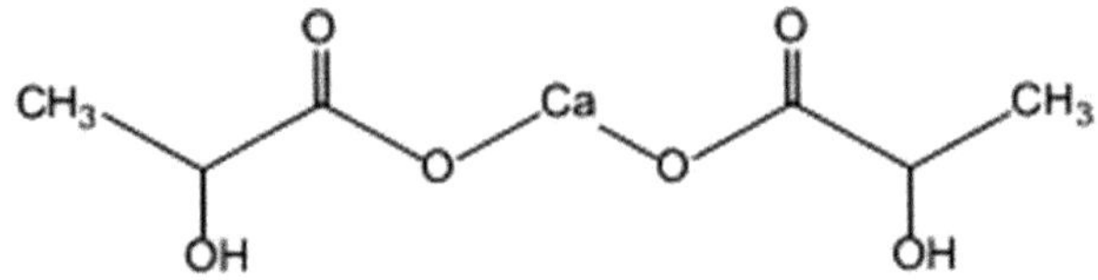

FIGURE 13.4 Structure of calcium lactate.

Calcium lactate is suitably used as coagulant because it is a completely nontoxic coagulant. It is commonly used as food additive. It can be used as baking powder. Besides that, it also can retain the firmness of table olives and other pickled vegetables. Calcium lactate is added to sugar-free foods so that it can prevent tooth decay. It also can be used as a medicine, for example, as an antacid and can treat calcium deficiency problem.[58]

Palm oil industry is among the key agricultural-based industries in Malaysia. The production of crude palm oil (CPO) leads to huge quantities of wastes, particularly palm oil mill effluent (POME). The biogas generation from anaerobic digestion of POME is predicted to become the future trend for the palm oil millers. However, the biogas-producing process effluent or anaerobically treated wastewater (AnPOMW) appearance is black-dark brown in color which indicates a water pollution by the public.[65] The AnPOMW contains particulates (0.32–0.39% w/w), that is, bioflocs, anaerobic microorganisms, and macrofibrils. The soluble fraction consists of carbohydrate, pectin, lignin, tannin, humic, and fulvic acid-like substance, melanoidin, and phenolics compounds[66–68] such as gallic, protocatechuic, 4-hydroxybenzoic, 4-hydroxylphenylacetic, caffeic, syringic acids, p-coumaric, and ferulic acids.[69]

The POMW from Aerobic Pond No. 1 was collected from the Beaufort, Sabah. The physicochemical characteristics of AnPOMW in this study are shown in Table 13.1. The wastewater in this study is similar to other studies but with higher content of ammonia nitrogen (Table 13.3).

During the jar tests, the appropriate volume of AnPOMW was transferred into the round jar. Predetermined dosage of calcium lactate (molecular mass 218 g/mol) (Merck, Germany) stock solution (75 g/L) was added to the solution in the jar, making the total volume of 500 mL. A standard flocculator apparatus (Phipps & Birds) equipped with stainless steel paddles and stirrer was used for the coagulation/flocculation tests. The aqueous solution was then rapidly mixed at a paddle speed of 258 rpm for 3 min followed by slow mixing for 20 min at 39 rpm. After allowing settling to occur (10–240 min), about 25 mL of the liquid was withdrawn using a pipette from a height of about 3 cm below the liquid surface in each jar.[39]

TABLE 13.3 The Characteristics of Palm Oil Mill Wastewater from Aerobic Pond No. 1 (AnPOMW).

Parameters	AnPOMW (this study)	AnPOMW from other studies [66, 71–74]
pH	8.0–8.4	7.2–8.3
Electrical conductivity, μS/cm	4686–5010	–
Ammonia nitrogen, mg/L	127–407	45–100
Total chemical oxygen demand (COD), mg/L	400–11600	1003–13532
Suspended solids, mg/L	140–870	290–12750
Lignin-tannin, mg/L	1173–1517	–
Zeta potential, mV	−25	–

The COD was determined using the Spectrophotometer HACH DR 2010. Lignin and ammonia nitrogen were analyzed at the absorbance of λ_{max} = 290 nm with a V-650 UV/Vis Spectrophotometer (Jasco). The lignin and nitrogen ammonia content were tested via the Tyrosine and Nessler method, respectively.[70-73] The absorbance at λ_{max}=290 nm represented the low molecular mass colored compounds (LMMCC). The pH and conductivity was measured by using meter HI 9611-5, Hanna Instrument. The zeta potential was determined by using Malvern Zetasizer Nano Series model ZS machine.

Figure 13.5 shows the effect of sedimentation time during coagulation of lignin–tannin in the AnPOMW. During the first 10 min, the lignin–tannin removal is 86% and then, the lignin–tannin removal increases steadily afterwards. However, after 2 h of sedimentation time, there was no increment for the lignin–tannin removal. In the 2-h sedimentation time, the lignin–tannin and LMMCC removal is 92% and 70%, respectively. Lower removal of LMMCC is expected due to the small size of colorant particles as well as high solubility particles compared to lignin.[74] Based on this result, the next experiment will be carried out by using 2-h sedimentation time.

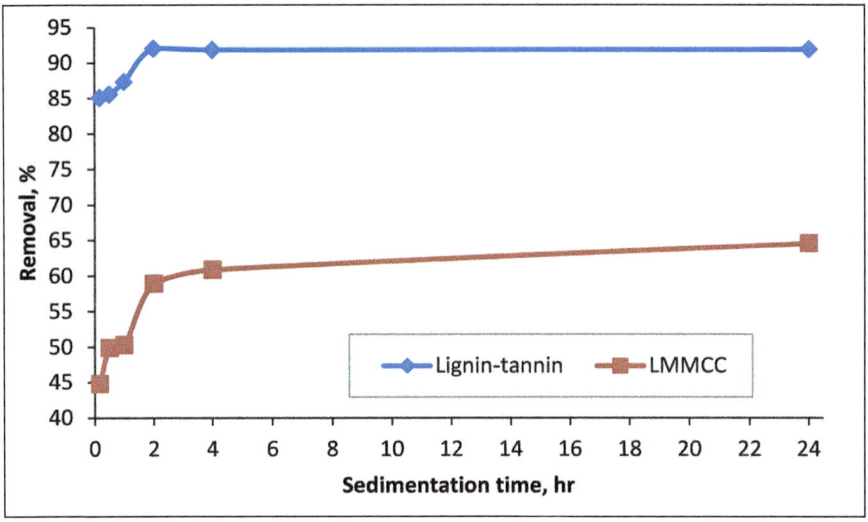

FIGURE 13.5 Effect of sedimentation time during coagulation of AnPOMW (dosage of calcium lactate: 0.1 g/L).

Figure 13.6 shows the effect of calcium lactate for the LMMCC and lignin removal. It can be seen that the lignin removal is unaffected by the calcium lactate dosage (0.1–4.0 g/L) with average of 92% (Fig. 13.6) since the major mechanism might be precipitation.[34] The LMMCC removal increases rapidly until it achieves the maximum removal (~70%) at the dosage of 0.5 g/L. Coagulation of the LMMCC is probably due to attraction between surfaces bearing a mosaic of positive and negative sites.[75] However, as the dosage increases to greater than 0.5 g/L, the LMMCC removal starts to decrease. The decrease in LMMCC reduction might be due to restabilization of the system by excess dosage of calcium lactate as shown by constant value of zeta potential (Fig. 13.7).

Figure 13.8 shows the effect of the calcium lactate on COD and NH_3–N removal (Fig. 13.8). The COD removal at the dosage of 0.1–2.0 g/L is ~90%. This might be due to the effect of precipitation of the dissolved organic compounds, adsorption by electrostatic attraction and physical entrapment.[76] Since the NH_3–N removal is due to the entrapment to the floc produced,[76] the NH_3–N removal trend is quite similar with LMMCC removal trend (Fig. 13.6). The maximum NH_3–N removal in this study is 90%.

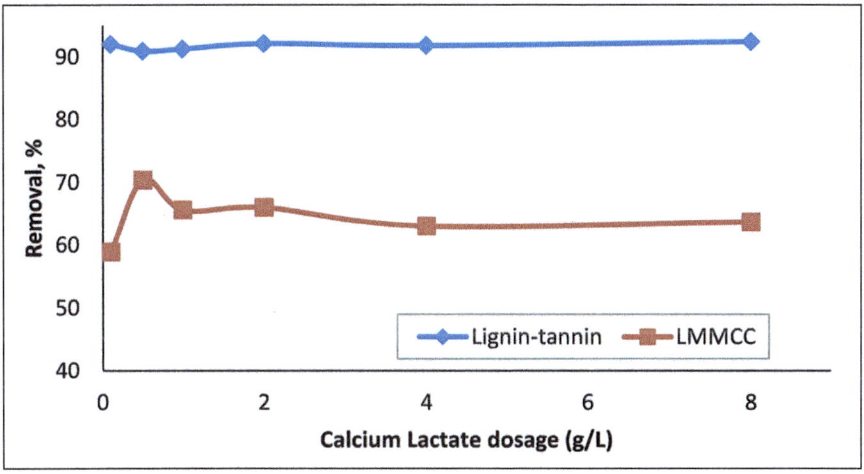

FIGURE 13.6 Effect of calcium lactate on the LMMCC and lignin removal during coagulation of AnPOMW (sedimentation time: 2 h).

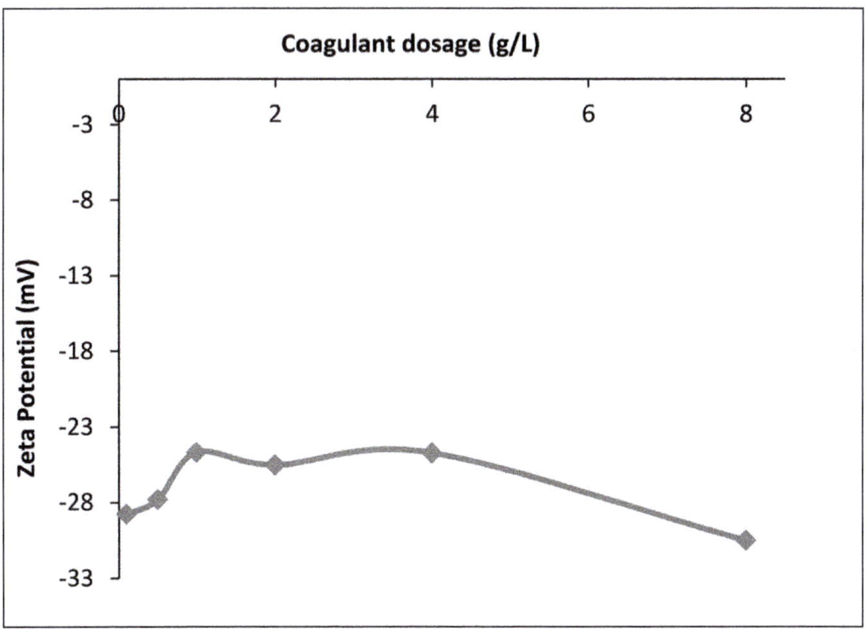

FIGURE 13.7 Effect of calcium lactate on the zeta potential during coagulation of AnPOMW.

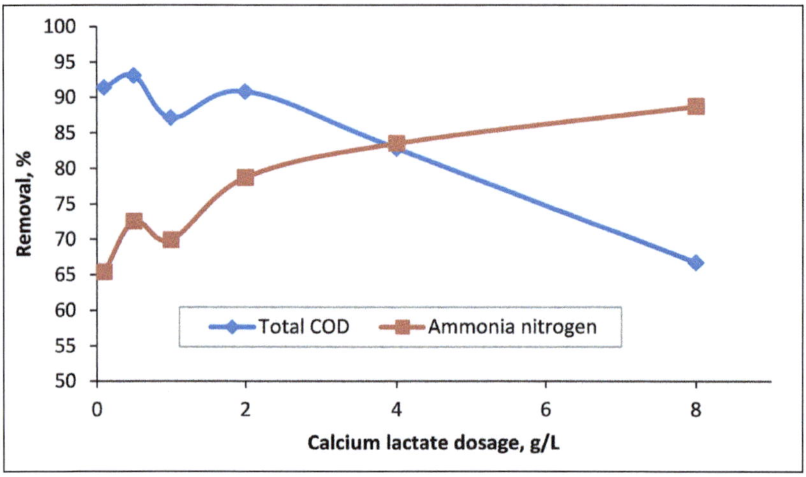

FIGURE 13.8 Effect of calcium lactate on the COD and NH_3–N removal during coagulation of AnPOMW (sedimentation time: 2 h, initial COD: 11,600 mg/L, initial ammonia nitrogen concentration: 349 mg/L).

As being expected, as the calcium dosage increases from 0.1 g/L to 4.0 g/L, the conductivity also increases from 5751 to 6419 μS/cm. changing calcium lactate dosage, the pH of treated AnPOMW also changes slightly.

13.6 CONCLUSION

Lignin is a recalcitrant; organic coagulation of lignin particles is important to ensure the effectiveness of biological wastewater treatment plant. Various studies have shown that composite coagulants were found to be more effective in reducing lignin particles than its individual coagulant. Over the years, the treatment of lignin-containing wastewater was dominated by aluminum- and iron-based coagulants. However, the application of these conventional coagulants should be minimized, especially if the posttreatment is a biological one. Calcium-based coagulant could produce more biodegradable treated water than the conventional aluminum- and iron-based coagulants. However, published literatures on this matter are lacking. In addition, the calcium-based power of coagulation can be increased by the addition of suitable additive(s). We also report a case study on the application of calcium lactate, a nonhazardous coagulant for treating POME. The application of calcium lactate could remove 92%

lignin–tannin, 70% LMMCC, 90% COD, and 90% NH3–N at optimum conditions.

ACKNOWLEDGMENT

The author would like to thank Ministry of Education, Malaysia for awarding FRGS grant, FRGS/2/2013/TK05/UMS/02/1 and SBK0075-TK-2013. Thanks to Hillery and Yasmi, for technical assistantship.

KEYWORDS

- **lignin**
- **coagulation**
- **agro-based industry wastewater**
- **calcium-based salt coagulant**
- **aluminum- and iron-based coagulant**

REFERENCES

1. Shak, K. P. Y.; Wu, T. Y. Coagulation–Flocculation Treatment of High-strength Agro-industrial Wastewater Using Natural Cassia Obtusifolia Seed Gum: Treatment Efficiencies and Flocs Characterization. *Chem. Eng. J.* **2014,** *256*, 293–305.
2. Anjaneyulu, Y.; Chary, N. S.; Raj, D. S. S. Decolourization of Industrial Effluents—Available Methods and Emerging Technologies: A Review. *Rev. Environ. Sci. Bio/ Technol.* **2005,** *4*(4), 245–273.
3. Ayeche, R. Treatment by Coagulation–Flocculation of Dairy Wastewater with the Residual Lime of National Algerian Industrial Gases Company (NIGC-Annaba). *Energy Procedia* **2012,** *18*, 147–156.
4. Zahrim, A. Y.; Hilal, N. Treatment of Highly Concentrated Dye Solution by Coagulation/Flocculation–sand Filtration and Nanofiltration. *Water Resour. Ind.* **2013,** *3*, 23–34.
5. Zahrim, A. Y.; Mariani, R. Diluted Biologically Digested Palm Oil Mill Effluent as a Nutrient Source for *Eichornia* crassipes. *Curr. Env. Eng.* **2014,** *1*(1), 45–50.
6. Neoh, C.; et al. Optimization of Decolorization of Palm Oil Mill Effluent (POME) by Growing Cultures of *Aspergillus fumigatus* Using Response Surface Methodology. *Environ. Sci. Poll. Res.* **2012,** *20*(5), 1–12.

7. Wong, Y. S.; Kadir, M. O. A. B.; Teng, T. T. Biological Kinetics Evaluation of Anaerobic Stabilization Pond Treatment of Palm Oil Mill Effluent. *Bioresour. Technol.* **2009,** *100*(21), 4969–4975.
8. Rupani, P. F.; et al. Review of Current Palm Oil Mill Effluent (POME)Treatment Methods: Vermicomposting as a Sustainable Practice. *World Appl. Sci. J.* **2010,** *10*(10), 1190–1201.
9. Leiviska, T.; Ramo, J. Coagulation of Wood Extractives in Chemical Pulp Bleaching Filtrate by Cationic Polyelectrolytes. *J. Hazard. Mater.* **2008,** *153*, 525–531.
10. Lewis, R.; et al. Colour Formation from Pre and Post-coagulation Treatment of *Pinus radiata* Sulfite Pulp Mill Wastewater Using Nutrient Limited Aerated Stabilisation Basins. *Sep. Purif. Technol.* **2013,** *114*, 1–10.
11. Karthik, M.; et al. Biodegradability Enhancement of Purified Terephthalic Acid Wastewater by Coagulation–Flocculation Process as Pretreatment. *J. Hazard. Mater.* **2008,** *154*, 721–730.
12. Gosselink, R. J. A.; et al. Co-ordination Network for Lignin—Standardisation, Production and Applications Adapted to Market Requirements (EUROLIGNIN). *Ind. Crops Prod.* **2004,** *20*, 121–129.
13. Lora, J. H.; Glasser, W. G. Recent Industrial Applications of Lignin: A Sustainable Alternative to Nonrenewable Materials. *J. Polym. Environ.* **2002,** *10*, 39–48.
14. Suhas, P. J. M.; Carrott, M. M. L.; Carrott, R. Lignin—From Natural Adsorbent to Activated Carbon: A Review. *Bioresour. Technol.* **2007,** *98*, 2301–2312.
15. Boeriu, C. G.; et al. Characterisation of Structure-dependent Functional Properties of Lignin with Infrared Spectroscopy. *Ind. Crops Prod.* **2004,** *20*, 205–218.
16. Chakar, F. S.; Ragauskas, A. J. Review of Current and Future Softwood Kraft Lignin Process Chemistry. *Ind. Crops Prod.* **2004,** *20*, 131–141.
17. Norgren, M. and Edlund, H.. Lignin: Recent advances and emerging applications. *Curr. Opin. Colloid Interface Sci*, 2014 *19*, 409-416.
18. Tan, J.; et al. In *Chemical Precipitation of Palm Oil Mill Effluent (POME).* Proceedings of the 1st International Conference on Natural Resources Engineering and Technology, Putrajaya, Malasia, July 24–25, 2006; 2006, pp 400–407.
19. Devesa-Rey, R.; et al. Evaluation of Non-conventional Coagulants to Remove Turbidity from Water. *Water Air Soil Pollut.* **2012,** *223*(2), 591–598.
20. Zahrim, A. Y.; Tizaoui, C.; Hilal, N. Coagulation with Polymers for Nanofiltration Pre-treatment of Highly Concentrated Dyes: A Review. *Desalination* **2011,** *266*, 1–16.
21. Lee, K. E.; et al. Development, Characterization and the Application of Hybrid Materials in Coagulation/Flocculation of Wastewater: A Review. *Chem. Eng. J.* **2012,** *203*, 370–386.
22. Ozkan, A.; Yekeler, M. Coagulation and Flocculation Characteristics of Celestite with Different Inorganic Salts and Polymers. *Chem. Eng. Process.: Process Intensif.* **2004,** *43*(7), 873–879.
23. Ahmad, A. L.; Sumathi, S.; Hameed, B. H. Coagulation of Residue Oil and Suspended Solid in Palm Oil Mill Effluent by Chitosan, Alum and PAC. *Chemi. Eng. J.* **2006,** *118*, 99–105.
24. Zahrim, A.; Tizaoui, C.; Hilal, N. Evaluation of Several Commercial Synthetic Polymers as Flocculant Aids for Removal of Highly Concentrated CI Acid Black 210 Dye. *J. Hazard. Mater.* **2010,** *182*(1), 624–630.

25. Bolto, B.; Gregory, J. Organic Polyelectrolytes in Water Treatment. *Water Res.* **2007,** *41*(11), 2301–2324.
26. Lee, C. S.; Robinson, J.; Chong, M. F. A Review on Application of Flocculants in Wastewater Treatment. *Process Saf. Environ. Prot.* **2014,** *92*(6), 489–508.
27. Rohella, R.; et al. Removal of Colour and Turbidity in Pulp and Paper Mill Effluents Using Polyelectrolytes. *Ind. J. Environ. Health* **2001,** *43*(4), 159–163.
28. Irfan, M.; et al, The Removal of COD, TSS and Colour of Black Liquor by Coagulation–Flocculation Process at Optimized pH, Settling and Dosing Rate. *Arab. J. Chem.* **2013,** *10*, S2307–S2318.
29. Ganjidoust, H.; et al. Effect of Synthetic and Natural Coagulant on Lignin Removal from Pulp and Paper Wastewater. *Water Sci. Technol.* **1997,** *35*(2–3), 291–296.
30. Wang, J.-P.; et al. Optimization of the Coagulation–Flocculation Process for Pulp Mill Wastewater Treatment Using a Combination of Uniform Design and Response Surface Methodology. *Water Res.* **2011,** *45*(17), 5633–5640.
31. Amuda, O. S.; Amoo, I. A. Coagulation/Flocculation Process and Sludge Conditioning in Beverage Industrial Wastewater Treatment. *J. Hazard. Mater.* **2007,** *141*(3), 778–783.
32. Srivastava, V. C.; Mall, I. D.; Mishra, I. M. Treatment of Pulp and Paper Mill Wastewaters with Poly Aluminium Chloride and Bagasse Fly Ash. *Colloids Surf., A: Physicochem. Eng. Asp.* 2005, 260(1–3), 17–28.
33. Teh, C. Y.; Wu, T. Y.; Juan, J. C. Potential Use of Rice Starch in Coagulation–Flocculation Process of Agro-industrial Wastewater: Treatment performance and flocs characterization. *Ecol. Eng.* 2014, 71, 509–519.
34. Sundin, J. Precipitation of Kraft Lignin Under Alkaline Conditions. Ph.D. Thesis, Department of Pulp and Paper Chemistry and Technology, Royal Institute of Technology: Stockholm, 2000.
35. Dentel, S. K.; Gossett, J. M. Effect of Chemical Coagulation on Anaerobic Digestibility of Organic Materials. *Water Res.* **1982,** *16*(5), 707–718.
36. Lees, E. J.; et al. The Impact of Residual Coagulant on the Respiration Rate and Sludge Characteristics of an Activated Microbial Biomass. *Process Saf. Environ. Prot.* **2001,** *79*(5), 283–290.
37. Zahid, W. M.; El-Shafai, S. A. Impacts of Alum Addition on the Treatment Efficiency of Cloth-media MBR. *Desalination* **2012,** *301*, 53–58.
38. Tatsi, A. A.; et al. Coagulation–Flocculation Pretreatment of Sanitary Landfill Leachates. *Chemosphere* **2003,** *53*(7), 737–744.
39. Zahrim, A. Y.; Hilal, N.; Tizaoui, C. Evaluation of Several Commercial Synthetic Polymers as Flocculant Aids for Removal of Highly Concentrated C. I. Acid Black 210 Dye. *J. Hazard. Mater.* **2010,** *182*(1–3), 624–630.
40. Kumar, P.; et al. Treatment of Pulp and Paper Mill Effluent by Coagulation. *Int. J. Civ. Environ. Eng.* **2011,** *5*(8), 715–720.
41. Karim, M. I. A.; Hie, L. L. The Use of Coagulating and Polymeric Flocculating Agents in the Treatment of Palm Oil Mill Effluent (POME). *Biol. Wastes* **1987,** *22*(3), 209–218.
42. Ahmad, A. L.; et al. Improvement of Alum and PACl Coagulation by Polyacrylamides (PAMs) for the Treatment of Pulp and Paper Mill Wastewater. *Chem. Eng. J.* **2008,** *137*, 510–517.
43. Dilek, F.; Bese, S. Treatment of Pulping Effluents by Using Alum and Clay—Colour Removal and Sludge Characteristics. *AJOL* **2001,** *27*(3), 361-366.

44. Zahrim, A. Y.; et al. In *Towards Recycling of Palm Oil Mill Effluent: Coagulation/Precipitation of Anaerobically Digested Palm Oil Effluent as a Pre-treatment*. Third International Conference on Recycling and Reuse of Materials. Kottayam, Kerala, India, 2014.
45. LeiviskÃ, T.; RÃmÃ, J. Coagulation of Wood Extractives in Chemical Pulp Bleaching Filtrate by Cationic Polyelectrolytes. *J. Hazard. Mater.* **2008,** *153*(1–2), 525–531.
46. Zeng, Y.; Park, J. Characterization and Coagulation Performance of a Novel Inorganic Polymer Coagulant—Poly-zinc-silicate-sulfate. *Colloids Surf. A: Physicochem. Eng. Asp.* **2009,** *334*(1–3), 147–154.
47. Ariffin, A.; Razali, M. A. A.; Ahmad, Z. PolyDADMAC and Polyacrylamide as a Hybrid Flocculation System in the Treatment of Pulp and Paper Mills Waste Water. *Chem. Eng. J.* **2012,** *179*, 107–111.
48. Wang, J.-P.; et al. Synthesis and Characterization of a Novel Cationic Chitosan-based Flocculant with a High Water-solubility for Pulp Mill Wastewater Treatment. *Water Res.* **2009,** *43*(20), 5267–5275.
49. Brovkina, J.; Shulga, G.; Ozolins, J. In *Coagulation of Wood Pollutants from Model Wastewater by Aluminium Salts*, Proceedings of the 8th International Scientific and Practical Conference, Latvia, Rezekne, June 20–22, 2011, pp 63–67.
50. Shi, H.; et al. A Combined Acidification/PEO Flocculation Process to Improve the Lignin Removal from the Prehydrolysis Liquor of Kraft-based Dissolving Pulp Production Process. *Bioresour. Technol.* **2011,** *102*(8), 5177–5182.
51. Wong, S. S.; et al. Treatment of Pulp and Paper Mill Wastewater by Polyacrylamide (PAM) in Polymer Induced Flocculation. *J. Hazard. Mater.* **2006,** *135*, 378–388.
52. Ahmad, A. L.; Chong, M. F.; Bhatia, S. Population Balance Model (PBM) for Flocculation Process: Simulation and Experimental Studies of Palm Oil Mill Effluent (POME) Pretreatment. *Chem. Eng. J.* **2008,** *140*(1–3), 86–100.
53. Ahmad, A. L.; Sumathi, S.; Hameed, B. H. Coagulation of Residue Oil and Suspended Solid in Palm Oil Mill Effluent by Chitosan, Alum and PAC. *Chem. Eng. J.* **2006,** *118*(1–2), 99–105.
54. Jami, M. S., S. A. Muyibi, and M. I. Oseni, Comparative Study of the Use of Coagulants in Biologically Treated Palm Oil Mill Effluent (POME) *Adv. Nat. Appl. Sci.* **2012,** *6*(5), 646–650.
55. Ahmad, A. L.; Ismail, S.; Bhatia, S. Optimization of Coagulation−Flocculation Process for Palm Oil Mill Effluent Using Response Surface Methodology. *Environ. Sci. Technol.* **2005,** *39*(8), 2828–2834.
56. Ganjidoust, H.; et al., Effect of Synthetic and Natural Coagulant on Lignin Removal from Pulp and Paper Wastewater. *Water Sci. Technol.* **1997,** *35*(2–3), 291–296.
57. Pawels, R.; Bhole, A. G. Use of Synthetic Polyelectrolytes in Treatment of Kraft Paper Mill Wastewater. *Indian J. Environ. Health* **1997,** *39*(3), 177–181.
58. Devesa-Rey, R.; et al. Optimization of the Dose of Calcium Lactate as a New Coagulant for the Coagulation–Flocculation of Suspended Particles in Water. *Desalination* **2011,** *280*(1–3), 63–71.
59. Vazquez-Almazan, M. C.; et al. Use of Calcium Sulphate Dihydrate as an Alternative to the Conventional Use of Aluminium Sulphate in the Primary Treatment of Wastewater. *Water SA* **2012,** *38*(5), 813–818.

60. Georgiou, D.; et al. Treatment of Cotton Textile Wastewater Using Lime and Ferrous Sulfate. *Water Res.* **2003,** *37*(9), 2248–2250.
61. van Vuuren, L. R. J.; et al. Advanced Purification of Sewage Works Effluent Using a Combined System of Lime Softening and Flotation. *Water Res.* **1967,** *1*(7), 463–474.
62. Aktas, E. S.; Imre, S.; Ersoy, L. Characterization and Lime Treatment of Olive Mill Wastewater. *Water Res.* **2001,** *35*(9), 2336–2340.
63. Boukhoubza, F.; et al. Application of Lime and Calcium Hypochlorite in the Dephenolisation and Discolouration of Olive Mill Wastewater. *J. Environ. Managt.* **2009,** *91*(1), 124–132.
64. Devesa-Rey, R.; et al. Evaluation of Non-conventional Coagulants to Remove Turbidity from Water. *Water Air Soil Pollut.* **2012,** *223*, 591–598.
65. Zahrim, A. Y. Palm Oil Mill Biogas Producing Process Effluent Treatment: A Short Review. *J. Appl. Sci.* **2014,** *14*, 3149–3155.
66. Ho, C. C.; Tan, Y. K.; Wang, C. W. The Distribution of Chemical Constituents Between the Soluble and the Particulate Fractions of Palm Oil Mill Effluent and Its Significance on Its Utilisation/Treatment. *Agric. Wastes* **1984,** *11*(1), 61–71.
67. Kongnoo, A.; et al. Decolorization and Organic Removal from Palm Oil Mill Effluent by Fenton's Process. *Environ. Eng. Sci.* **2012,** *29*(9), 855–859.
68. Yaser, A. Z.; Nurmin, B.; Rosalam, S. Coagulation/Flocculation of Anaerobically Treated Palm Oil Mill Effluent (AnPOME): A Review. In *Developments in Sustainable Chemical and Bioprocess Technology*; Ravindra, P., Bono, A., Ming, C. C., Eds. Springer Science+Business Media: New York, 2013; pp 3–9.
69. Jamal, P.; Idris, Z. M.; Alam, M. Z. Effects of Physicochemical Parameters on the Production of Phenolic Acids from Palm Oil Mill Effluent Under Liquid-state Fermentation by *Aspergillus niger* IBS-103ZA. *Food Chem.* **2011,** *124*(4), 1595–1602.
70. Chan, Y. J.; Chong, M. F.; Law, C. L. Optimization on Thermophilic Aerobic Treatment of Anaerobically Digested Palm Oil Mill Effluent (POME). *Biochem. Eng. J.* **2011,** *55*(3), 193–198.
71. Ng, W. J.; Wong, K. K.; Chin, K. K. Two-phase Anaerobic Treatment Kinetics of Palm Oil Wastes. *Water Res.* **1985,** *19*(5), 667–669.
72. Zahrim, A. Y.; et al. Decolourisation of Anaerobic Palm Oil Mill Effluent via Activated Sludge-granular Activated Carbon. *World Appl. Sci. J.* **2009,** *5*, 126–129.
73. Bunrung, S.; Prasertsan, S.; Prasertsan, P. In *Decolourisation of Biogas Effluent of Palm Oil Mill Using Palm Ash*, TIChE International Conference 2011, Hatyai, Songkhla, Thailand, 2011.
74. Kim, T. H.; et al. Decolorization of Disperse and Reactive Dye Solutions Using Ferric Chloride. *Desalination* **2004,** *161*(1), 49–58.
75. Iler, R. K. Coagulation of Colloidal Silica by Calcium Ions, Mechanism, and Effect of Particle Size. *J. Colloid Interface Sci.* **1975,** *53*(3), 476–488.
76. Feng, J.-W.; et al. Treatment of Tannery Wastewater by Electrocoagulation. *J. Environ. Sci.* **2007,** *19*(12), 1409–1415.

CHAPTER 14

PREPARATION AND CHARACTERIZATION OF WOOD–PLASTIC COMPOSITE BY PLASTIC WASTE AND SAW DUST

NEENU GEORGE[1*], CINCY GEORGE[2], SONA JOHN[2], AJI JOSEPH[3], and IVY MATHEW[1]

[1]*Research and Post Graduate Department of Chemistry, St. Georges College, Aruvithura, Kerala, India*

[2]*Department of Chemistry, Newman College, Thodupuzha, Kerala, India*

[3]*Department of Chemistry, Bishop Kurialacherry College, Amalagiri, Kottayam, Kerala, India*

[*]*Corresponding author. E-mail: neenuathickal@gmail.com*

CONTENTS

ABSTRACT

Consumption of plastic products has increased dramatically over the past few decades. This trend results in the generation of a vast waste stream that needs to be properly managed to avoid environmental damage. This study designed to transform plastic waste into fabrication of value-added products such as particle boards. Polyethylene-saw dust (PE–SD) composites are prepared by melt blending and compression molding techniques from recycled PE and SD. The mechanical properties and contact angle analysis of PE–SD composites were investigated. The results showed that the addition of SD increased the tensile strength, elastic modulus of the PE–SD composites, and strain at break was decreased as per SD concentration. Contact angle measurements revealed that presence of SD increased the hydrophilic nature of the composites by reduced its contact angle. Plastic waste can be reduced though the fabrication of value-added product such as wood–plastic composite from the SD and recycled PE.

14.1 INTRODUCTION

As a result of the increase in the consumption of plastics, the wastes generated from their production, transportation, and consumption create various environmental problems. The problem of waste plastic management can be solved if economic, political, technological, energetic, material, and environmental dimensions are all considered.[1] Since plastics are generally high-calorific value products ranging approximately from 18,000 to 38,000 kcal/kg, utilization for their energy alone or for related chemical production may be an alternative option.[2] Some polymers such as; polystyrene and poly(methyl methacrylate) undergo to produce monomers and other monoaromatics besides other hydrocarbons.[3,4] However, polyethylene (PE) and polypropylene having 0% and 2% monomer yield should not be used for monomer production processes. These kinds of polymers undergo pyrolysis process to produce valuable hydrocarbons.[5] Recycling processes for waste plastics are classified into two categories: mechanical and feedstock recycling. The former covers a range of physical methods aimed at converting the polymeric residue into plastic pellets or directly into secondary plastic materials.[6,7]

The use of conventional wood-based panels, such as particleboard and medium-density fiberboard (MDF), is quite limited for exterior and moist applications, due to the strong tendency of such materials to absorb water. By contrast, wood–plastic composites (WPCs) show a considerably reduced affinity toward water, compared to conventional wood-based panels, what is caused by their relatively high-thermoplastic content. WPCs represent a growing class of materials used by the residential construction industry and the furniture industry. Further expansion into the residential construction industry and development of applications for the furniture industry require an understanding of the fire resistance of flat-pressed WPCs. Wood is typically impregnated with solutions of FRs, commonly salts, such as monoammonium phosphate (MAP) and diammonium phosphate (DAP), ammonium sulfate, zinc chloride, sodium tetraborate, boric acid, and guanylurea phosphate.[8–11] Polyolefin (low density polyethylene, LDPE; high density polyethylene, HDPE; and poly propylene, PP) are a major type of thermoplastic used throughout the world in such applications as bags, toys, containers, pipes (LDPE), housewares, industrial wrappings and film, gas pipes (HDPE), film, battery cases, automotive parts, electrical components (PP). Addition polymers (such as polyethylene (PE)) in contrast to condensation polymers [i.e., poly(ethylene terephthalate), PET] cannot be easily recycled by simple chemical methods.[12,13] Disposing of waste plastics in to the landfill is becoming undesirable due to legislation pressures (waste to landfill must be reduced by 35% over the period from 1995 to 2020), rising costs, and the poor biodegradability of commonly used polymers. The approaches that have been proposed for recycling of waste polymers include,[14,15] *Primary recycling* referring to the "in-plant" recycling of the scrap material of controlled history. In *Mechanical recycling*, the polymer is separated from its associated contaminants and it is reprocessed by melt extrusion. *Chemical recycling* leading in total depolymerization to the monomers or partial degradation to other secondary valuable materials. The recycling of model and waste products based on LDPE, HDPE, and PP was examined using two different methods: the traditional method of dissolution/reprecipitation and the more challenging technique of pyrolysis. The first belongs to the mechanical recycling techniques while the second to the chemical/feedstock recycling. During the first technique, the polymer can be separated and recycled using a solvent/nonsolvent system. For this purpose, different solvents/nonsolvents were examined at different weight percent amounts and temperatures using either model polymers as raw material or commercial waste products (packaging film, bags, pipes, and food-retail products). This technique has been widely used by Papaspyrides et al.[16–18] and other researchers.[19]

Saw dust (SD), a waste from wood-processing industries, also creates environmental hazard unless reprocessed for different applications such as particle board and pulp. The recycled PET and SD can be used to produce WPC by flat-press method, which might be good value-added products from waste and would help to minimize the waste. This technology possesses some advantages such as higher productivity with relatively lower pressure requirement and as a consequence naturally given wood structure lefts undestroyed. Thus, the density of WPCs reduces considerably and increases the moisture resistance properties compared to the conventional wood-based composites.

This study investigated the technical evaluation of fabrication of WPC by utilizing plastic waste and SD from different mixing ratios of SD with waste PE.

14.2 EXPERIMENTAL

14.2.1 MATERIALS

SD was obtained from the local saw mills in Kerala, India. SD was screened to remove the impurities. It was then dried in an oven at 103°C ± 2°C for 24 h for a moisture content of 2%. Clean consumer PE products wastes were collected locally and grind in a grinder for getting the recycled PE powder. The PE powder was sieved by 60 mesh size sieve to remove the oversized particles. The PE powder was then dried in an oven at 103 ± 2°C for 24 h for a moisture content of 3% or less. Sodium hydroxide pellets were of analytical grade obtained from Nice Chemicals, India.

14.2.2 PRETREATMENT OF SD (NAOH TREATMENT)

The SD is first subjected to mild chemical delignification process using 5% of NaOH solution using an autoclave at 125°C for 1 h. Pressure was released suddenly and the process is repeated three times. Then, it is washed well with water and then dried in an oven at 103 + 2°C for 24 h for removing the moisture content.

14.2.3 PREPARATION OF SD–PE COMPOSITE

Composite of PE and SD at various composition were prepared by the melt mixing techniques. The melt mixing of PE and SD in different formulation is performed in Bra bender twin-screw computer (GmbH and Co. KG, Duisburg, Germany) at temperature of 160°C for 9 min at a screw speed of 90 rpm. The different composition is used for the PE and SD blends are 100–0, 90–10, 80–20, and 70–30. PE–SD composite panels were manufactured using these extruded pellets by compression molding process. The samples were compression molded at 160°C for 3 min, in a specially designed mold, so that the mold with the sample inside could be cooled keeping the sample under pressure.

14.2.4 CHARACTERIZATION TECHNIQUES

14.2.4.1 MECHANICAL ANALYSIS

The static tensile properties (i.e., modulus, strength, and toughness) of PE–SD composites were measured at room temperature (25°C) and atmospheric conditions (relative humidity of 50% ± 5%) using Tinos Olsen Universal Testing Machine (Model #5985) equipped with a 10 kN load cell. The tests were carried at a crosshead speed of 10 mm per min. Five identical specimens were tested for each formulation.

14.2.4.2 CONTACT ANGLE MEASUREMENTS

Contact angle measurements of the PE–SD composites were conducted in a SEO Phoenix-10 by using double-distilled water. Measurements were carried out on samples of dimension 2 × 2 × 0.2 cm^3 at room temperature. The volume of the sessile drop was maintained as 5 μL in all cases using a micro syringe. For accuracy, measurements were repeated five times on different pieces of the same sample. Along with contact angle, work of adhesion (WA), wetting energy, and surface energy were calculated.

WA is the work required to separate the solid and liquid phases or the negative free energy associated with the adhesion of solid and liquid phases used to express the strength of the interaction between the two phases. It is given by the Young–Dupree equation as:

$$WA = (1 + \cos\theta)\,\gamma L \tag{14.1}$$

where γ_L is the surface tension of the liquid used for the contact angle measurement.

The interfacial free energy ($\gamma s1$) can be calculated using the Duper's equation given below:

$$\gamma s1 = \gamma s + \gamma_L - WA \tag{14.2}$$

where γs is the total solid surface free energy. Spreading coefficient (Sc) is a measurement of the ability of one liquid to spontaneously spread across another. The work done in spreading one liquid over a unit area of another is given by:

$$Sc = \gamma s - \gamma s_L - \gamma_L \tag{14.3}$$

14.3 RESULT AND DISCUSSION

14.3.1 MECHANICAL ANALYSIS

All mechanical testing results are summarized in Table 14.1. The tensile strength and elastic modulus of PE–SD composites increased with SD content up to 20% and it decline its properties at 30% SD content (Figs. 14.1 and 14.2). The tensile strength increased from 11.8 to 15 MPa with 20 wt% of SD. This variation might be due to the interaction between the NaOH treated SD and PE. Elongation-at-break shows a decreasing nature as the increase in SD content (Fig. 14.3). The difference in polarity and molecular weight, decrease the interaction between the SD and PE and leads to formation SD aggregates in the composites. As the load is applied to the composite, the lack of interfacial adhesion between the PE and SD will limit the load-transfer process. These aggregates can lead to failure at lower stresses and modulus. Also failure may be due to the formation of cavities in the composite and due to the weak interactions.

TABLE 14.1 Mechanical Testing Results of PE–SD Composites.

Sample name	Tensile strength (MPa)	Elongation at break (%)	Modulus (MPa)
100 PE	11.81	11.85	661
90 PE 10 SD	12.94	15.1	760
80 PE 20 SD	14.99	12.98	853
70 PE 30 SD	12.84	10.41	757

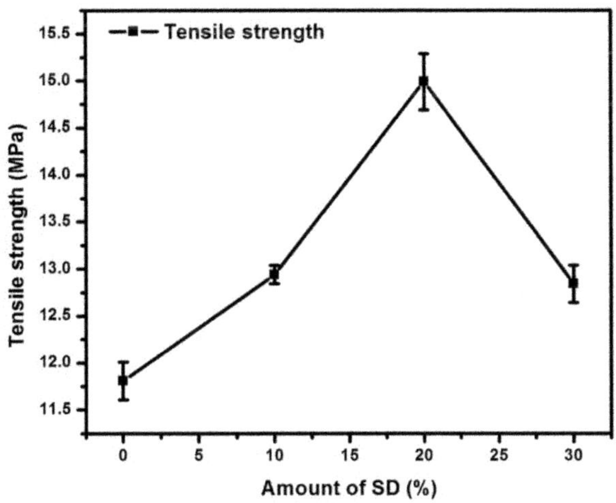

FIGURE 14.1 Tensile strength of PE–SD composites.

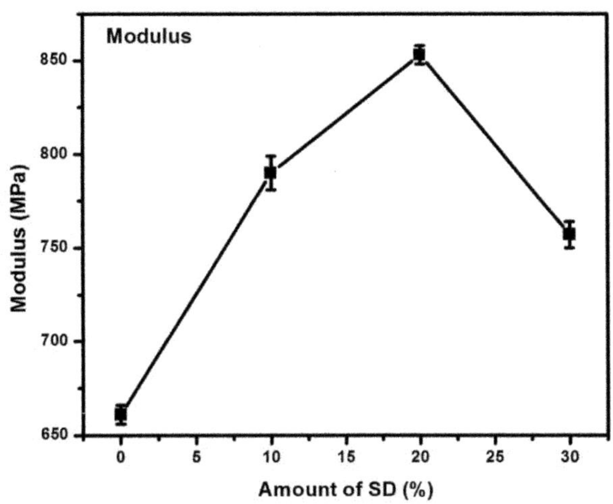

FIGURE 14.2 Tensile modulus of PE and PE–SD composites.

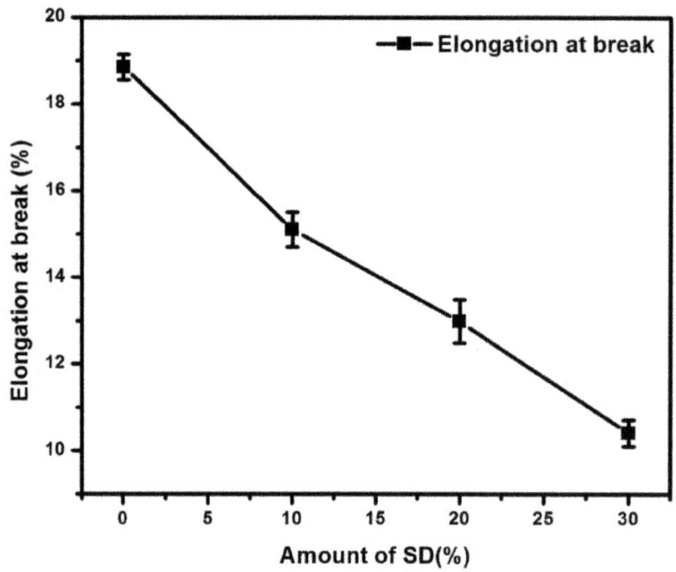

FIGURE 14.3 Elongation at break of PE–TSD and PE–ASD composites.

14.3.2 CONTACT ANGLE ANALYSIS

The wetting behavior of the composite with respect to water is analyzed, which focuses on the effect of the filler concentration on wetting characteristics such as WA, total surface free energy, interfacial free energy, and spreading coefficient. PE shows a wetting behavior by the addition of TSD and the decrease in contact angle value is in perfect correlation with increase in filler concentration. The hydrophilic nature of the composite is found to increase with the addition of SD. Figure 14.4 presents the representative figures of contact angle measurements of PE and PE–SD composite with water. The increase in hydrophilicity is very well understood from the images. All parameters of contact angle studies are summarized in Table 14.2. The nature of solvents affects the contact angle measurements. PE–SD composite shows decrease in contact angles with increase in SD concentration. By the addition of SD, polymer surface becomes more polar, that is, hydrophilic and the contact angle decreases accordingly. With water, the contact angle of PE (73°) decreases to 40.7° for PE-15–TSD composite, that is, it becomes a wetting surface.

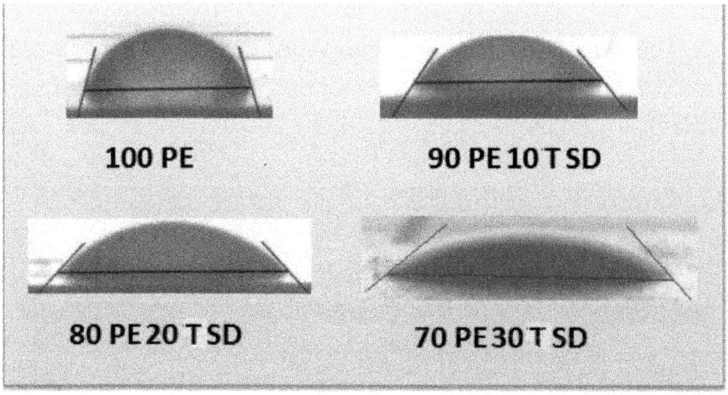

FIGURE 14.4 Contact angle measurements of PE–SD composites.

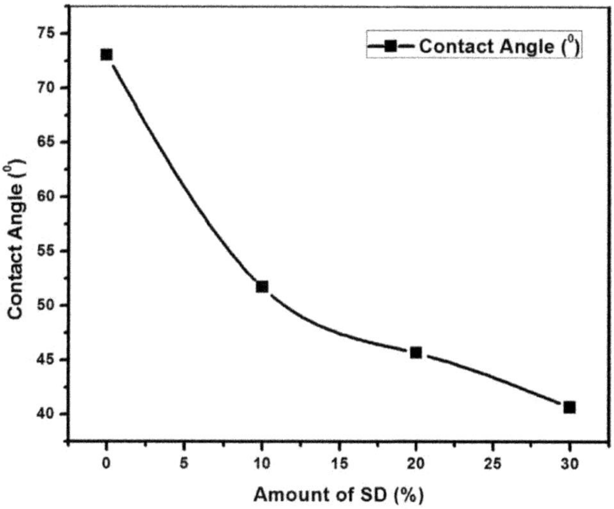

FIGURE 14.5 Contact angle curve of PE–SD composites.

TABLE 14.2 Wetting Properties of PE–TSD Composite on Water System.

Sample name	Contact angle (°C)	Work of adhesion WA = (1 + cos θ) γ_l (mJ/m^2)	Wetting energy $\gamma_{sl} = \gamma_s + \gamma_L -$ WA (mJ/m^2)	Spreading coefficient $Sc = \gamma_s - \gamma_{sL} - \gamma_L$ (mJ/m^2)
100 PE	73.02	94.04	21.24	51.56
90 PE 10 TSD	51.71	117.91	45.11	27.69
80 PE 20 TSD	45.68	123.13	50.33	22.47
70 PE 30 TSD	40.70	127.99	55.19	17.61

Surface free energy characteristics of PE–SD composite are given in Table 14.2. The total surface free energy is showing an increasing trend with SD concentration. According to the principle of wetting process, if the solid–vapor interfacial energy is low, the tendency for spreading to eliminate the interface will be less. In that case, the system exhibits more hydrophilic nature. Lowering the free energy of the system, the polymer chain must preferentially interact with the SD surface, where wettability plays a dominant role in successfully achieving the desired structure. Surface energy is the energy associated with the interface between two phases. The solid surface is rich in hydrocarbon molecule. The forces that hold between hydrocarbons together are much weaker than the force that acts between water molecules and consequently water on a hydrocarbon surface remain in nonwetting foam. Composite with SD, however, shows significantly decreased contact angles, indicating that the hydrophilic nature of the composites was significantly enhanced, that is, the surface free energy of the nanoparticle was reduced and by which it affects the total surface energy of the entire composite system. The *Sc* indicates that a liquid will spontaneously wet and spread on the solid surface if the value is positive, whereas a negative value of *Sc* implies the lack of spontaneous

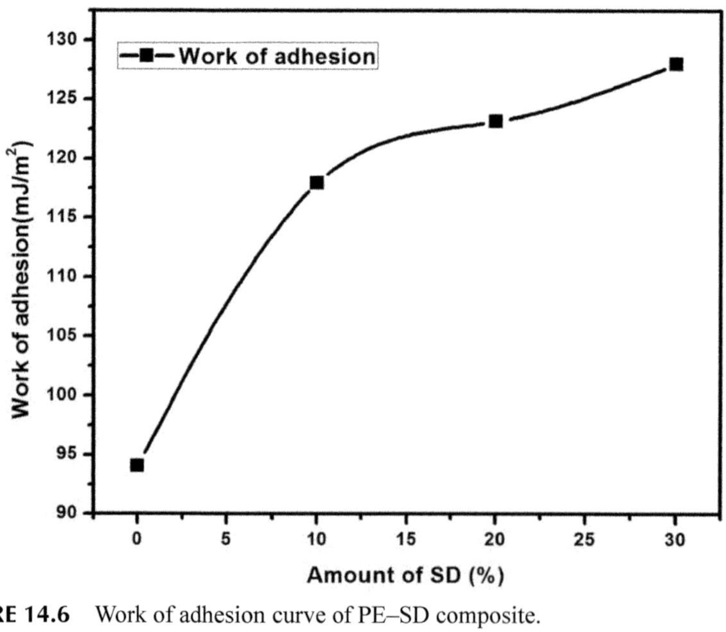

FIGURE 14.6 Work of adhesion curve of PE–SD composite.

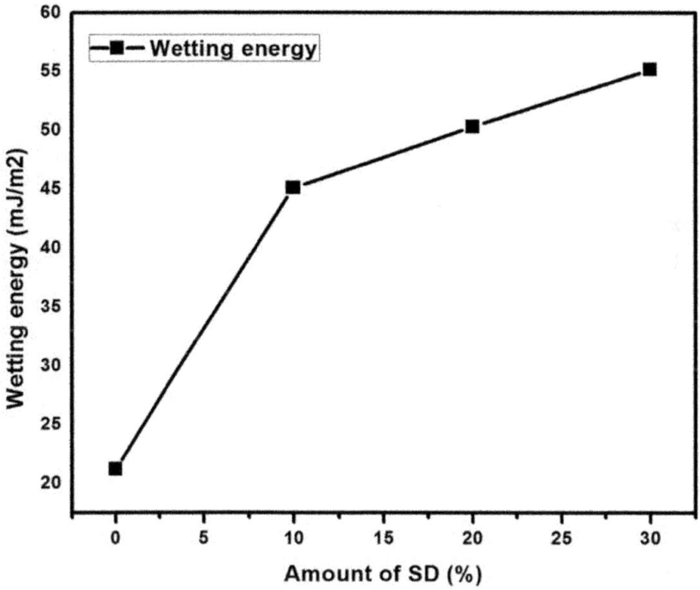

FIGURE 14.7 Wetting energy of PE–SD composites.

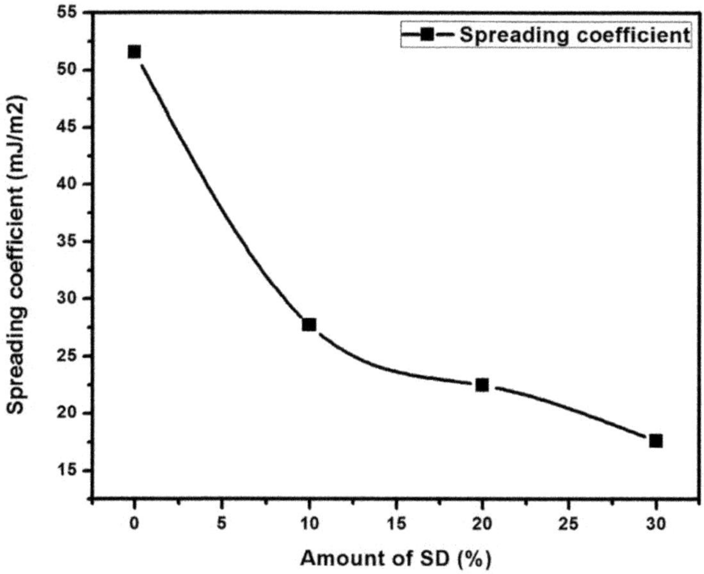

FIGURE 14.8 Spreading coefficient curve of PE–SD composite.

wetting. This means the existence of a finite contact angle. Wetting characteristics of water increase upon filler addition. The polar–polar interaction across the interface is a measure of wetting. The WA, which is the work required to separate the composite surface and the liquid drop, increases with filler concentration. WA shows an increasing tendency as the interfacial bonding is decreased. It is clear from the Table 14.2 that SD addition increases the polar behavior of the surface and by which it shows an increasing tendency of WA toward polar solvents. Generally, WA can be correlated to the filler matrix interaction.

14.4 CONCLUSIONS

This study investigated the technical evaluation of fabrication of WPC by utilizing plastic waste and SD from different mixing ratios of SD with waste PE. On the basis of mechanical properties, it appears that fabrication of PE–SD composites with using a Brabender twin-screw compounder and compression molding machine process is technically feasible for various structural purposes. The addition of SD increased the tensile strength, elastic modulus of the PE–SD composites, and strain at break was decreased as per SD concentration. Contact angle measurements revealed that presence of SD increased the hydrophilic nature of the composites by reducing its contact angle. Reduction of plastic waste production though the WPC fabrication from the SD and PE is technically feasible; it would be better to mix additives such as coupling agents to enhance interaction between SD and PE by reducing the melting temperature of PE, and thus, could ensure the adequate physical and mechanical properties of composites.

KEYWORDS

- **recycling of plastic**
- **plastic pollution**
- **wood–plastic composites**
- **saw dust–PE composite**
- **plastic waste**

REFERENCES

1. Li, J.; Zeng, X.; Stevels, A. Ecodesign in Consumer Electronics: Past, Present, and Future. *Crit. Rev. Env. Sci. Tec.* **2015,** *45*, 840–860.
2. Laurent, A.; Bakas, I.; Clavreul, J.; Bernstad, A.; Niero, M.; Gentil, E.; Christensen, T. H. Review of LCA Studies of Solid Waste Management Systems: Part I. Lessons Learned and Perspectives. *Waste Manage.* **2014,** *34*, 573–588.
3. Yang, X.; Sun, L.; Xiang, J.; Hu, S.; Su, S. Pyrolysis and Dehalogenation of Plastics from Waste Electrical and Electronic Equipment (WEEE): A Review. *Waste Manage.* **2013,** *33*, 462–473.
4. Oyake-Ombis, L.; van Vliet, B. J.; Mol, A. P. Managing Plastic Waste in East Africa: Niche Innovations in Plastic Production and Solid Waste. *Habitat Int.* **2015,** *48*, 188–197.
5. Ignatyev, I. A.; Thielemans, W.; Vander Beke, B. Recycling of Polymers: A Review. *Chem. Sus. Chem.* **2014,** *7*, 1579–1593.
6. Ajay, V. R.; Abitha, V. K.; Vinayak, K.; Jayaja, P.; Gaikwad, S. Sustainable Development Through Feedstock Recycling of Plastic Wastes. *Macromol. Symp.* **2016,** *362*, 39–51.
7. Tukker, A. Plastic Waste Feedstock Recycling, Chemical Recycling and Incineration. *Rapra Rev. Rep.* **2002,** *13*, 1–10.
8. Laufenberg, T.; Ayrilmis, N.; White, R. Fire and Bending Properties of Block Board with Fire Retardant Treated Veneers. *Holz als Roh- und Werkstoff.* **2006,** *64*, 137–143.
9. Winandy, J. E.; Wang, Q.; White, R. H. Fire-retardant-treated Strandboard: Properties and Fire Performance. *Wood Fiber Sci.* **2008,** *40*, 62–71.
10. Ayrilmis, N.; Kartal, S. N.; Laufenberg, T. L.; Winandy, J. E.; White, R. H. Physical and Mechanical Properties and Fire, Decay, and Termite Resistance of Treated Oriented Strandboard. *Forest Prod. J.* **2005,** *55*, 74.
11. Wang, Q.; Jian L.; Jerrold E. W. Chemical Mechanism of Fire Retardance of Boric Acid on Wood. *Wood Sci. Technol.* **2004,** *38*, 375–389.
12. Achilias, D. S.; Roupakias, C.; Megalokonomos, P.; Lappas, A. A.; Antonakou, E. V. Chemical Recycling of Plastic Wastes Made from Polyethylene (LDPE and HDPE) and Polypropylene (PP). *J. Hazard. Mater.* **2007,** *149*, 536–542.
13. Karayannidis, G. P.; Achilias, D. S. Chemical Recycling of Poly (Ethylene Terephthalate). *Macromol. Mater. Eng.* **2007,** *292*, 128–146.
14. Wang, R.; Xu, Z. Recycling of Non-metallic Fractions from Waste Electrical and Electronic Equipment (WEEE): A Review. *Waste Manage.* **2014,** *34*, 1455–1469.
15. Achilias, D. S.; Karayannidis, G. P. The Chemical Recycling of PET in the Framework of Sustainable Development. *Water Air Soil Pollut. Focus.* **2004,** *4*, 385–396.
16. Papaspyrides, C. D.; Poulakis, J. G.; Varelides, P. C. A Model Recycling Process for Low Density Polyethylene. *Res. Conserv. Recycl.* **1994,** *12*, 177–184.
17. Poulakis, J. G.; Papaspyrides, C. D. Recycling of Polypropylene by the Dissolution/Reprecipitation Technique: I. A Model Study. *Res. Conserv. Recycl.* **1997,** *20*, 31–41.
18. Poulakis, J. G.; Papaspyrides, C. D. The Dissolution/Reprecipitation Technique Applied on High-density Polyethylene: I. Model Recycling Experiments. *Adv. Polym. Technol.* **1995,** *14*, 237–242.
19. Pappa, G.; Boukouvalas, C.; Giannaris, C.; Ntaras, N.; Zografos, V.; Magoulas, K.; Lygeros, A.; Tassios, D. The Selective Dissolution/Precipitation Technique for Polymer Recycling: A Pilot Unit Application. *Res. Conserv. Recycl.* **2001,** *34*, 33–43.

CHAPTER 15

RECYCLING OF PVC WASTE BY FABRICATION OF A NBR–PVC BLEND

NEENU GEORGE[1*], JITHIN JOY[2], CINTIL JOSE[2], and IVY MATHEW[1]

[1]*Research and Post Graduate Department of Chemistry, St. Georges College, Aruvithura, Kerala, India*

[2]*Department of Chemistry, Newman College, Thodupuzha, Kerala, India*

[*]*Corresponding author. E-mail: neenuathickal@gmail.com*

CONTENTS

ABSTRACT

Elastomeric copolymer of acrylonitrile and 1,3 butadiene have been blended with poly(vinyl chloride) (PVC) to achieve a number of purposes. Small amount of nitrile elastomers improve the mechanical strength of rigid PVC composition. Recycling of plastic wastes through blending, by solution casting of acrylonitrile-butadiene rubber (NBR) and PVC, where replacing of virgin PVC to some extend by waste PVC is used. Cyclohexanone is used as the solvent to prepare the blends. The mechanical properties such as tensile stress–strain measurements, elongation at break, and modulus of NBR with pure and waste PVC were investigated. Blends can be compactable or incompatible. NBR/PVC blend is compactable one, and loss of compatibility if any, due to the presence of waste PVC.

15.1 INTRODUCTION

As a result of the increase in the consumption of plastics, the wastes generated from their production, transportation, and consumption create various environmental problems. The problem of waste plastic management can be solved if economic, political, technological, energetic, material, and environmental dimensions are all considered. Since plastics are generally high calorific value products ranging approximately from 18,000 to 38,000 kcal/kg, utilization for their energy alone or for related chemical production may be an alternative option.[1,2] Some polymers such as polystyrene and poly(methyl methacrylate) undergo to produce monomers and other monoaromatics besides other hydrocarbon.[3,4] However, polyethylene and polypropylene having 0% and 2% monomer yield should not be used for monomer production processes. These kinds of polymers undergo pyrolysis process to produce valuable hydrocarbons.[5] Recycling processes for waste plastics are classified into two categories: mechanical and feedstock recycling. The former covers a range of physical methods aimed at converting the polymeric residue into plastic pellets or directly into secondary plastic materials.[6,7]

Nowadays, considerable research interest is focused on new polymeric materials obtained by blending two or more polymers.[8–10] Mixing of different polymers have revealed a new realm of technically important materials. Their properties can be altered by varying the composition of the polymer blends. Although a large number of combinations of polymers are possible, there are relatively few that lead to a totally miscible system. A blend of two components is classified[11] as miscible, thermodynamically, if the Gibbs free energy of mixing is less than zero and the second derivative of the Gibbs free energy of mixing is zero or positive. The major feature of such process is that the intermediate properties are in some cases better than those exhibited by either of the single components.[12–14] In addition, some modifications in terms of processing characteristics, durability, and cost can be achieved via polymer blending.[15] In recent years, the blends of acrylonitrile-butadiene rubber (NBR) and poly(vinyl chloride) (PVC) have been widely used in industry.[16] Major applications of these blends include conveyor belt covers, cable jackets, hose cover linings, gaskets, footwear, and cellular products.[17] It is worth noting that NBR acts as a permanent plasticizer for PVC in applications such as wire and cable insulation in which PVC improves the chemical resistance, thermal ageing, and abrasion resistance of NBR.[18,19] PVC is miscible with NBR (23–45% acrylonitrile content) at all composition ranges.[20] The aim in blending plastic and rubber is to improve the physical, thermal, and mechanical properties as well as to modify the processing characteristics and cost reduction of the final product. Some blends, however, are incompatible. In a blend of rubber–plastic, such as NBR–PVC, the rubber must be vulcanized in the final form.[16] One way to improve the final performance of this blend is by means of interfacial modifier or compatibilizing agents acting from the matrix side.[21] In general, these interfacial modifications have generated great interest in materials based on polymers as polymer blend or poly blends, because these agents are able to enhance the interaction level between the material components, such interactions take place through the interphase.[16,22] Recycling plastic wastes through blending, by solution casting of NBR and PVC where the extend of replacing virgin PVC by waste PVC is used. Studies were conducted on PVC/NBR blends. Three different compositions are chosen for the present study, that is, 20%, 50%, and 80% both waste PVC and pure PVC with NBR. Solution blending was used as the technique for the preparation of blends of various compositions. Cyclohexanone is used as the solvent to prepare the blends.

15.2 EXPERIMENTAL

15.2.1 MATERIALS

Butadiene-acrylonitrile used in the study are of acrylonitrile content is 42% and the mooney viscosity is 50. PVC used was extrusion grade suspension polymer in powder form with a K value of 67. Cyclohexane is used as the solvent to prepare blend, and the waste PVC pipes were collected from a small-scale industry nearby.

15.2.2 SYNTHESIS OF NBR–PVC BLEND WITH PURE AND WASTE PVC

NBR/PVC blends with PVC content ranging from 20 to 80 wt% were prepared. Another set of blends with waste PVC in the range 20–80 wt% was also prepared. NBR and PVC are dissolved in cyclohexane solvent separately by using magnetic stirrer. It is then mixed and blend is prepared.

15.2.3 SCANNING ELECTRON MICROSCOPIC ANALYSIS

Studies on morphology of the fractured surfaces of blends were carried out using a Cambridge Stereoscan 360 scanning electron microscope (SEM); surfaces of the sample were coated with a thin gold layer.

15.2.4 MECHANICAL STRENGTH MEASUREMENTS

Test specimens were cut from the sheet. Tensile testing was carried out in Universal Tensile Testing machine using a load cell of 500 N with a gauge length of 30 mm and cross head speed of 500 mm/min. Dumbbell specimens for testing were punched from molded sheets parallel to the grain direction using a dumbbell die. The sample was held tight on the grips of the machine, the upper grip of which being fixed. Different types of grips were used according to test materials. The computer attached to machine plots load versus displacement curves and gives ultimate load, maximum deformation, and deformation at maximum force. From these curves, tensile strength, elongation at break, and modulus at 100% can be calculated.

15.3 RESULT AND DISCUSSIONS

Figure 15.1 shows the SEM micrographs of tensile fracture surfaces of (a) 50/50 NBR/PVC and (b) 70/30 NBR/PVC blends. Phase morphology of ensile fractured surface of blend sample based on 50/50 NBR/PVC and 70/30 NBR/PVC composition with various mixing procedure and NBR forms are shown in Figure 15.1. In blends, which are prepared with NBR powder, the size of the PVC domains is usually smaller than blend with NBR bale. The effect of increased interaction between NBR powder and PVC blends which can be seen from the presence of many tear lines on the tensile fractured surfaces.

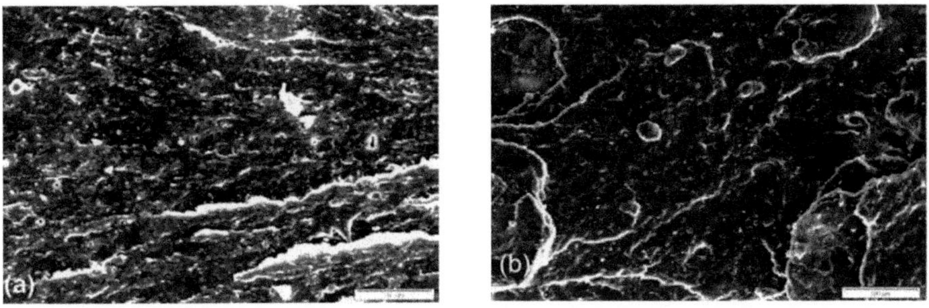

FIGURE 15.1 SEM micrographs of tensile fracture surfaces of (a) 50/50 NBR/PVC and (b) 70/30 NBR/PVC blends.

Three different compositions are of 30%, 50%, 70%, and 100% both waste PVC and pure PVC with NBR was prepared by solution blending. Cyclohexane is an excellent solvent which dissolves both NBR and PVC. Figure 15.2 shows tensile stress–strain of NBR/PVC blends with various PVC content. The plastic phase is rich for these blends and the nature of the curves are of typical plastics which exhibit a property called yielding, that is, taking higher force at longer elongation and then brittle failure with low elongation. The blends are not rich in rubber phase; hence lower will be the elongation. Blend with waste PVC also shows comparable tensile strength with pure PVC tensile strength which is reduced only slightly, showing no deterioration during second process. Comparison of the curves shows only slight reduction in the tensile strength between waste PVC and pure PVC (Fig. 15.3). Elongation is higher for 70/30 blends compared to 30/70 NBR/PVC blends due to the increase in the rubber content (Fig. 15.4). Addition of waste PVC is not reducing the mechanical properties of the blends to large extent. Elongation break values are highest for these blends because of richness in the rubber phase. All the blend compositions show only slight reduction in tensile strength.

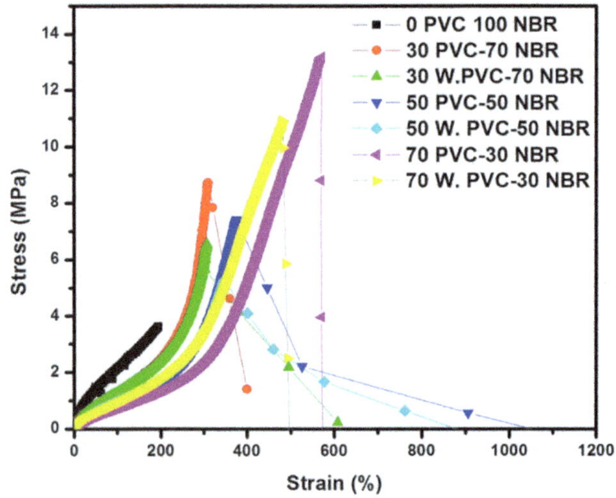

FIGURE 15.2 Stress–strain curve of PVC–NBR blends with pure PVC and waste PVC.

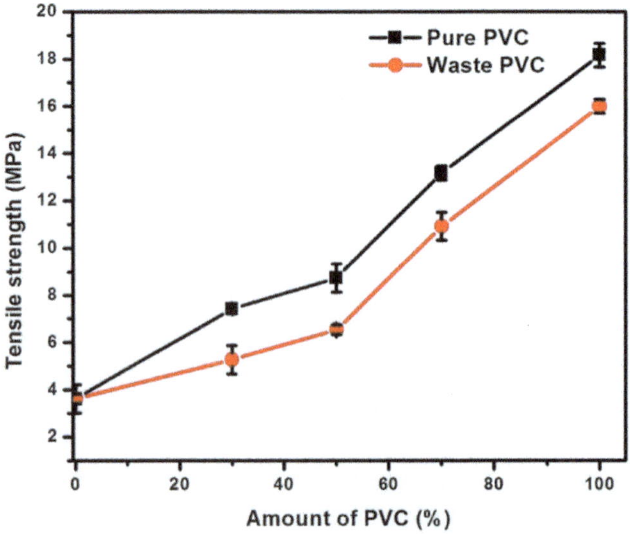

FIGURE 15.3 Tensile strength of PVC–NBR blends with pure PVC and waste PVC.

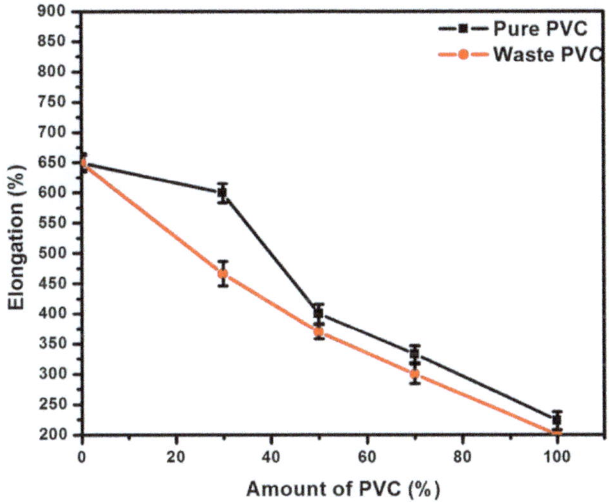

FIGURE 15.4 Elongation of PVC–NBR blends with pure PVC and waste PVC.

Figure 15.5 shows the modulus of PVC–NBR blends with pure PVC and waste PVC. Modulus is slightly enhanced with increase in PVC content. This is more pronounced in the case of plastic-rich blends due to the stiffness of the plastic phase. Modulus values are almost same or

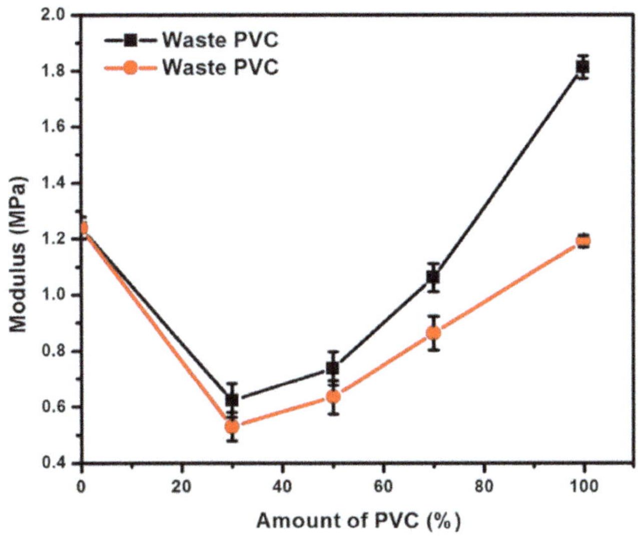

FIGURE 15.5 Modulus of PVC–NBR blends with pure PVC and waste PVC.

slightly increasing for the 50% blend with pure and waste PVC. Here, the plastic phase is strong enough to withstand the applied stress. For 30% PVC blends, the rich rubber phase is unable to withstand high stress showing lower modulus values. All these results show uniform increase in the modulus values from 100% to 200% to 300% elongation. All these results show that it is safe to recycle plastic waste for specific application without much deterioration in mechanical properties.

15.4 CONCLUSIONS

The expansion of polymers which is detectable all over the world, has also brought with it a huge problem, the decomposition of considerable quantities of potentially useful materials discarded from the industry as well as domestic source annually. At present, recycling or reprocessing provides a solution for the problem, but research has been going on all over the world to find stirs facing solution for this problem. Studies were conducted on PVC/NBR blends. Three different compositions are chosen for the present study, that is, 20%, 50%, and 80%, both waste PVC and pure PVC with NBR. Solution blending was used as the technique for the preparation of blends of various compositions. Selection of the proper solvent for both phases was done by trial and error. Cyclohexanone an excellent solvent which dissolves both NBR and PVC was selected for present study.

Tensile properties of the blends of different composition show an increasing effect when the PVC content increases due to the increase in the rigidity of plastic phase. Modulus values increase when the PVC phase increases both when pure and waste PVCs are used. This is again due to increased stiffness of the PVC phase. Elongations at break values are found to decrease as the PVC content increases. Elongation of the blend occurs mainly due the presence of rubber phase. So, as the rubber phase decrease, elongation at break also decreases. Waste PVC was used to prepare blends of NBR and PVC. Any deterioration in the mechanical properties was studied and found that mechanical properties decreased only slightly. Compactability of the blend also was not found to be decreasing for blends with waste PVC. It was found that the incorporation of NBR powder has a good effect on the improvement of mechanical properties and swelling behavior of these blends. It is worth noting that the increasing effect of NBR powder on mechanical properties could be due to fine particle size and high surface area that is well covered with PVC. In conclusion, we can see that waste plastic can be recycled for specific application without deteriorating their properties to some extent.

KEYWORDS

- **NBR**
- **pure and waste PVC**
- **tensile strength–strain**
- **elongation**
- **modulus**

REFERENCES

1. Li, J.; Zeng, X.; Stevels, A. Ecodesign in Consumer Electronics: Past, Present, and Future. *Crit. Rev. Env. Sci. Tec.* **2015,** *45*, 840–860.
2. Laurent, A.; Bakas, I.; Clavreul, J.; Bernstad, A.; Niero, M.; Gentil, E.; Christensen, T. H. Review of LCA Studies of Solid Waste Management Systems–Part I: Lessons Learned and Perspectives. *Waste Manage.* **2014,** *34*, 573–588.
3. Yang, X.; Sun, L.; Xiang, J.; Hu, S.; Su, S. Pyrolysis and Dehalogenation of Plastics from Waste Electrical and Electronic Equipment (WEEE): A Review. *Waste Manage.* **2013,** *33*, 462–473.
4. Oyake-Ombis, L.; van Vliet, B. J.; Mol, A. P. Managing Plastic Waste in East Africa: Niche Innovations in Plastic Production and Solid Waste. *Habitat Int.* **2015,** *48*, 188–197.
5. Ignatyev, I. A.; Thielemans, W.; Vander Beke, B. Recycling of Polymers: A Review. *Chem. Sus. Chem.* **2014,** *7*, 1579–1593.
6. Ajay, V. R.; Abitha, V. K.; Vinayak, K.; Jayaja, P.; Gaikwad, S. Sustainable Development Through Feedstock Recycling of Plastic Wastes. *Macromol. Symp.* **2016,** *362*, 39–51.
7. Tukker, A. Plastic Waste Feedstock Recycling, Chemical Recycling and Incineration. *Rapra Rev. Rep.* **2002,** *13*, 1–10.
8. Ghaisas, S. S.; Kale, D. D.; Kim, J. G.; Jo, B. W. Blends of Plasticized Poly (Vinyl Chloride) and Waste Flexible Poly(Vinyl Chloride) with Waste Nitrile Rubber Powder. *J. Appl. Polym. Sci.* **2004,** *91*, 1552–1558.
9. Manoj, N. R.; De, P. P. Hot Air and Fuel Ageing of Poly (Vinyl Chloride)/Nitrile Rubber and Poly (Vinyl Chloride)/Hydrogenated Nitrile Rubber Blends. *Polym. Degrad. Stab.* **1994,** *44*, 43–47.

10. Watanabe, N. *Thermoplastic Elastomers from Rubber Plastic Blends*; De, S. K., Bhowmick, A. K., Eds.; Ellis Horwood: London, 1990.
11. Douwel, C. H. K.; Maas, W. E. J. R.; Veeman, W. S.; Buning, G. H. W.; Vankan, J. M. J. Miscibility in PMMA/Poly(Vinylidene Fluoride) Blends, Studied by Fluorine-19-enhanced Carbon 13 CPMAS NMR. *Macromolecules* **1990,** *23*, 406.
12. Mousa, A.; Ishiaku, U. S.; Mohd. I. Z. A. Rheological and Mechanical Properties of Dynamically Cured Poly(Vinyl Chloride)/Nitrile-butadiene Rubber Thermoplastic Elastomers. *Polym. Int.* **2003,** *52*, 120–125.
13. Manoj, N. R.; De, P. P. An Investigation of the Chemical Interaction in Blends of Poly(Vinyl Chloride) and Nitrile Rubber During Processing. *Polymer* **1998,** *39*, 733–741.
14. Hardiman, C. J.; McKenzie. G. T. Nitrile Rubber/Poly(Vinyl Chloride) Blends. U.S. Patent 6,043,318, March 28, 2000.
15. Manoj, N. R.; De, P. P.; De, S. K. Self-crosslinkable Plastic–Rubber Blend System Based on Poly(Vinyl Chloride) and Acrylonitrile-co-butadiene Rubber. *J. Appl. Polym. Sci.* **1993,** *49*, 133–142.
16. Supri, I. H.; Yusof A. M. M. Blend of Waste Poly(Vinyl Chloride) (PVCw)/Acrylonitrile Butadiene Rubber (NBR): The Effect of Maleic Anhydride (MAH). *Polym. Test.* **2004,** *23*, 675–683.
17. Huang, J. C. Analysis of the Thermodynamic Compatibility of Poly(Vinyl Chloride) and Nitrile Rubbers from Inverse Gas Chromatography. *J. Appl. Polym. Sci.* **2003,** *89*, 1242–1249.
18. Ishiaku, U. S.; Lim, F. S.; Mohd, I. Z. A. Mechanical Properties and Thermooxidative Aging of a Ternary Blend, PVC/ENR/NBR, Compared with the Binary Blends of PVC. *Polym. Plast. Technol. Eng.* **1999,** *38*, 939–954.
19. Liu, Z.; Zhu, X.; Wu, L.; Li, Y. Effects of Interfacial Adhesion on the Rubber Toughening of Poly(Vinyl Chloride): Part 1. Impact Tests. *Polymer* **2001,** *42*, 737–746.
20. Zakrzewski, G. A. Investigation of the Compatibility of Butadiene–Acrylonitrile Copolymers with Poly(Vinyl Chloride). *Polymer* **1973,** *14*, 347–351.
21. Wu, S. *Polymer Interface and Adhesion*; Marcel and Dekker: New York, 1982.
22. Mascia, L. *Thermoplastics: Materials Engineering*; Elsevier: London, 1989.

INDEX

S

T

V

W

Y